William H. Burr

A Course on the Stresses in Bridge

and roof trusses, arched ribs and suspension bridges

William H. Burr

A Course on the Stresses in Bridge
and roof trusses, arched ribs and suspension bridges

ISBN/EAN: 9783337349325

Printed in Europe, USA, Canada, Australia, Japan

Cover: Foto ©berggeist007 / pixelio.de

More available books at **www.hansebooks.com**

A COURSE

ON

THE STRESSES

IN

BRIDGE AND ROOF TRUSSES, ARCHED RIBS AND SUSPENSION BRIDGES,

PREPARED FOR THE DEPARTMENT OF CIVIL ENGINEERING AT THE
RENSSELAER POLYTECHNIC INSTITUTE.

By WILLIAM H. BURR, C.E.,

ASST. TO THE CHIEF ENGINEER OF THE PHŒNIX BRIDGE CO., MEMBER OF THE
AMERICAN SOCIETY OF CIVIL ENGINEERS. FORMERLY WILLIAM HOWARD
HART PROFESSOR OF RATIONAL AND TECHNICAL MECHANICS
AT THE RENSSELAER POLYTECHNIC INSTITUTE.

FOURTH EDITION.

NEW YORK:
JOHN WILEY & SONS.
1888.

PREFACE TO THIRD EDITION.

SINCE the publication of the first edition of this book engineering practice in iron and steel construction, especially in the department of bridge building, has made very material progress. The distribution of metal in pier structures has been considerably modified so as to produce concentrations in larger members; but chiefly the treatment of moving loads has experienced such a radical transformation as to bring it to a thoroughly rational basis. Hence, portions of the book as originally written have been cancelled and replaced by entirely new matter, so amplified and extended as to bring the work in all its details abreast of the best practice of the present day.

My indebtedness to the published papers of Prof. H. T. Eddy, of the University of Cincinnati, on the arched rib, will be evident to any reader even slightly acquainted with his valuable work entitled " Researches in Graphical Statics."

Certain matters are of such common occurrence in the following pages that it may conduce to clearness to mention them here.

The word " ton " signifies a ton of 2,000 pounds, unless it is otherwise specifically stated.

The word " stress " means the force acting in any member of a structure, while "strain " is the distortion which accompanies the stress.

The sign $+$ indicates a tensile stress, and the sign $-$, a compressive one.

Unless otherwise stated, the stress in any member of a structure will be represented by inclosing with a parenthesis the letter or letters which belong to it in the diagrams or plates. Thus (A B), or (a), signifies " stress in the member A B," or " stress in the member a."

As a matter of convenience to those who may be familiar with the first and second editions, it is well to state that pages 19 to 60, Arts. 72, 74 and 85 are entirely new. Portions of pages at several other places in the book have also been re-written, but it is unnecessary to name them here.

For convenience in swing bridge computations, Appendix IV. has been inserted. Formulæ for moments and reactions are there collected in the simplest form for application.

W. H. B.

PHŒNIXVILLE, PA ,
Feb. 24th, 1886.

CONTENTS.

CHAPTER I.

General Consideration of the Laws Governing the Action of Stresses in Trusses.

PAGE

Art. 1.—The Truss Element... 1

"   2.—General Case... 2

"   3.—Web and Chord Stresses in General............................ 4

"   4.—Overhanging Truss—Parallel Chords—Bracing with two Inclinations... 8

"   5.—Overhanging Truss—Parallel Chords—Uniform Bracing—Vertical and Diagonal Bracing................................... 11

CHAPTER II.

Special Non-continuous Trusses with Parallel Chords.

Art. 6.—Distribution of Fixed and Moving Loads...................... 18

"   7.—Position of Moving Load for Greatest Shear and Greatest Bending. 20

"   8.—Fixed Weight 36

"   9.—Single System of Bracing with two Inclinations............ 38

"   10.—Single System of Vertical and Diagonal Bracing—Verticals in Tension... 50

"   11.—Single System of Vertical and Diagonal Bracing—Verticals in Compression 54

"   12.—Two Systems of Vertical and Diagonal Bracing—Verticals in Compression 59

"   13.—Truss with Uniform Diagonal Bracing—Two Systems of Triangulations 66

"   14.—Compound Triangular Truss................................. 69

"   15.—Methods of Obtaining Stresses—Stress Sheets................ 74

"   16.—Ambiguity caused by Counterbraces......................... 74

CHAPTER III.

Non-continuous Trusses with Chords not Parallel.

PAGE

ART. 17.—General Methods... 75
" 18.—Curved Upper Chord—Two Systems of Vertical and Diagonal
Bracing 78
" 19.—General Considerations..................................... 88
" 20.—Position of Moving Load for Greatest Stress in any Web Member. 89
" 21.—Position of Moving Load for Greatest Stresses in Chords........ 92
" 22.—Horizontal Component of Greatest Stress in any Web Member—
Constant Value of the Same for Vertical and Diagonal Bracing
with Parabolic Chord..................................... 93
" 23.—Bowstring Truss—Diagonal Bracing—Example............... 93
" 24.—Bowstring Truss—Vertical and Diagonal Bracing with Counters.. 102
" 25.—Bowstring Truss—Vertical and Diagonal Bracing without Counters.. 105
" 26.—Deck Truss with Curved Lower Chord, Concave Downward—Example 108
" 27.—Crane Trusses... 111
" 28.—Preliminary to the Treatment of Roof Trusses—Wind Pressure—
Notation... 115
" 29.—First Example... 117
" 30.—Second Example.. 118
" 31.—Third Example... 121
" 32.—Fourth Example.. 122
" 33.—General Considerations.................................... 123

CHAPTER IV.

Swing Bridges. Ends Simply Supported.

ART. 34.—General Considerations.................................... 125
" 35.—General Formulæ for the Case of Ends Simply Supported—Two
Points of Support at Pivot Pier—One Point of Support at Pivot
Pier... 126
" 36.—Ends Simply Supported—Four Points of Support at Centre—Example 133
" 37.—Ends Simply Supported—Four Points of Support at Centre—Partial Continuity—Example................................. 167
" 38.—Ends Simply Supported—One Support at Centre—Example..... 187

CHAPTER V.

Swing Bridges. Ends Latched to Supports.

ART. 39.—General Considerations 203
" 40.—Ends Latched Down—One Point of Support between Extremities
of Truss—Example..................................... 203

CHAPTER VI.

Swing Bridges. Ends Lifted.

PAGE

ART. 41.—General Considerations..................................... 212
" 42.—Ends Lifted—One Point of Support between Extremities—Example.. 212
" 43.—Final Observations on the preceding Methods 222

CHAPTER VII.

Continuous Trusses other than Swing Bridges.

ART. 44.—Formulæ for Ordinary Cases—Reactions—Methods of Procedure. 224

CHAPTER VIII.

Arched Ribs.

ART. 45.—Equilibrium Polygons 229
" 46.—Bending Moments... 234
" 47.—General Formulæ 236
" 48.—Arched Rib with Ends Fixed............................... 237
" 49.—Arched Rib with Free Ends................................ 247
" 50.—Thermal Stresses in the Arched Rib with Ends Fixed 250
" 51.—Thermal Stresses in the Arched Rib with Ends Free.......... 256
" 52.—Arched Rib with the Fixed Ends—I and n Variable............ 259
" 53.—Determination of the Stresses in the Members of an Arched Rib—Example—Fixed Ends—Consideration of Details............ 267
" 54.—Arched Rib Free at Ends and Jointed at the Crown. 278

CHAPTER IX.

Suspension Bridges.

ART. 55.—Curve of Cable for Uniform Load per Unit of Span—Suspension Rods Vertical—Heights of Towers Equal or Unequal—Generalization 279
" 56.—Parameter of Curve—Distance of Lowest Point of Cable from either Extremity of Span—Inclination of Cable at any Point... 281
" 57.—Resultant Tension at any Point of Cable.................... 282
" 58.—Length of Curve between Vertex and any Point whose Co-ordinates are x and y, or at which the Inclination to a Horizontal Line is i.. 283
" 59.—Deflection of Cable for Change in Length, the Span remaining the same 285
" 60.—Suspension Canti-Levers................................... 287

PAGE

ART. 61.—Suspension Bridge with Inclined Suspension Rods—Inclination of Cable to a Horizontal Line—Cable Tension—Direct Stress on Platform—Length of Cable............................... 288

" 62.—Suspension Rods; Lengths and Stresses...................... 292

" 63.—Pressure on the Tower—Stability of the Latter—Anchorage..... 294

" 64.—Theory of the Stiffening Truss—Ends Anchored—Continuous Load—Single Weight................................. 296

" 65.—Theory of the Stiffening Truss—Ends Free—Continuous Load—Single Weight... 307

" 66.—Approximate Character of the Preceding Investigations—Deflection of the Truss 311

CHAPTER X.

Details of Construction.

ART. 67.—Classes of Bridges—Forms of Compression Members—Chords Continuous or Non-Continuous............................ 313

" 68.—Cumulative Stresses 315

" 69.—Direct Stress Combined with Bending in Chords.............. 316

" 70.—Riveted Joints and Pressure on Rivets....................... 323

" 71.—Riveted Connections between Web Members and Chords....... 325

" 72.—Floor-Beams and Stringers—Plate Girders 326

" 73.—Eye-Bars or Links.. 343

" 74.—Size of Pins... 345

" 75.—Camber.. 350

" 76.—Economic Depth of Trusses with Parallel Chords............. 353

" 77.—Fixed and Moving Loads 354

" 78.—Safety Factors and Working Stresses........................ 358

" 79.—General Observations....................................... 365

CHAPTER XI.

Wind Stresses and Braced Piers.

ART. 80.—Wind Pressure 366

" 81.—Sway Bracing 370

" 82.—Transverse Bracing for Transferring Wind Stresses from One Chord to Another—Concentrated Reaction.................. 379

" 83.—Transverse Bracing with Distributed Reactions................ 384

" 84.—Stresses in Braced Piers 387

" 85.—Complete Design of a Railway Bridge....................... 395

APPENDIX I.

The Theorem of Three Moments................................... 431

APPENDIX II.

The Resistance of Solid Metallic Rollers............................ 443

APPENDIX III.

The Schwedler Truss 447

APPENDIX IV.

Reactions and Moments for Continuous Beams 453

CHAPTER I.

GENERAL CONSIDERATION OF THE LAWS GOVERNING THE ACTION OF STRESSES IN TRUSSES.

Art. 1.—The Truss Element.

A TRUSS may be defined to be a structure so composed of individual pieces that, if all the externally applied forces called loading are parallel in direction, the other external forces called reactions will be parallel both to each other and the loading.

The simplest of all trusses is a triangle, and all trusses, however complicated, containing no superfluous members, are, and may be considered, assemblages of triangles simply. That the triangle is the truss element arises from the fact that it is the only geometrical figure whose form may not be changed without varying the lengths of its sides.

In the elementary truss of the figure, let any force act vertically downwards at B, and consider the two triangles ABD and BDC having the common side BD. Since all external forces are parallel, the reaction at A is to the reaction at C as DC is to AD.

For if BD be taken to represent the vertical force acting at B, and DF be drawn parallel to AB as well as EF parallel to AC, then will DF represent the stress in AB, and BF that in BC. But by the construction $ED : EB = DC : AD$, but ED is the vertical component of the stress in AB as well as the reaction at A, while BE is the same component of the stress in BC, and, similarly, the reaction at C. It is to be particularly noticed that EF is the common horizontal component in each of the members AB and BC, and also the resultant stress in AC.

I   I

When, therefore, the truss is horizontal, as is supposed in the figure, the vertical component in each of the members *AB* and *BC* is equal to the reaction at its foot; also the horizontal component of stress in each of these members is equal to the horizontal component in the other, as well as to the resultant stress in the third horizontal member.

These simple principles constitute the foundation of all stress analyses in trusses.

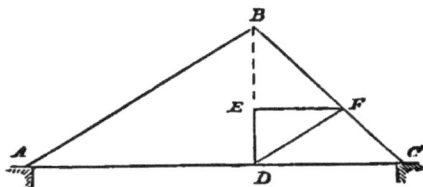

Fig. 1.

Art. 2.—General Case.

Again, consider any truss whatever, as that in Fig. 1, in which the supports are not on the same level, nor are any two of the triangles of which it is composed similar. Suppose a vertical load to act at any apex, as *A*, the reactions will be vertical. Let the truss be cut by any plane which divides the line *GH* (*AH* is the trace of such a plane), then the part of the truss which is found on the left of *AH* is held in equilibrium by a component of the vertical force at *A*, the vertical reaction at *C*, and the induced stress in *GH*. Since there is equilibrium, the lines of action of those forces must intersect in a point; and since the forces acting through *C* and *GH* have lines of action intersecting at *D*, the line of action of the component of the vertical force at *A* must pass through the same point. Thus the line of action *DA* for one component is established.

In precisely the same manner *BE* is erected and produced until it intersects *GH*, produced, in *E* and the line of action *AE* of the other component established. Connect *D* and *E*, then, so far as the reactions are concerned, the case will not

be changed if the actual truss be supposed displaced by the
simple truss DEA. Let AN represent the vertical load at A,
then make NO, parallel to AE, and DA, produced, intersect
at O. If MO be drawn parallel to DE, AM will evidently
represent the reaction at C, and MN the reaction at B. Pro-

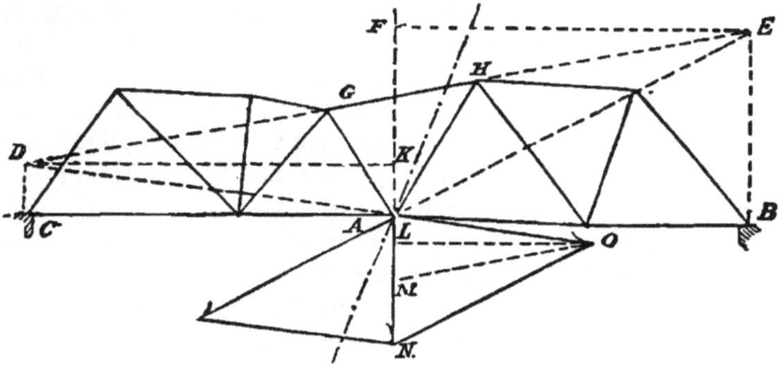

FIG. 1.

duce AN until it intersects GH in G, and draw DK and EF
in a horizontal direction, then, from similar triangles:

$$\frac{OM}{MA} = \frac{DG}{GA}, \quad \text{and} \quad \frac{OM}{MN} = \frac{EG}{GA}.$$

Dividing one by the other,

$$\frac{MN}{MA} = \frac{DG}{EG} = \frac{DK}{EF}.$$

But $DK + EF$ is equal to the span, hence the reactions are
inversely as the segments of the span, or any load or system
of loading, vertical in direction, to which any truss whatever
is subjected, is divided into reactions according to the law of
the lever. Farther, whatever the internal stresses of the truss
may be, at the ends the sum of the vertical components must
be equal to the reactions.

Art. 3.—Web and Chord Stresses in General.

In the preceding case no account was taken of the stresses to which the individual members of the truss were subjected, but it will now be necessary to consider them. For this purpose Pl. I., Fig. 1, will be used, in which the points of support will be taken in the same level, the loading, vertical; and the truss will be considered as made up of similar triangles. The first and last suppositions in nowise affect the generality of the conclusions, but the operations are thereby simplified and given a character approaching more nearly to that of the ordinary operations of the engineer.

In Pl. I., Fig. 1, is the representation of a truss placed upon two supports A and L in the same horizontal line. CH and AL are parallel, and the oblique members included between them may have any inclinations whatever, only they are made symmetrical in reference to a vertical centre line through O. Let any weight W act at any point, as N, and let t and t' represent the tangents of the greater and less inclinations, respectively, of the oblique members to a vertical. Erect verticals at A and L which will intersect the prolongations of CH in B and K; then, as has been shown, BR and UK drawn through N will be the lines of action of the components of W which act on the two parts of the truss. The force parallelogram $WRNU$ can then be drawn, in which RT and UV are to be drawn parallel to AL. $NT = VW$ is the reaction at A, and NV that at L.

Resolve NR in the direction of FN and NM' by drawing $M'R$ parallel to FN, then will $M'R$ represent the stress in FN. The stress in FN will induce the stresses Fb and Fb'', in FG and FO, and in the same manner the stresses shown in the figure will be induced at all the points on the left of FN.

It is to be noticed that all the inclined stresses at the points A, C, Q, D, P, E, O, F, as well as $M'R$, have the same vertical component, NT; also, that all the horizontal stresses at C, D, E, and F are equal to each other and act in the same direction, each one having the value $NT \times (t + t')$. Let n be the number of the points C, D, E, and F, then the total horizontal stress acting along CH from left to right will be

$$NT \times (t + t') \times n.$$

At N there is the horizontal force NM' acting from left to right.   Let d equal the depth of the truss, or AB, and l and l' the segments AN and NL respectively of the span AL or s; then from the figure it is seen that

$$NM' = NT\left(\frac{l}{d} - t'\right).$$

At the points O, P, Q, A, there are horizontal stresses acting from right to left.   From the diagram it is seen that the stress at O is $(2\,t \times NT)$; at P, $(t + t') \times NT$; at Q, $(t + t') \times NT$; at A, $t' \times NT$; hence the total stress on the left of N acting from right to left is

$$NT(nt + (n - 1)t') = NT\left(\frac{n(t + t')d}{d} - t'\right) = NT\left(\frac{l}{d} - t'\right).$$

Hence there is deduced the important result that NM' is just equal to the total horizontal stress on the left of N, and possesses the same line of action but is opposite in direction, therefore the two forces balance each other.

Next prolong GN until it cuts UV in Y, then will NY represent the stress induced in GN by the component UN. The stress at N, acting from right to left, is therefore UY, or

$$UY = NV \times \left(\frac{l''}{d} - t\right).$$

Although the diagrams are not drawn, it is plain that the horizontal stresses at M and L are $NV \times (t + t')$ and $NV \times t'$ respectively; also, that their directions are from left to right; hence the total horizontal stresses, on the right of N, which act from left to right, are

$$NV \times (t + 2\,t') = NV\left(\frac{2\,(t + t')\,d}{d} - t\right) = NV \times \left(\frac{l'}{d} - t\right).$$

Hence the stress UY is balanced by the horizontal stresses on the right of N. All the internal horizontal stresses acting along AL are, therefore, balanced.

According to the two force parallelograms drawn from G and H, it is seen that all the horizontal stresses on the right of N, which act from right to left along CH, are

$$n' \times NV \times (t + t') = NV \times \frac{l'}{d},$$

in which n' is the number of apices G and H.

But $NV = NT\dfrac{l}{l'}$; hence

$$NV\frac{l'}{d} = NT\frac{l}{d}.$$

But it has already been shown that all the stresses which act along CH from left to right are

$$NT \times (t + t') \times n = NT\frac{l}{d}.$$

Hence all the horizontal internal stresses of the truss are perfectly balanced among themselves.

This important characteristic belongs only to the "truss" proper, and distinguishes it from all other bridge superstructures.

If an irregular truss, like that in Fig. 1 above, were treated, precisely the same result would be reached, but the resultant horizontal stress would be expressed by ΣPt or $\Sigma' Pt$, in which P is a variable portion of W; and d would be a " mean " depth, such that $\Sigma t\,d = l$, and $\Sigma' t\,d = l'$.

The portion FG is subjected to all the stress induced at the points C, D, E, and F; EF to all that induced at C, D, and E; etc. It is important to notice this accumulation of stresses from panel to panel in the horizontal lines CH and AL, for it shows that the stresses in those portions are not uniform from end to end. A stress induced at one point may be felt at any distance from that point.

The upper and lower portions of the truss, *CDEFGH* and *AQPON.ML*, are called the top and bottom "chords," and all members included between the chords, whether inclined or vertical, are called " braces " or web members. The various portions into which the chords are divided, usually equal to each other, are called panels.

From the figure it is seen that the vertical components of the stresses in the braces or web members on one side of *N* are equal to each other ; also, that the chord stresses have no vertical component, being horizontal. Farther, the vertical component in any brace or web member is equal to the reaction found on the same side of the load as itself. In a truss provided with horizontal chords, therefore, the office of the web members is solely to transfer, so to speak, the load from its point of application to the abutments or piers of the bridge ; their duty, therefore, is precisely the same as that of the web in a flanged girder, hence their name " web members." In other words, the braces or web members take up the shearing stress at any section.

Let *sec i* and *sec i'* be the secants of the angles of inclination of the web members, corresponding to the tangents *t* and *t'*, and let *S* and *S'* be the shearing stresses in the two segments of the span ; then the web stresses in the left-hand segment will be *S sec i* and *S sec i'*, and those for the right-hand segment *S' sec i* and *S' sec i'*.

The general principles brought out in the preceding results, therefore, are these : *With horizontal chords the web stresses are products of the shears by the secants of the inclinations, and the chord stresses are functions of the tangents of the inclinations of the braces or web members from vertical lines.*

From an inspection of the force parallelogram at *D*, for instance, it is seen that *the increment of chord stress at any panel point is equal to the algebraic sum of the horizontal components of the stresses in the web members intersecting at that point.* The sum is numerical when, as in the figure, the braces slope on different sides of the vertical line passing through the panel point, but the numerical difference is to be taken when both braces are found on the same side.

Two web members intersecting at any panel point have stresses of opposite kinds induced by the same shearing stress.

The stress in *CH* is of course compressive, while that in *AL* is tensile.

The preceding general results have been deduced on the supposition of the application of but one weight, but they are equally true for any system of loading. For the effect of any system of loading is simply the summation of the effects of the individual loads of which it is composed, hence only those principles which are true for the individuals can be true for the system, and *those* at least must hold, for the action of each load is independent of all the others.

Art. 4.—Overhanging Truss.—Parallel Chords.—Bracing with Two Inclinations.

Probably the simplest case of a truss subjected to the action of external loading occurring in the practice of the engineer is a simple truss fixed at one end, and is the case with one arm of a swing-bridge when open and subjected to its own weight as load.

Now, in all cases of actual trusses, the load will be supposed divided in its application to the truss between the upper and lower chords. It will not, however, be equally divided, because the floor system of the bridge will rest wholly on one chord or the other.

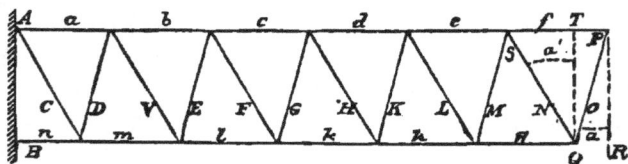

FIG. 1.

In the figure, let the truss be fixed at *AB*, and let *W* and

W' be the panel loads on the upper and lower chords respectively, except at the extremity P of the upper chord, where $\frac{1}{2}W$ will rest. Let α represent the angle QPR, and α' the angle SQT; the line PR is vertical, and AP horizontal.

A simple and direct application of the principles and formulæ of Art. 3 gives the following results :

Stress in O ....$\frac{1}{2}W$ *sec* α.
"   " M........$(1\frac{1}{2}W + W')$ "
"   " K ....$(2\frac{1}{2}W + 2W')$ "
"   " G ....$(3\frac{1}{2}W + 3W')$ "
"   " E ....$(4\frac{1}{2}W + 4W')$ "
"   " D ....$(5\frac{1}{2}W + 5W')$ "
"   " N ....$(\frac{1}{2}W + W')$ *sec* α'
"   " L ....$(1\frac{1}{2}W + 2W')$ "
"   " H ....$(2\frac{1}{2}W + 3W')$ "
"   " F ....$(3\frac{1}{2}W + 4W')$ "
"   " V ....$(4\frac{1}{2}W + 5W')$ "
"   " C ....$(5\frac{1}{2}W + 6W')$ "

It will be observed that in every instance the stress in any brace is the "shearing stress" in the section to which the brace belongs, multiplied by the secant of the inclination to a vertical line of the brace in question. For instance, if the brace L be divided by a vertical plane, the weights on the right of it, or the shearing stress, are $(1\frac{1}{2}W + 2W')$, and this multiplied by the secant of α' is the stress desired.

The chord stresses are also determined by a direct and simple application of the principles and formulæ of the preceding article.

Stress in f is $\frac{1}{2}W$ *tan* α
"   " e " $1\frac{1}{2}W$ *tan* $\alpha + (W' + \frac{1}{2}W)$ $(tan\ \alpha + tan\ \alpha')$
"   " d " $2\frac{1}{2}W$ *tan* $\alpha + (2W + 3W')$ $(tan\ \alpha + tan\ \alpha')$
"   " c " $3\frac{1}{2}W$ *tan* $\alpha + (4\frac{1}{2}W + 6W')$ $(tan\ \alpha + tan\ \alpha')$
"   " b " $4\frac{1}{2}W$ *tan* $\alpha + (8W + 10W')$ $(tan\ \alpha + tan\ \alpha')$
"   " a " $5\frac{1}{2}W$ *tan* $\alpha + (12\frac{1}{2}W + 15W')$ $(tan\ \alpha + tan\ \alpha')$

(1).

In the lower chord the stresses are as follows :

Stress in g is $W'\ tan\ \alpha' + \frac{1}{2} W\ (tan\ \alpha + tan\ \alpha')$

"   " h " $2 W'\ tan\ \alpha' + (2 W + W')\ (tan\ \alpha + tan\ \alpha')$

"   " k " $3 W'\ tan\ \alpha' + (4\frac{1}{2} W + 3 W')\ (tan\ \alpha + tan\ \alpha')$

"   " l " $4 W'\ tan\ \alpha' + (8 W + 6 W')\ (tan\ \alpha + tan\ \alpha')$ $\Big\}(2).$

"   " m " $5 W'\ tan\ \alpha' + (12\frac{1}{2} W + 10 W')(tan\ \alpha + tan\ \alpha')$

"   " n " $6 W'\ tan\ \alpha' + (18 W + 15 W')\ (tan\ \alpha + tan\ \alpha')$

In determining these quantities, it is to be remembered that the stresses cumulate from the free end to the fixed ; *i.e.*, the stress developed at any panel point is felt throughout those portions of the chords included between that point and the fixed end of the truss.

General formulæ for the Eqs. (1) and (2) may easily be found. Let n be the number of the panel, from the free end, in the chord AP (f is number 1 ; c, 2 ; d, 3 ; etc.), then the formula expressing the results in Eq. (1) is the following :

$$\textit{Stress in any panel} = (n - \tfrac{1}{2}) W\ tan\ \alpha + \left\{ \frac{(n - 1)^2}{2}\ W + \right.$$
$$\left. \frac{n(n - 1)}{2}\ W' \right\} \{tan\ \alpha + tan\ \alpha'\}\quad . \quad . \quad . \quad (3).$$

This expression gives the stress in any panel of AP.

The formula which expresses the results shown in Eq. (2) is the following :

$$\textit{Stress in any panel} = n W'\ tan\ \alpha + \left\{ \frac{n^2}{2}\ W + \right.$$
$$\left. \frac{n(n - 1)}{2} W' \right\} \{tan\ \alpha + tan\ \alpha'\}\quad . \quad . \quad . \quad (4).$$

In which n denotes the number of the panel in the chord BQ starting from the free end ; *i.e.*, g is number 1, h number 2, etc.

The weight at P has been taken at half that applied at other panel points in the same chord. In the case of a swing-bridge, however, it is greater than that, since some of the details of the locking apparatus, etc., are hung from that point. Yet the Equations (1) to (4) may still be used, only a simple term is to be added to each of those equations. Let p be the panel length, d the depth of the truss, and W_1 the actual weight hung from P. Also, let $W_1 - \frac{1}{2}W = w'$. In order to find the additional stress produced in any panel d of the chord AP, let the moment of w be taken about the intersection of H and K in the lower chord; this moment is $w\{(n-1)p + d\,tan\,\alpha\}$. Consequently the additional stress desired is

$$s = w \left\{ (n-1)\frac{p}{d} + tan\,\alpha \right\} \quad . \quad . \quad . \quad . \quad (5).$$

The stress s is to be added to each of equations (1) and (3) if $W_1 > \frac{1}{2}W$, otherwise it is to be subtracted.

In precisely the same manner, the additional stress for the lower chord BQ is

$$s' = wn\frac{p}{d} \quad . \quad . \quad . \quad . \quad . \quad . \quad (6).$$

The stress s' is to be added to equations (2) and (4) if $W_1 > \frac{1}{2}W$, otherwise it is to be subtracted.

If, as in Fig. 1, AP is the upper chord, the stress in QP and all members parallel to it will be compressive; while the stress in QS and all braces parallel to it will be tensile. Likewise the stress in AP is tensile, and that in BQ compressive.

If the truss were turned over so that BQ would become the top chord, the expressions for the stresses in equations (1) to (6) would remain exactly as they are, only the signs of the stresses would change. The condition of stress would be exactly represented in the preceding paragraph by simply changing "compressive" to "tensile," and "tensile" to "compressive."

Art. 5.—Overhanging Truss—Parallel Chords—Uniform Bracing—Vertical and Diagonal Bracing.

The two most frequent cases of Fig. 1, Art. 4, are, first, that in which $\alpha = \alpha'$, and, second, that in which $\alpha' = 0$. The first of these cases is represented in Fig. 1, and the second in Fig. 2.

The web stresses for this case will be precisely the same in general form as those given in Art. 4, but *sec α* will be written for *sec α'*.

Very simple general formulæ can be written for these web stresses. Let n' denote the number of any brace starting from O, which is called 1; then observing the general values in Art. 4, the stress in any brace n' parallel to O will be

$$+ b = \left\{ \frac{n'}{2} W + \frac{(n' - 1)}{2} W' \right\} \sec \alpha + w \sec \alpha. \quad . \quad . \quad (1).$$

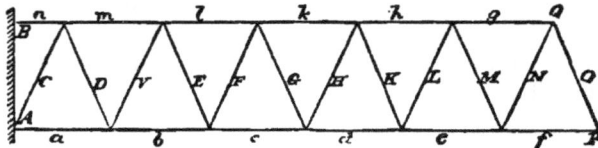

FIG. 1.

The expression $(+ b)$, of course, denotes tensile stress in any brace parallel to O.

In precisely the same manner, the compressive stress in any brace parallel to N (n' possessing the same signification as before; *i. e.*, n' for N is 2; for L, 4, etc.) is

$$- b = \left\{ \frac{(n' - 1)}{2} W + \frac{n'}{2} W' \right\} \sec \alpha + w \sec \alpha \quad . \quad . \quad (2).$$

In determining the chord stresses, it is to be remembered that the weights W rest on the lower chord AP. Making *tan α = tan α'* in Eq. (3) of Art. 4, the stress in any lower-chord panel is

$$C = (n - \tfrac{1}{2}) W \tan \alpha + \left\{ (n - 1)^2 W + n(n - 1) W' \right\} \tan \alpha + s \quad . \quad (3).$$

Making the same change in Eq. (4) of the previous Article, the upper-chord tensile stresses will be found to be

$$T = n W' \tan \alpha + \left\{ n^2 W + n(n - 1) W' \right\} \tan \alpha + s' \quad . \quad . \quad . \quad (4).$$

Some of the results given by the formulæ should always be checked by the method of moments.

Let it be desired to determine the stress in k by the method of moments. Let the origin of moments be taken at the intersection of G and H. The moments which balance each other about that point are that of the stress in k acting with the lever-arm d, the depth of the truss, and those of the weights applied to the truss on the right of the panel point in question ; these latter act against the former. Calling the panel length p, and taking the moments mentioned :

$$Td = 3W' \cdot \tfrac{3}{2}p + 2W \cdot \tfrac{3}{2}p + \tfrac{1}{2}W \cdot 3p + w \cdot 3p.$$

$$\therefore \ T = 4\tfrac{1}{2}W'\frac{p}{d} + 4\tfrac{1}{2}W\frac{p}{d} + 3w\frac{p}{d} \ \cdot \ \cdot \ \cdot \ (5).$$

The result of Eq. (5) ought to be the same as that of Eq. (4). Two or three panels in each chord ought to be treated in the same manner.

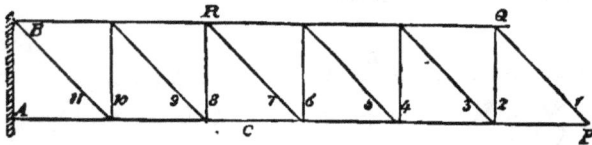

FIG. 2.

The cantilever truss represented in Fig. 2 shows the case in which α' of Fig. 1, Art. 4, is equal to zero. The notation is precisely the same as that used before.

The general expression for the stress in any inclined brace is simply Eq. (1) repeated—that is :

$$+ b = \left\{ \frac{n'}{2}W + \frac{(n'-1)}{2}W' \right\} \sec \alpha + w \sec \alpha \ \cdot \ \cdot \ (6).$$

Making $\sec \alpha' = 1$, there results for the compressive stress in any of the verticals 2, 4, 6, etc. :

$$- b = \left\{ \frac{(n' - 1)}{2} W + \frac{n'}{2} W' \right\} + w \quad . \quad . \quad . \quad . \quad . \quad (7).$$

Making *tan α' = 0*, in Eq. (3) of Art. 4, gives the compressive chord stress in any panel of the lower chord *AP*. Hence, for that chord:

$$C = \left\{ (n - \tfrac{1}{2}) W + \frac{(n - 1)^2}{2} W + \frac{n (n - 1)}{2} W' \right\} tan \, \alpha +$$

$$wn \frac{p}{d} \quad . \quad . \quad . \quad (8).$$

In a similar manner, from Eq. (4) of the previous article, for the tensile stress in any panel of the upper chord, there results the equation:

$$T = \left\{ \frac{n^2}{2} W + \frac{n (n - 1)}{2} W' \right\} tan \, \alpha + wn \frac{p}{d} \quad . \quad . \quad (9).$$

If W_1 should be less than $\frac{1}{2} W$, the term which expresses the additional stress, whether in braces or chords, will be subtractive, as will be indicated by the sign of *w*.

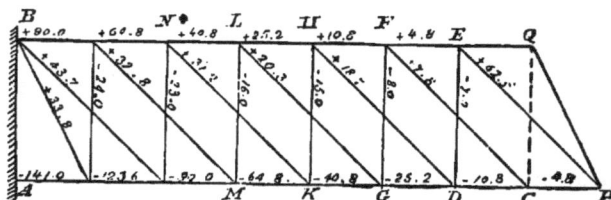

FIG. 3.

Again, applying the moment test to any panel, as *c*, by taking the origin of moments at *R*, the notation remaining the same as before, there results:

$$Cd = 3 (W + W') \cdot 2p + \tfrac{1}{2} W \cdot 4p + w \cdot 4p$$

$$\therefore C = (4 W + 3 W') \frac{2p}{d} + 4w \frac{p}{d} \quad . \quad . \quad . \quad (10).$$

This result ought to agree with that shown by Equation (8), and several panels in each chord should be tested.

In the great majority of cases it is not convenient to apply a general formula, but the numerical values are usually determined directly from the diagram, and the stress in each member written along it, as shown in Fig. 3.

Fig. 3 shows a truss which frequently occurs in the practice of the American engineer; it is in reality one arm of an open swing-bridge.

Let the panel length $= p = 12$ ft., and the depth of the truss $= d = 20$ ft. The tangent of $CQP = 12 \div 20 = 0.6$, and the secant of $CQP = 1.166$. The tangent of $DEP = 1.2$, and the secant of $DEP = 1.562$. The panel loads at E, F, H, etc. $= W' = 3.00$ tons; at C, D, G, K, etc., $W = 5.00$ tons, at P, $W_1 = 4.00$ tons. No load is taken at Q.

In the figure there are two systems of right-angled triangulation; P, D, E, H, K, etc., is one system, and C, F, G, L, M, etc., is the other. This does not, however, complicate the matter in the least, for *each system of triangulation is regarded as an individual truss carrying its own weights only.* Calculations are therefore made for each system of triangulation as if they were independent trusses, and then the two are added.

The weight of the portion EQP is supposed divided between E and P, thus showing $W_1 > \frac{1}{2}W$. The vertical braces are evidently in compression, while the inclined ones are in tension.

The figures in the diagram denote tons (2000 lbs.) of stress; + indicates tension, while − indicates compression.

$$
\begin{aligned}
\text{Stress in } PE &= W_1 \sec DEP = 4 \times 1.562 = 6.248 \text{ tons.} \\
\text{``} \quad \text{``} \quad CF &= W \sec DEP = 5 \times 1.562 = 7.810 \text{ ``} \\
\text{``} \quad \text{``} \quad DH &= (W_1 + W' + W) \times 1.562 = 18.744 \text{ ``} \\
\text{``} \quad \text{``} \quad GL &= 2W + W' \times 1.562 = 20.306 \text{ ``}
\end{aligned}
$$

The other brace or web stresses are found in precisely the same manner.

Stress in $CP = W_1 \tan DEP$ $= 4 \times 1.2$
                                                        $= 4.8$ tons.
"   " $DC = W \tan DEP + 4.8$ $= 6.0 + 4.8$
                                                        $= 10.8$ tons.
"   " $GD = (W_1 + W' + W) \times 1.2 + 10.8 = 14.4 + 10.8$
                                                        $= 25.2$ tons.
"   " $KG = (W + W' + W) \times 1.2 + 25.2$ $= 40.8$ tons.

Other lower-chord stresses are found in exactly the same manner.

By an inspection of the diagram it is seen that the general expression for the stress in FE is precisely the same as that for CP; the same can be said of HF in reference to DC; LH in reference to DG; etc. The explanation of this is simple. If the truss be divided by a plane normal to the paper and parallel to the inclined braces, only vertical and horizontal members will be cut. But the truss is in equilibrium, and since the loading is wholly vertical, the sum of the horizontal stresses must be zero; or, the stress in the lower-chord panel cut must be equal and opposite to the stress in the upper-chord panel cut.

Let the moment test be applied to the stress in the panel MK. The origin of moments for the loads applied to the system $PEDH$, etc., is N, and the origin for the other system is L. Taking moments about those points:

$$C'd + C''d = 16 \times 36 + 4 \times 72 + 8 \times 24 + 5 \times 48$$

$$\therefore C = C' + C'' = 64.8 \text{ tons.}$$

Again, for the lower-chord panel adjacent to A, B is the moment origin for the whole load.

$$C = (48 \times 42 + 5 \times 84 + 4 \times 96) \div 20 = 141 \text{ tons.}$$

Thus the numerical results are verified.

According to one of the principles of Art. 3, the horizontal component of the stress in any inclined web member ought to be equal to the increment of chord stress at either of its extremities, and such will be found to be the case. If, for example, 20.3 be multiplied by the cosine of the angle *GLH*, the result will be 15.6 = 40.8 — 25.2.

This last is a verification of the web stresses, and both methods of checking are perfectly general and may be applied to all trusses, as should be done in actual cases.

2

CHAPTER II.

SPECIAL NON-CONTINUOUS TRUSSES WITH PARALLEL CHORDS.

Art. 6.—Distribution of Fixed and Moving Loads.

THE trusses treated heretofore have been of rather an elementary character, and general principles have been considered instead of special and practical applications. Before taking up the technical treatment of trusses it will be necessary to consider some preliminary matters.

The total load on a bridge-truss always consists of two parts, the *fixed* load and the *moving* load. The fixed load consists of the entire weight of the bridge, including tracks, flooring, etc. The moving load, as its name indicates, consists of that load (whether single or continuous) which moves over the bridge.

The truss is, of course, always subjected to the action of the fixed load.

If the truss is of uniform depth the panel-fixed loads will be uniform in amount for one chord; but if the depth is variable it may be necessary to make a varying distribution of the weight of the trusses and lateral bracing. The amount and rate of this variation can only be determined by the circumstances of each particular case.

The moving load on a railway bridge may be taken as continuous or as a series of single weights as actually applied at the wheels under the locomotives and cars. The assumption of continuity of moving load was formerly always made, a larger amount per lineal foot being taken to represent the extra locomotive weight. In such a case, if the moving load extends from the end of the bridge to the centre of any panel, or to the end of that panel, the panel point immediately in front of the train will not sustain a full panel load; but if it be assumed that this panel point does sustain the *full* load, then a small error on the side of safety will be committed. Such an assumption was formerly made, and the consequent method of computation will be given in some of the Arts. which follow in this chapter.

At the present time (1885), however, the demands of the best practice require the moving load to be taken at the actual points of application of locomotive and car wheels. This method of computation will be given in several of the first cases taken.

If the span is short, or less than 125 feet, the moving load should be taken entirely of locomotives, as two will nearly cover the structure. The amount and character of the moving load, however, is usually indicated by specifications.

The moving load of a bridge may pass along the upper chord or the lower chord. In the first case the bridge is called a "*deck*" bridge, and in the latter case a "*through*" bridge. The methods employed in the determination of stresses in the various truss members are exactly the same in both cases.

"Pony" trusses are through trusses not sufficiently high or deep to need overhead cross-bracing.

Every truss-bridge is composed of the following parts:

Upper and lower chords,

Upper sway-bracing,

Web members,

Floor system, including beams, stringers, ties, floor-hangers, lower sway-bracing, and rails.

The sum of the weights of the parts is the "fixed" load of the bridge.

In the case of highway bridges the calculations are precisely the same as for railway bridges, except that the moving load is assumed to be uniform per lineal foot of bridge. The greatest load that can ordinarily pass on a bridge is a dense crowd of people, the greatest weight of which can be taken at eighty-five pounds per square foot. The late Mr.

Hatfield, of New York City, found by experiment that it was scarcely possible to exceed seventy pounds per square foot. The moving panel load of a highway bridge may then be found by multiplying the width of the clear way, including sidewalks, by the product of the panel length with the load per square foot.

If the span is not over 125 feet, or about that value, the moving load for the truss members may be taken at eighty-five pounds per square foot, or sixty pounds for greater lengths. In all cases, however, the floor beams and joists should be designed for a moving load of 100 pounds per square foot, in order to provide for the increased fatigue of those members due to shocks and sudden application of loads.

In some cases highway bridges are subjected to enormous concentrated loads of a special character. Such loads can only be known from local considerations, and the bridges must be built with a view to sustaining such special weights.

All the methods or principles used, then, in the following cases, which will be those of railway bridges, are equally applicable to highway structures, and no further special attention will be given to the latter.

Art. 7.—Position of Moving Load for Greatest Shear and Greatest Bending.

That method of computation which treats the moving load as composed of a system of isolated weights requires some simple method of finding the greatest possible shear in a given panel for a given system of loading. Among the first to use such a method was Theodore Cooper, C. E.; and the results of the following investigation are the same as those determined by him.

The method and formula first developed apply to *any* single system of triangulation so far as the web stresses are concerned, but for the chord stresses they only apply to such a system when composed of alternately vertical and inclined members. Subsequent modifications for web members all inclined will be made for the chord stresses.

CASE I.

Let the moving load consist of the advancing weights W_1, W_2, W_3, ... W_n separated by the distances a, b, c, d, etc.; then let the weights W_1, W_2, ... Wn' be

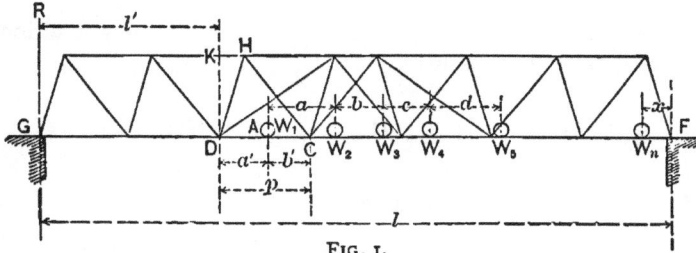

FIG. 1.

found in the panel DC, in which it is desired to find the greatest possible shear, and at the distances b', b'', . . . from C; it being understood that the moving load advances from F toward G. The last load W_n is found at the distance x from F. The length of span FG is l, while the length of panel DC is p. With the assumed position of loading, the reaction R at G will be:

$$R = \begin{cases} W_1 \dfrac{a + b + c + \ . \ . \ . \ . \ . \ + x}{l} \\[2mm] + W_2 \dfrac{b + c + d + \ . \ . \ . \ . \ . \ + x}{l} \\[2mm] + W_3 \dfrac{c + d + \ \ \ . \ . \ . \ . \ . \ + x}{l} \quad . \ . \ . \ (1) \\[2mm] . \ . \ . \ . \ . \ . \ . \ . \ . \ . \ . \ . \\[2mm] + W_n \dfrac{x}{l} \end{cases}$$

The parts of the weights W_1, W_2, ... resting on

DC, which pass to D, are $W_1, \dfrac{b'}{p}$, $W_2 \dfrac{b''}{p}$, ... Hence the shear in the panel DC will be

$$S = R - (W_1 \frac{b'}{p} + W_2 \frac{b''}{p} + \ . \ . \ . \ . \ . \ ) \ . \ . \ . \ . \ (2)$$

If the train advance by the amount $\triangle x$, the new reaction R', at G, takes the value:

$$R' = R + (W_1 + W_2 + W_3 + \ . \ . \ . \ . + W_n) \frac{\triangle x}{l} ; \ . \ (3)$$

and the new shear will become:

$$S' = R' - (W_1 \frac{b'}{b} + W_2 \frac{b''}{p} + \ . \ . \ . \ ) - (W_1 + W_2 + \ . \ . \ . \ ) \frac{\triangle x}{p}. \ (4)$$

$$\therefore \quad S' = S + (W_1 + W_2 + W_3 + \ . \ . \ . \ + W_n) \frac{\triangle x}{l}$$

$$- (W_1 + W_2 + \ . \ . \ . \ ) \frac{\triangle x}{p}. \ . \ . \ . \ . \ . \ . \ . \ (5)$$

Or, $S' - S = \dfrac{\triangle x}{l} \Big\{ (W_1 + W_2 + W_3 + \ . \ . \ . \ + W_n)$

$$- (W_1 + W_2 \ . \ . \ . \ ) (\frac{l}{p} = n) \Big\} \ . \ . \ . \ . \ . \ . \ (6)$$

Whenever $S' - S$ becomes equal to zero, S' will be either a maximum or a minimum. If the difference is positive just before it becomes equal to zero, S' will be a maximum, and that is the only case of interest in the present connection. Hence, by placing $S - S$ equal to zero, the following condition is obtained:

$$n(W_1 + W_2 + \ . \ . \ . \ ) = W_1 + W_2 + W_3 + \ . \ . \ . \ + W_n \ . \ . \ (7)$$

The shear in the panel in question will therefore take its greatest value *when n times the moving load which it contains is equal to, or most nearly equal to, the entire moving load on the bridge.*

That equality will seldom or never exist unless one of the weights W is placed on a panel point, since $W_1 + W_2 + \ldots$ is seldom or never an exact divisor of the entire load on the bridge. If a weight rests on a panel point, any part of such a weight may be taken as acting in one adjacent panel and the remainder in the other; the desired equality may thus be obtained.

In case Eq. (7) should hold, the position of the moving load is a matter of indifference so long as the panel in question contains the same number of loads $W_1 + W_2 + \ldots$, as there is no trace of b', b'', etc., in that equation. A load may then always be taken as resting at the rear extremity of the panel where the greatest shear in it exists.

These considerations frequently essentially simplify computations.

When the value of x has been found for the position of the greatest shear, the latter being determined by the preceding method, Eq. (2) may be put in the following convenient shape by the aid of Eq. (1):

$$S = \frac{1}{l} \left[W_1\, a + (W_1 + W_2)\, b + (W_1 + W_2 + W_3)\, c + \ \ldots \right.$$

$$+ (W_1 + W_2 + \ldots W_n)\, x] - \frac{1}{p} [W_1\, a + (W_1 + W_2)\, b$$

$$+ \ \ldots \ + (W_1 + W_2 + \ldots + W_{n'-1})\, ?] \ . \ (8)$$

The sign (?) stands for the distance between the wheel concentrations $W_{n'-1}$ and $W_{n'}$, since the latter rests directly at the panel point in question.

It is thus seen that all the parts of Eq. (8) may be taken at once from tables, except that term involving x.

CASE II.

In the preceding case it has been supposed that for the greatest shear in DC, the front weight W_1 is found between D and C; but let $W_1 + W_2 +$ etc., be supposed between D and G.

With the notation remaining the same as before, the shear S will become:

$$S = R - (W_1 + W_2 + \text{etc.}) - (W_3\frac{b'}{p} + W_4\frac{b''}{p} + \quad . \quad . \quad . \quad) ; (9).$$

while S' takes the value:

$$S' = S + (W_1 + W_2 + W_3 + \quad . \quad . \quad . + W_n)\frac{\Delta x}{l}$$
$$- (W_3 + W_4 + \quad . \quad . \quad . \quad)\frac{\Delta x}{p}.$$

Hence for a maximum, the following expression must never become negative:

$$S' - S = \frac{\Delta x}{l}\left\{ (W_1 + W_2 + W_3 + \quad . \quad . \quad . \quad + W_n)\right.$$
$$\left. - (W_3 + W_4 + \quad . \quad . \quad . \quad)\left(\frac{l}{p} = n\right)\right\} = o. \quad . \quad . (10).$$

But Eq. (10) is identical with Eq. (7). Hence, the same conditions for a maximum obtain wherever may be the head of the moving load. The second member of Eq. (8), however, must contain the negative sum of all the weights between D and G.

EXAMPLE.

If each one of the weights W_1, W_2, etc., is equal to any other, *i.e.*, if they are all uniform, and if $a = b = c = d = \quad . \quad . \quad . = p$, Eq. (7) shows that the front weight W_1 must be taken at the first extremity of the panel in question. The same result holds if the first weight W_1 is not exceeded in amount by any that follows it, provided that a, b, c, etc., still equal p.

In cases where the same system of concentrated loads is to be used for a number of spans, it will be shown that the tabulation of the products of the sums of the weights W_1, W_2, etc., by the distances a, b, c, etc., can be advantageously used

to shorten and simplify computation. In other cases, however, the quickest and simplest method is partly graphical;
it is as follows:

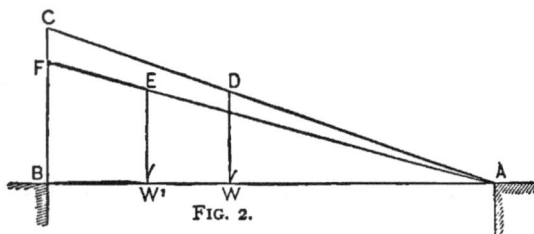

FIG. 2.

In Fig. 2 let AB be the length of span, and W any weight
resting anywhere in the span. Erect a vertical at B and let
BC represent W by any convenient scale; then draw the
straight line AC. The vertical intercept WD will represent
the reaction at B due to W, by the
same scale on which BC represents
that weight.

In the same manner if BF represents W^1, then W^1E will represent
the reaction at B due to W^1. Thus
there must be as many verticals
BF, BC, etc., as there are different
weights resting in the span, and the

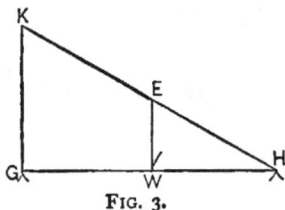

FIG. 3.

total reaction at B, for any given position of the moving
load will be the sum of the vertical intercepts WD, W^1E,
etc., erected at each load W for that position.

The negative shears $W_1\dfrac{b'}{p}$, $W_2\dfrac{b''}{p}$, etc., appearing in Eq. (2)
are most readily found in the same manner. If GH, Fig. 3,
is a panel length and W any weight represented by the vertical line GK, drawn to any convenient scale, while WH is
equal to b', b'', etc., then the vertical intercept WE between
GH and KH will represent $W\dfrac{b'}{p}$, $W\dfrac{b''}{p}$, etc., $i. e.$, the reaction
at G due to W.

If the reaction at B, Fig. 2, is then given by ΣWD, and $W_1\dfrac{b'}{p} + W_2\dfrac{b''}{p} +$ etc. $= \Sigma WE$, Fig. 3, the shear (see Eq. (2)) will be:

$$S = \Sigma WD - \Sigma WE.$$

If Figs. 2 and 3 are drawn on profile, or cross-section paper, the shears for any span can be found with great ease and rapidity.

Position of Moving Load for Greatest Bending Moment.

The Fig. and notation of the preceding cases will be used in connection with this. Moments will be taken about the panel point C, horizontally distant l' from G. There will be supposed to be n' weights in front of C (*i. e.*, between C and G), and the weight $W_{n'}$ will be taken at the distance x' from C towards G. The bending moment M will then take the value:

$$M = Rl' - \begin{cases} W_1\,(a + b + c + \quad . \quad . \quad . \quad + x') \\ + W_2\,(\qquad b + c + \quad . \quad . \quad . \quad + x') \\ . \quad . \quad . \quad . \quad . \quad . \quad . \quad . \\ + W_{n'}x'. \end{cases}$$

Or, after taking the value of R from the preceding cases:

$$M = \frac{l'}{l}\,[\,W_1 a + (W_1 + W_2)\,b + (W_1 + W_2 + W_3)\,c + \quad . \quad . \quad .$$
$$+ (W_1 + W_2 + W_3 + \quad . \quad . \quad . \quad + W_n)\,x\,] - W_1 a - (W_1 + W_2)\,b$$
$$- (W_1 + W_2 + W_3)\,c - \quad . \quad . \quad . \quad - (W_1 + W_2 + W_3 + \quad . \quad . \quad .$$
$$+ W_{n'})\,x' \quad . \quad . \quad . \quad . \quad . \quad . \quad . \quad . \quad . \quad . \quad . \quad . \quad (11).$$

If the train advances by the amount Δx, the moment becomes:

$$M' = M + \frac{l'}{l}(W_1 + W_2 + W_3 + \quad . \quad . \quad . \quad + W_n)\,\Delta x - (W_1 + W_2$$
$$+ \quad . \quad . \quad . \quad + W_{n'})\,\Delta x, \quad . \quad . \quad . \quad . \quad . \quad . \quad (12).$$

Hence, for a maximum, the following value must never be negative:

$$M' - M = \triangle \; x \left\{ \frac{l'}{l}(W_1 + W_2 + W_3 + \; \ldots \; + W_n) - (W_1 + W_2 \right.$$
$$+ \; \ldots \; \left. + W_{n'}) \right\} = 0. \; \ldots \ldots \ldots \; (13).$$

Or, the desired condition for a maximum takes the form:

$$\frac{l'}{l} = \frac{W_1 + W_2 + \; \ldots \; + W_{n'}}{W_1 + W_2 + W_3 + \; \ldots \; + W_n} \; \cdot \; (14).$$

It will seldom or never occur that this ratio will exactly exist if $W_{n'}$ is supposed to be a *whole* weight; hence, $W_{n'}$ will usually be that part of a whole weight at C which is necessary to be taken in order that the equality (14) may hold.

It is to be observed that if the moving load is very irregular, so that there is great and arbitrary diversity among the weights W, there may be a number of positions of the train which will fulfil Eq. (14), some one of which will give a value greater than any other; this is the absolute maximum desired.

Since $W_{n'}$ will always rest at a panel point for the greatest bending moment, x' in Eq. (11) may always be put equal to zero when that equation expresses the greatest value of the moment. The latter then becomes:

$$M = \frac{l'}{l} \left[W_1 a + (W_1 + W_2) \, b + \; \ldots \; + (W_1 + W_2 + \; \ldots \right.$$
$$\left. + W_n) \, x \right] - W_1 a - (W_1 + W_2) \, b - \; \ldots \; - (W_1 + W_2$$
$$+ \; \ldots \; + W_{n'-1}) ? \; \ldots \ldots \ldots \; (15).$$

In this equation, of course x corresponds to the position of maximum bending, while the sign (?) represents the distance between the wheel concentrations $W_{n'-1}$ and $W_{n'}$.

It is known that for any given condition of loading the greatest bending moment in the beam or truss, will occur at that section for which the shear is zero. But if the shear is zero at that section, the reaction R must be equal to the sum of the weights $(W_1 + W_2 + W_3 + \; \ldots \; + W_{n'})$ between G and C; the latter now being that section at which the greatest

moment in the span exists.   Hence for that section Eq. (14).
will take the form :

$$\frac{l'}{l} = \frac{R}{W_1 + W_2 + W_3 + \quad . \quad . \quad . \quad + W_n};$$

or, the centre of gravity of the load is at the same distance
from one end of the truss as the section or point of greatest
bending is from the other.   In other words, *the distance between
the point of greatest bending for any given system of loading,
and the centre of gravity of that loading is bisected by the centre of span.*

If the load is uniform, therefore, it must cover the whole
span.

It will be observed that Eq. (15) is composed of the sums
of W_1, $W_1 + W_2$, etc., multiplied by the distances a, b, c, etc.,
precisely as in Eq. (8), hence the same tabulation as there
indicated may be used to advantage.

Limitations of the preceding methods.

The preceding methods are limited to a single system of
triangulation.   By the use of certain assumptions in reference to the distribution of the loading between two or more
systems of triangulations in the same truss, a somewhat similar investigation might be made for such cases, but such
analysis would not be rigorously exact.   Hence it is as well
to pursue the usual method and assume that each system
acts as an independent truss, then place the moving load in
such a position for each system that the front panel load for
that system will be the greatest possible.   This panel concentration will then be the forward panel-moving load, and
the succeeding ones may either be those concentrations
which actually correspond to the forward one, or may be
supposed to be composed of a uniform load equivalent to
the concentrated one.   The latter plan will be employed
hereafter.

Application of the preceding method to an all-inclined web system.

As was observed at the beginning of this Art. the analysis for chord stresses, as already given, is directly applicable

to a single system of triangulation in which a vertical web member is found in each panel. The general demonstration, however, is easy.

If the web members are *all* inclined, the formulæ, as already given, are directly applicable in any case to the determination of stress in that chord which does *not* carry the moving load, since moments are taken about the panel points of the chord traversed by the moving load.

But let it be required to determine the position of the moving load for the maximum stress in *DC*, Fig. 1, and the expression for the corresponding moment. As before, let the load move from *F* toward *G*. Let q represent the horizontal distance of *D* from *H*, i. e., $q = KH$; evidently q is constant for the same span. Let x_1 represent the *horizontal* distance from *H* of the first load to the left of *D*, and let that load be represented by W^1_n. That portion of the loads resting in *DC*, which is transferred to *D*, is $\Sigma W \frac{b'}{p}$. l' will now represent $GD + q$.

By taking moments about *H*, Eq. (11) will take the form:

$$M = \frac{l'}{l} [W_1 a + (W_1 + W_2) b + \ldots + (W_1 + W_2 + \ldots + W_n) x]$$
$$- W_1 a - (W_1 + W_2) b - \ldots - (W_1 + W_2 + \ldots + W^1_n) x_1$$
$$- \frac{q}{p} \Sigma W b' \quad . \quad . \quad . \quad . \quad . \quad . \quad . \quad . \quad . \quad (16).$$

By advancing the train $\triangle x$, since $\triangle x = \triangle x_1 = \triangle b'$, Eq. (13) will become:

$$M - M = \triangle x \left\{ \frac{l'}{l} (W_1 + W_2 + \ldots + W_n) - (W_1 + W_2 + \ldots + W^1_n) \right.$$
$$\left. - \frac{q}{p} \Sigma W \right\} = 0. \quad . \quad . \quad . \quad . \quad . \quad . \quad (17).$$

The condition for a maximum or minimum then takes the shape:

$$\frac{l'}{l} = \frac{W^1_1 + W_2 + \ldots + W^1_n + \frac{q}{p} \Sigma W}{W_1 + W_2 + W_3 + \ldots + W_n} \cdot \quad . \quad . \quad . \quad . \quad (18).$$

Eqs. (16) and (18) are the general expressions of which Eqs. (15) and (14) are special forms.

After x and, hence, x_1 have been determined by the aid of Eq. (18), the maximum moment will be given by Eq. (16), in which the tabulations already indicated can be advantageously employed.

Application of preceding methods to a system of concentrations followed by a uniform load.

If the uniform load does *not* reach to the panel under consideration, which is usually the case, W_n in Eqs. (7), (14) and (18) represents the total uniform load on the bridge, but the formulæ are in no wise changed. In Eqs. (8), (15) and (16), however, it is to be observed that while W_n again represents the total uniform load, x will represent the distance from its centre of gravity to the end of the span (*i. e.*, half the length covered by the uniform load), also that the distance between W_n and W_{n-1} will be equal to x plus the space which separates W_{n-1} from the front of the uniform load.

In the case of the existence of this uniform load it will happen that $W_{n'}$ will not rest at a panel point. The last term in the negative expressions of the second members of Eqs. (8) and (15) will then be $(W_1 + W_2 + \ . \ . \ . \ + W_{n'})\, x'$; x' being the distance of $W_{n'}$ in front of the panel point C. Eq. (16) is general, and needs no change on this account.

If the concentrations are so few that the uniform load extends over a portion of the panel in question, the observations made above still hold. But in addition to them, $W_{n'}$ or W^1_n will represent the amount of uniform load in the panel, and x' or x_1 will represent the distance from its centre of gravity to the panel point. The interval or space between $W_{n'-1}$ or W^1_{n-1} and $W_{n'}$ or W^1_n will then be the distance from either of the former to the centre of gravity of the uniform load.

Finally, in Eq. (17) or (18) ΣW will be either wholly or partly composed of uniform load.

Modifications for Skew Spans.

If a skew bridge is under treatment, the preceding methods

apply in all respects, so far as the general principles are concerned.

It will be sufficiently accurate in all cases to treat the

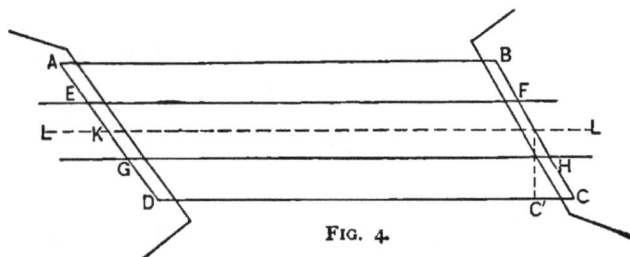

FIG. 4.

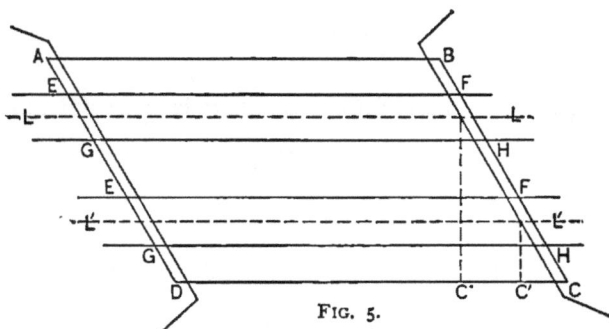

FIG. 5.

moving load as if it were passing along the centre line LL $L'L'$ of each track.

In the case of the single track bridge, Fig. 4, if the load is passing from right to left, the moving load does not rest on the truss CD until it passes the point C', and continues to act on that truss *until it passes to the same distance to the left of D*, it being borne in mind that all moving load is transferred to the trusses by transverse floor beams placed normal to the axis of the bridge. It results from these considerations that if the load passes from right to left in Fig. 4 and along LL, the reactions at D will be greater than a half of those at K by the amount of the half products of the total load corresponding to those reactions by the ratio

$$\frac{CC'}{CD}$$

Hence, if R is any reaction at K and ΣW the total load, the corresponding reaction at D will be:

$$\frac{R}{2} + \tfrac{1}{2}\frac{CC'}{CD}\Sigma W.$$

On the other hand, with the load moving in the same direction, the reaction at A will be:

$$\frac{R}{2} - \tfrac{1}{2}\frac{CC'}{CD}\Sigma W.$$

In Eqs. (8), (15), and (16), then, there is to be written for

$$(W_1 + W_2 + \quad . \quad . \quad . \quad + W_n)\,x$$

the expression,

$$(W_1 + W_2 + \quad . \quad . \quad . \quad + W_n)\,(x \pm CC'),$$

according to the direction in which the train is moving; *but the negative portions are to remain unchanged.* It is to be remembered that the quantity x is to be measured on the centre line.

In the case of the double track skew bridge of Fig. 5, in which there are the two trusses AB and CD only, precisely the same observations hold. For one track, however, CC'' is to be used and CC' for the other. Separate computations are to be made for each track for each truss.

If there are three trusses in Fig. 5, each pair of trusses constitutes a single track bridge for the track between, and is to be treated precisely as Fig. 4.

If the skew is so great that one or more floor beams have their end or ends resting on the masonry, obvious modifica-

tions must be made according to the preceding general principles.

The Graphical Method.

With convenient means for constructing an accurate equilibrium polygon with a large number of loads, this method is
a very rapid one for either shears or moments. A perfect
familiarity with the principles and operations of Art. 45 is
here supposed.

Let the *entire* moving load for a given truss be represented
by the system of forces 1, 2, 3, 4, and 5, in Fig. 6. They are
given in actual position under the polygon *PKO*. *P* is the

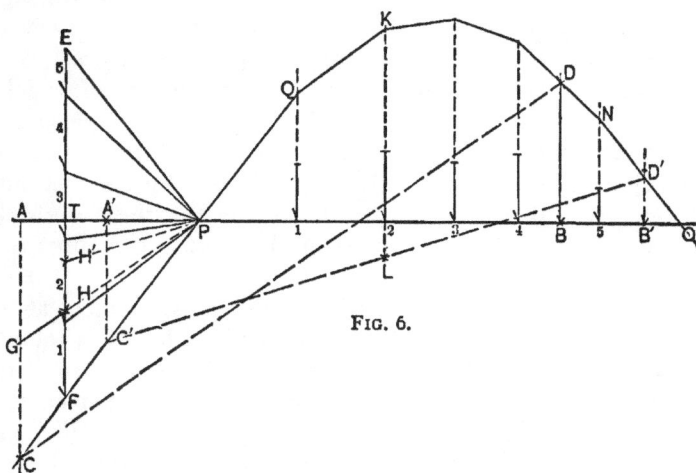

FIG. 6.

pole, *EF* the load line, and *PO* an indefinite horizontal line
normal to *EF*. As usual, the moving load is supposed to
move from *O* toward *P*. The polygon *PQK* . . . . *NO* is
formed in the ordinary manner by making its sides parallel
to the lines radiating from *P*. For reasons that will presently be evident, *QF* should be continued considerably
beyond *C*, while *NO* should be carried somewhat below
the horizontal line through *P*.

After the constructions indicated have been made to any

convenient scale, let the span under consideration be laid off to the same scale by which the horizontal separations between the loads, 1, 2, 3 and 4, etc., are laid down, and let indefinite vertical lines be drawn through the panel points, but let all this latter construction be made on tracing cloth. In order to avoid confusion, neither the panel points nor the vertical lines through them will be shown in the Fig.

In the first place, let the position of the moving load for the maximum shear in some web member be determined by Eq. (7), and in this position let the loads, 1, 2, 3 and 4, be supposed to rest on the truss ; and let $B4$ represent the distance between the last concentration and the right hand of the span (*i. e.*, $B4$ is x of Eq. (8)). Now let the tracing cloth be superimposed on the equilibrium polygon in such a manner that AB shall represent the span ; then erect the verticals BD and AC, and draw CD, to the latter of which PH is drawn parallel. FH will then represent the reaction at the left end of the span, *i. e.*, at A. If from this reaction the negative shear shown in Eq. (8), or found by the method of Fig. 3 be subtracted, the result will be the shear desired in the web member under consideration. In this manner all the maximum shears may be found.

Again, let it be required to find the greatest moment at a given panel point, for which the moving load has been found by Eqs. (14) or (18), to occupy such a position that the distance from the right end of the span to the last concentration (*i. e.*, x in Eqs. (15) and (16)) is represented by $B'5$ in Fig. 6. Also with the position of moving load thus determined, let it be supposed that the load 2 rests at the panel point considered. Now, let the tracing cloth be so superimposed on the equilibrium polygon that $A'B'$ will represent the span, then erect the vertical lines $B'D'$ and $A'C'$, and join $C'D'$. Since load 2 was found at the panel point, at which the moment is to be determined, KL will represent the maximum moment in question. Each linear unit in KL, measured by the same scale to which the span and distances, 1–2, 2–3, etc., are laid down, will represent as many moment units as there are force units in the pole distance PT. If PT is in pounds and *KL* in feet, the *PT* × *KL* will be the moment desired in foot-pounds. In the same manner all maximum moments may be determined. In every position of moving load required, the vertical intercept between the closing line and equilibrium polygon, drawn through the panel point considered, will represent the maximum moment at that point. *PH'* drawn parallel to *C'D'* will give *FH'* as the reaction at *P*, though it is of no special value in this connection. The application of this method to the different panels of a truss will give all the greatest chord stresses.

This method is, of course, subject to all the modifications that have been outlined for the various special cases and conditions. It cannot, however, be applied to the moving load on skew bridges. The reaction for the centre line of the track must be reduced to the trusses, while the negative moments of the loads in advance of the panel point which serves as the moment origin remain unchanged. The generality of this method is not, therefore, complete.

The Maximum Floor-beam Reaction.

The moving load is carried to each transverse floor beam by the adjacent stringers. Hence each floor beam is a pier for two adjacent spans of stringers, and it becomes necessary to determine that position of the moving load on those two spans which will subject the floor beam to its greatest load.

In the Fig. let a section of the beam be shown at *R*, while *l* and *l'* are the two adjacent stringer spans traversed by the moving load; then let the *x*'s be measured from the right and left ends of *l* and *l'*, while *W*, *W¹*, etc., *W₁*, *W₂*, etc., represent the weights or wheel concentrations resting in the two spans. the reaction *R* will then have the value:

$$R = \frac{W_1 x^1 + W_2 x_2 + \text{etc.}}{l_1} + \frac{W x + W^1 x^1 + \text{etc.}}{l} \quad . \quad (19).$$

If the whole system of loading move to the left by the distance $\triangle x$, the new reaction will be:

$$R^1 = R - \frac{(W_1 + W_2 + \text{etc.})\ \Delta x}{l_1} + \frac{(W + W^1 + \text{etc.})\ \Delta x}{l}.$$

In that position which gives a maximum or minimum, $R_1 - R = 0$; hence :

$$(W_1 + W_2 + W_3 + \text{etc.})\ \frac{l}{l_1} = (W + W^1 + W'' + \text{etc.})\ (20).$$

It will seldom happen that Eq. (20) will be satisfied unless

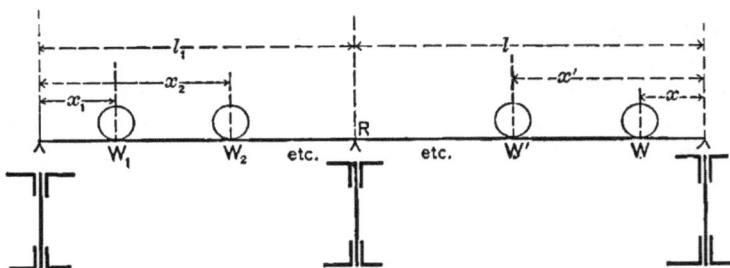

FIG. 7.

a concentration rest on the point R, so that the proper portion of it may be taken for one span or the other, precisely as in the problems of maximum shear and maximum moments.

Ordinarily the two adjacent spans are equal, or

$$l = l_1 . \therefore \qquad W_1 + W_2 + \text{etc.} = W + W' + \text{etc.} . . . (21).$$

Eq. (21) shows that when the two spans are equal, the amounts of load each side of R must also be equal.

After the proper position of loading has been determined, Eq. (19) will give the maximum reaction desired.

Article 8.—Fixed Weight.

As the weight of a structure forms a very considerable portion of its total load, it becomes a matter of importance to find at least an approximate value for it, when the moving load has been once assumed.

The weight of ties, guard timbers or rails, rails, spikes, etc., may be taken at 350 to 475 pounds per lineal foot of single track, and forms an invariable part of the fixed weight, (*i. e.*, weight of structure), since it is independent of span, length of panel, or depth of truss. With ordinarily heavy traffic and standard gauge, 400 pounds per lineal foot is usually taken.

After having fixed the weight of track, the stringers are to be designed. A very little experience will enable their weight per lineal foot to be assigned in advance, so that the total load resting upon them may be used in making computations. These track stringers are almost invariably plate girders, and the main object of the computation is to determine the area of flange section. If that area, as computed, makes the weight of the stringer very different from that assumed, it will be necessary to again assume a weight, guided by the results of the first computation, and re-calculate for the flange area ; and then repeat the operation until sufficient accuracy is obtained.

The weight of the track stringer thus obtained becomes a part of the load sustained by the floor beam. The weight per lineal foot of the latter is then to be assumed, and calculations made and repeated, if necessary, precisely as in the case of the stringers. With a very little practice the weights of the stringers and floor beams may be so accurately assigned in advance, that a re-calculation is seldom or never necessary.

The lateral and transverse systems of bracing should next be subject to computation, and as the wind pressure is their only load, their own weight does not affect the operations; hence no re-calculation will ever be required.

The remaining calculations are those of the truss proper, and the accurate assignment of their own weights presents more difficulty than any other part of the operation. Experience, however, enables even this weight to be quite closely taken in advance. For each single track of standard gauge the weight of two through trusses, designed for the moving load taken in Art. 9, together with the lateral system, in pounds per lineal foot, may be approximately taken at five times the length of the span in feet, or double that amount for a two truss double track structure. If this is not sufficiently close, one or more re-calculations will be required. For spans over two hundred and fifty feet in length, this rule gives too small results.

By thus designing the floor system, and lateral and transverse bracing before the computations are made for the trusses, it is only necessary to assign before each step, the weight of that part immediately under consideration, and, hence, enables the actual weight to agree very closely with the assumed, without re-calculation.

If the bridge is of the deck variety and carries the ties directly on the upper chord, so that they act as a transverse load on the latter, thus obviating a system of track stringers and floor beams, the total fixed weight will be reduced about 125 pounds per lineal foot. If, on the other hand, such a bridge carries a regular system of stringers and floor beams, the truss weight may be the same as if it were a through bridge.

If the span is less than about eighty or more than about two hundred and fifty feet, the weight per lineal foot will somewhat exceed the value given by the preceding rule.

If the bridge is a highway structure, the same general method of operations (and taken in the same order) is to be followed as for a railway bridge. A rule for truss weights cannot, however, be so easily given, because the moving load is so very variable ; they may equal or exceed those of railway structures of the same span, or may not exceed a third of that value. The weight of a highway floor, if of plank and timber joists, will be from twenty to twenty-five pounds per square foot.

Art. 9.—Single System of Bracing with Two Inclinations.

The first case taken will be that shown in Fig. 2 of Pl. I. The span s is 120 feet ; depth d 20 feet ; panel length p 20 feet; angle NAR, 18° 30′, and angle NAM, 33° 40′. The moving load will be taken as two coupled consolidation loco-motives, each with weights distributed as shown in Fig. 1, Art. 77, followed by a uniform load of 1.5 tons per lineal foot. The bridge will be supposed to be a single track "through" structure. Each truss will be taken to weigh approximately 240 pounds per lineal foot. The upper lateral bracing will be taken at fifteen pounds per lineal foot for each truss. The total weight concentrated in each of the upper chord panel points will then be 10 × 270 = 2,700 pounds. The fixed weight concentrated in the lower panel points will be taken at 9,300 pounds. Hence, if W' is the upper chord fixed panel load and W the same for the lower:

$$W' = 2700 \text{ pounds} = 1.35 \text{ tons, for one truss.}$$
$$W = 9300 \text{ " } = 4.65 \text{ " " " "}$$
$$tan\, NAR = 0.333 \qquad\qquad tan\, NAM = 0.666$$
$$sec \text{ " } = 1.054 \qquad\qquad sec \text{ " } = 1.202$$

In the above fixed weight, the ties, rails, guard timbers, etc., were taken at 400 pounds per lineal foot.

The stresses due to the fixed load only, in each of the truss members will first be found, and it will be convenient to begin by determining those in the web members.

On page 7 it is shown that the stress in any web member of a truss with horizontal chords is the vertical shear, multi-plied by the secant of its inclination to a vertical line. But the vertical shear in any web member of such a truss is simply *the algebraic sum of all the vertical forces or weights (includ-ing the end reaction) between the end and the web member in question.*

In web member 5, for example, the fixed load shear will be the difference between the reaction R and the weights at A, B, M and L; or, since the truss is symmetrical with the centre, the shear will be the weight at C added to half the weight at K. In fact, as a general principle, when the truss and its load are symmetrical with the centre, *the shear in any web member will be the load between that member and the centre.* Hence, the shear in the web member of half the truss will possess the following values:

$$
\left.
\begin{aligned}
\text{Shear in } 6 &= \tfrac{1}{2}W &&= 2.325 \text{ tons}\\
\text{`` `` } 5 &= \tfrac{1}{2}W + W' &&= 3.675 \text{ ``}\\
\text{`` `` } 4 &= 1\tfrac{1}{2}W + W' &&= 8.325 \text{ ``}\\
\text{`` `` } 3 &= 1\tfrac{1}{2}W + 2W' &&= 9.675 \text{ ``}\\
\text{`` `` } 2 &= 2\tfrac{1}{2}W + 2W' &&= 14.325 \text{ ``}\\
\text{`` `` } 1 &= 2\tfrac{1}{2}W + 3W' &&= 15.675 \text{ ``}
\end{aligned}
\right\} \quad \text{. . (1).}
$$

If the plus sign indicates tension and the minus sign compression, the web stresses will take the following values:

$$
\left.
\begin{aligned}
\text{Stress in } 6 &= + \ 2.325 \times 1.202 &&= + \ 2.79 \text{ tons}\\
\text{`` `` } 4 &= + \ 8.325 \times \text{`` } &&= + 10.00 \text{ ``}\\
\text{`` `` } 2 &= + 14.325 \times \text{`` } &&= + 17.19 \text{ ``}\\
\text{`` `` } 5 &= - \ 3.675 \times 1.054 &&= - \ 3.87 \text{ ``}\\
\text{`` `` } 3 &= - \ 9.675 \times \text{`` } &&= - 10.18 \text{ ``}\\
\text{`` `` } 1 &= - 15.675 \times \text{`` } &&= - 16.52 \text{ ``}
\end{aligned}
\right\} \quad \text{. (2).}
$$

On the same page, 7, it was shown that the increment of chord stress at any panel point is equal to the algebraic sum of the horizontal components of the web stresses intersecting at that point; *but those horizontal components are the vertical shears multiplied by the tangents of the respective inclinations to a vertical line.* Hence:

The chord increment at:

$$
\left.
\begin{aligned}
A &= 15.675 \times \tfrac{1}{3} + 14.325 \times \tfrac{2}{3} &&= 14.79 \text{ tons}\\
B &= 9.675 \times \tfrac{1}{3} + 8.325 \times \tfrac{2}{3} &&= 8.79 \text{ ``}\\
C &= 3.675 \times \tfrac{1}{3} + 2.325 \times \tfrac{2}{3} &&= 2.79 \text{ ``}\\
R &= 15.675 \times \tfrac{1}{3} + 0 &&= 5.24 \text{ ``}\\
M &= 9.675 \times \tfrac{1}{3} + 14.325 \times \tfrac{2}{3} &&= 12.79 \text{ ``}\\
L &= 3.675 \times \tfrac{1}{3} + 8.325 \times \tfrac{2}{3} &&= 6.79 \text{ ``}\\
K &= 0 + 2.325 \times \tfrac{2}{3} &&= 1.55 \text{ ``}
\end{aligned}
\right\} \cdot (3).
$$

Hence the following are the upper chord stresses:

$$
\left.
\begin{aligned}
(1) &= - 14.79 &&= - 14.79 \text{ tons}\\
(2) &= - (14.79 + 8.79) &&= - 23.58 \text{ ``}\\
(3) &= - (23.58 + 2.79) &&= - 26.37 \text{ ``}
\end{aligned}
\right\} \cdot (4).
$$

The lower chord stresses will take the values:

$$
\left.
\begin{aligned}
(1) \quad &= \quad\quad\quad\quad\quad 5.24 \ \text{tons} \\
(2) \quad &= \ 5.24 + 12.79 = 18.03 \ \text{``} \\
(3) \quad &= 18.03 + \ 6.79 = 24.72 \ \text{``}
\end{aligned}
\right\} \quad \cdots \quad (5).
$$

When the position of the moving load is once determined for a required maximum stress, the latter is found from the re-action and panel loads by precisely the same general methods used with the fixed loads. Hence the proper position of the moving load is first to be found for each of the web stresses.

Eq. (7) of Art. 7 is an expression of the condition which obtains with the greatest shear in any web member. The number of panels, n, is 6. W_1, W_2, W_3, etc., are the single locomotive weights given in Art. 77; W_1 has the value 3.75 tons; W_2, W_3, W_4 and W_5, 6 tons, etc., etc.

If the moving load passes on the bridge from the right, and W_2 rests at the foot of web member 10, W_1 . . . W_5 will be on the bridge, and n times 3.75 (*i.e.*, $6 \times 3.75 = 22.50$ tons) will be less than $W_1 + W_2 + \ . \ . \ . \ + W_5 = 27.75$ tons. But $6 \times (3.75 + 6)$ tons is much greater than the total moving load on the bridge. Hence, W_2 at G is the position desired for web member 10; W_1 will then be 8.083 feet from G toward H. At this point the tabulation mentioned in connection with Eqs. (8) and (15) of Art. 7, may be used.

The tabulation for the two locomotives is given herewith.

1	2	3	4
1	15,000 × 8.08	121,200	121,200
2	39,000 × 5.75	224,250	345,450
3	63,000 × 4.50	283,500	628,950
4	87,000 × 4.50	391,500	1,020,450
5	111,000 × 7.08	785,880	1,806,330
6	126,000 × 4.83	608,580	2,414,910
7	141,000 × 5.67	799,470	3,214,380
8	156,000 × 4.83	753,480	3,967,860
9	171,000 × 9.00	1,539,000	5,506,860
10	186,000 × 8.08	1,502,880	7,009,740
11	210,000 × 5.75	1,207,500	8,217,240
12	234,000 × 4.50	1,053,000	9,270,240
13	258,000 × 4.50	1,161,000	10,431,240
14	282,000 × 7.08	1,996,560	12,427,800
15	297,000 × 4.83	1,434,510	13,862,310
16	312,000 × 5.67	1,769,040	15,631,350
17	327,000 × 4.83	1,579,410	17,210,760
18	342,000 × 4.00	1,368,000	18,578,760

A little consideration of the table will make its composition evident. It is reproduced from a table in actual use, and is given in pounds and foot pounds.

When W_2 rests at G, the x of Eq. (8) will be 5.25 feet.

It should be explained that each quantity in column 4 is the sum of all the preceding and opposite numbers in column 3. As an example, 5,506,860 is the sum of all the numbers in column 3 from the top down to and including the ninth. Column 4, then, represents the positive parenthesis in Eq. (8) of Art. 7, less the term multiplied by x.

When W_2 rests at G, the x of Eq. (8), Art. 7, will be 5.25 feet. Hence that equation in connection with the preceding table gives:

$$Shear\ in\ 10 = \frac{1}{120}\left(\frac{111000}{2} \times 5.25 + \frac{1020450}{2}\right) - \frac{1}{20} \times \frac{121200}{2}.$$

$$= 3650\ pounds = 1.825\ tons. \quad . \quad . \quad . \quad . \quad (6).$$

This quantity will shortly be needed again.

If the same locomotive weight W_2 rest at H, W_1 . . . W_9, or one complete locomotive, will be found on the bridge, and n times $(3.75 + 6)$ or $6 \times 9.75 = 58.50$ tons is still in excess of 42.75 tons, *i. e.*, half the entire locomotive weight. Hence, W_2 at H is the position of moving load, which gives the greatest shear to web member 8, and x, in Eq. (8) of Art. 7, becomes 2.83 feet. That equation then gives:

$$Shear\ in\ 8 = \frac{1}{120}\left(\frac{171000}{2} \times 2.83 + \frac{3967860}{2}\right) - \frac{1}{20} \times \frac{121200}{2}.$$

$$= 15520\ pounds = 7.76\ tons. \quad . \quad . \quad . \quad . \quad (7).$$

If the same weight, W_2, be placed at K, it will be found that W_{12} will rest at the right extremity of the span, there will therefore be eleven weights on the bridge. Since $6 \times (12000 + 7500) = 117000 > 105000$, this position of the moving load gives the greatest shear in brace 6. Eq. (8) of Art. 7 then gives, since $x = 5.75$ feet:

$$Shear\ in\ 6 = \frac{8217240}{2 \times 120} - \frac{1}{20} \times \frac{121200}{2}.$$

$$= 31400\ pounds = 15.7\ tons. \quad . \quad . \quad . \quad . \quad (8).$$

If W_2 be placed at L, it will be found that six times $(W_1 + W_2)$ is less than W_1 . . . . W_{15}, which will then rest on the bridge; but $6(W_1 + W_2 + W_3) > (W_1$ . . . . $W_{16})$, hence W_3 must rest at L for the greatest shear in 4. The value of x will then be 4.83 feet. Hence:

$$Shear\ in\ 4 = \frac{1}{120}\left(\frac{312000}{2} \times 4.83 + \frac{13862310}{2}\right) - \frac{1}{20} \times \frac{345450}{2}$$

$$= 55480 \text{ pounds} = 27.74 \text{ tons.} \quad . \quad . \quad . \quad . \quad (9).$$

In the case of brace 2, the condition of maximum shear obtains with W_3 at M. This position of moving load places 10.5 feet of the uniform load of 1.5 tons per lineal foot on the bridge. The value of x in Eq. (8) of Art. 7, consequently, becomes 5.25 feet. The tabulation given above is not quite sufficient to completely cover this case; although the result in line 18 forms by far the greater portion of the moment product which must be divided by the span, in order to obtain the shear. The application of Eq. (8) of Art. 7, to this particular case becomes, then:

$$Shear\ in\ 2 = \frac{1}{120}\left(\frac{342000}{2} \times 10.5 + \frac{3000}{2} \times 10.5 \times 5.25 + \frac{18578760}{2}\right) - \frac{345450}{2 \times 20}.$$

$$= 84430 \text{ pounds} = 42.2 \text{ tons.} \quad . \quad . \quad . \quad . \quad (10).$$

The moving load shears in web members 1, 3, 5, 7 and 9, will be precisely the same as those in 2, 4, 6, 8 and 10 respectively, because each pair, as 10 and 9, 8 and 7, etc., intersect in a chord which carries no moving load. The web stresses due to the moving load will then take the following values:

$$
\left.
\begin{array}{llll}
\text{Stress in } 10 = + 1.825 \times 1.054 = + & 1.92 \text{ tons.} \\
\text{``} \quad \text{``} \quad 8 = + 7.760 \times \text{``} = + & 8.1 \text{ ``} \\
\text{``} \quad \text{``} \quad 6 = + \ 15.7 \times 1.202 = + & 18.87 \text{ ``} \\
\text{``} \quad \text{``} \quad 4 = + 27.74 \times \text{``} = + & 33.35 \text{ ``} \\
\text{``} \quad \text{``} \quad 2 = + 42.20 \times \text{``} = + & 50.72 \text{ ``}
\end{array}
\right\} . \ (11).
$$

$$
\left.
\begin{array}{l}
\text{Stress in} \quad 9 = -1.825 \times 1.202 = -2.19 \text{ tons.} \\
\quad\text{``}\quad\text{``}\quad 7 = -7.762 \times \quad\text{``}\quad = -9.33 \quad\text{``} \\
\quad\text{``}\quad\text{``}\quad 5 = -15.7 \times 1.054 = -16.49 \quad\text{``} \\
\quad\text{``}\quad\text{``}\quad 3 = -27.74 \times \quad\text{``}\quad = -29.13 \quad\text{``} \\
\quad\text{``}\quad\text{``}\quad 1 = -42.20 \times \quad\text{``}\quad = -44.30 \quad\text{``}
\end{array}
\right\} \quad . \ (12).
$$

By comparing (:1) and (2), it is seen that the moving load stress in web member (10) is of a kind opposite to that caused by the fixed load in web member 3, while those two members are, in reality, identical; the latter is compression and the former tension. Since it is evident that no piece of material can be compressed and extended at the same time, it is clear that the resultant stress will be the numerical difference or algebraic sum of the induced stresses. In other words :

If the action of forces external to a piece of material tend to subject that portion of material to stresses of opposite kinds, the resultant stress will be equal to the numerical difference of the opposite stresses, and will be of the same kind as the greater.

By comparing (11) and (12) with (2), it will be seen that in *all the web members of that half of the truss first traversed by the train, the fixed and moving load produce stresses of opposite kinds.* The moving load stresses predominate in the members near the centre of the span, but near the ends the fixed load stresses are the greatest. In web member 8, the resultant stress is $8.1 - 3.87 = 4.23$ tons of tension ; but if the bridge carries no moving load, it is subjected to 3.87 tons of compression. Now, that member may be so designed and constructed that it can resist tension or compression, according to the demands upon it; in such a case it is said to be *counterbraced.* If, however, it is formed to resist compression only, the member 14 must be introduced between H and C to take the moving load shear in tension. In order to provide for the movement of the load in the opposite direction, 13 must be introduced between L and D. Such web members as 13 and 14 are called *counterbraces.* Other web members than counterbraces are called *main braces* or *main web members.* The duty of *counterbraces,* then, or of counterbraced main web members, is to transfer moving load shear to the farther abutment or pier.

It is now clear that the extent to which a main web member must be counterbraced is found by taking the excess of the moving load stress over that caused by the fixed load. At first sight it would appear that the same method should hold in determining counterbrace stresses ; since it may be supposed that the main brace will carry shear until its fixed load stress is neutralized. This presupposes, however, that there is such an exact adjustment of members that each will perform just the amount of duty assigned to it. As that is an end that can never be confidently realized, it is only prudent to suppose that *the counterbrace takes all the moving load shear*, and this will be assumed in all that follows. This procedure appears the more advisable when one reflects that counterbraces are subject to the greatest fatigue of all truss members, and that the amount of metal concerned is trifling.

It now becomes necessary to determine where the counterbraces are to begin. Since the fixed and moving load stresses, or shears, neutralize each other in equal amounts, it at once results that *counterbraces or counterbraced web members must begin at that point, or with that web member, in which the stress or shear produced by the moving load is greater than that of the opposite kind produced by the fixed load*

" Stress " or " shear " is used indifferently, as the former is simply the product of the latter by the secant of the inclination to a vertical.

In order, therefore, to find the stress in a counterbrace, it is only necessary to ascertain the moving load shear, and multiply it by the secant of the inclination. The secant of the angle between counterbrace 13 or 14, and a vertical line is 1.944 ; and since the shear it has to carry is by Eq. (11) 7.76 tons:

Stress in counterbrace $(14) = 7.762 \times 1.944 = +15.09$ tons. (13).

Again, by comparing (11) with (2) it is seen that the fixed load shear in 10 (or 3) is -9.675 tons, while the moving load shear is $+ 1.825$ tons. Hence 13 and 14 are the only counter-braces needed.

It is farther seen that in the half of the truss traversed last in order by the train, the stresses produced by the fixed and moving loads are the same in kind. Hence, *in the main braces the fixed and moving loads induce stresses of the same kind and the resultant is simply the numerical sum.*

A tabulation of all the resultant web stresses, then, gives the following values:

$$
\left.
\begin{array}{l}
\text{Web member } 1 = -60.82 \text{ tons.} \\
\quad\text{``}\qquad\text{``}\quad 3 = -39.31 \quad\text{``} \\
\quad\text{``}\qquad\text{``}\quad 5 = -20.36 \quad\text{``} \\
\\
\quad\text{``}\qquad\text{``}\quad 2 = +67.91 \quad\text{``} \\
\quad\text{``}\qquad\text{``}\quad 4 = +43.35 \quad\text{``} \\
\quad\text{``}\qquad\text{``}\quad 6 = +21.66 \quad\text{``} \\
\\
\quad\text{``}\qquad\text{``}\quad 14 = +15.09 \quad\text{``}
\end{array}
\right\} \quad . \quad . \quad . \quad . \quad . \quad . \quad (14).
$$

The moving load chord stresses remain to be found, and it will be necessary to resort to the methods of Art. 7 in order to determine the proper positions for the stresses in the various panels.

Since none of the web members are vertical, the positions of moving load for the greatest stresses in the lower chord panels will be given by Eq. (18) of Art. 7. For this case, q in that equation will be one-third the panel length.

In finding the stresses in lower chord panels 2 and 3, it is necessary to bring 10.4 and 7 feet, respectively, of the uniform load on the bridge. The condition of greatest stress in the lower chord end panel coincides with that for brace 1, and it will only be necessary to multiply the greatest shear in that brace (already determined) by the tangent of its inclination to a vertical line.

In determining the greatest stresses in the upper chord, moments are taken about the lower chord points, and Eq. (14) of Art. 7 will be used.

The application of Eqs. (18) and (14) of Art. 7 give the following results:

Lower ch'd 2 .. $\dfrac{l'}{7} = \tfrac{4}{18}$.. $W_n^1 = W_3$ .. $x_1 = 6.67$ ft .. $2\,x = 10.4$ ft.

" " 3 .. " $= \tfrac{7}{18}$.. " $= W_6$ .." $= 7.2$ " .. $2\,x = 7.0$ "

Upper " 1 .. " $= \tfrac{1}{8}$ .. $W_{n'-1} = W_2$ .... $2\,x = 10.4$ "

" " 2 .. " $= \tfrac{1}{3}$ .. " $= W_5$ .... $2\,x = 6.4$ "

" " 3 .. " $= \tfrac{1}{2}$ .. " $= W_{11}$ .... $2\,x = 19.0$ "

Eq. (16) of Art. 7, applied to the lower chord, gives by the aid of the tabulation already employed the moments:

In lower chord 2; $\quad M = \tfrac{4}{18}\left[\dfrac{18{,}578{,}760}{2} + (2 \times 171000 + 10.4\right.$

$\times 1500)\,5.2\bigg] - \dfrac{345450}{2} - \dfrac{63000}{2} \times 6.67 - \tfrac{1}{3}(12000 \times 15.5$

$+ 12000 \times 11 + 7500 \times 3.92)$,

$$\therefore\ M = 1{,}936{,}400 \text{ ft. lbs.}$$

In lower chord 3; $\quad M = \tfrac{7}{18}\left[\dfrac{18{,}578{,}760}{2} + (2 \times 171000 + 7\right.$

$\times 1500)3.5\bigg] - \dfrac{1{,}806{,}330}{2} - \dfrac{126000}{2} \times 7.2 - \tfrac{1}{3}(7500 \times 15.67$

$+ 2 \times 7500 \times 7.58)$,

$$\therefore\ M = 2{,}658{,}490 \text{ ft. lbs.}$$

Eq. (15) gives for the upper chord moments:

In upper chord 1; $\quad M = \tfrac{1}{8}\left[\dfrac{18{,}578{,}760}{2} + (2 \times 171000 + 10.4\right.$

$\times 1500)\,5.2\bigg] - \dfrac{345450}{2}$,

$$\therefore\ M = 1{,}685{,}425 \text{ ft. lbs.}$$

In upper chord 2 ; $M = \frac{1}{3}\left[\dfrac{18,578,760}{2} + (2 \times 171000 + 6.4\right.$

$$\left. \times 1500)\, 3.2\right] - \dfrac{1,806,330}{2}$$

$$\therefore\ M = 2,568,335 \text{ ft. lbs.}$$

In upper chord 3 ; $M = \frac{1}{2}\left[\dfrac{18,578,760}{2} + (2 \times 171000 + 19\right.$

$$\left. \times 1500)\, 9.5 - 7500 \times 122.5\right] - \left(\dfrac{8,217,240}{2} - 7500 \times 62.5\right)$$

$$\therefore\ M = 2,305,315 \text{ ft. lbs.}$$

The negative moments (-7500×122.5) and (-7500×62.5) occur in the last moment above, for the reason that W_1 is found 2.5 feet off the span to the left when upper chord 3 takes the maximum bending.

It is notable that the moment in upper chord panel 3 is less than that in panel 2, while it is yet more important to observe that the uniform load of 1.5 tons per lineal foot gives the greatest bending moment in the centre panel of the upper chord. A panel uniform load is $10 \times 1.5 = 15$ tons $= 30000$ pounds, and the reaction with this load on the whole bridge is $2.5 \times 15 = 37.5$ tons $= 75000$ pounds. Hence, making K the origin of moments :

In upper chord 3 ; $M = 75,000 \times 60 - 2 \times 30,000 \times 30 = 2,700,000$
ft. lbs.

These operations verify a previous observation to the effect that with concentrated loads there may be several maxima.

Eq. (10) shows that the greatest moving load shear in braces 1 and 2 is 42.2 tons ; hence, multiplying that result by its tangent, 0.333, and dividing the preceding greatest moments by the depth of truss, *i.e.*, 20 feet, the following moving load chord stresses are found :

$$\text{In upper ch'd } 1 ; - \frac{1,685,425}{20} = 84,271 \text{ lbs.} = 42.14 \text{ tons.}$$

$$\text{`` `` `` } 2 ; - \frac{2,568,335}{20} = 128,417 \text{ `` } = 64.21 \text{ ``}$$

$$\text{`` `` `` } 3 ; - \frac{2,700,000}{20} = 135,000 \text{ `` } = 67.50 \text{ ``}$$

$$\text{In lower ch'd } 1 ; - \quad 42.2 \quad \times \quad 0.333 \quad = 14.06 \text{ ``}$$

$$\text{`` `` `` } 2 ; - \frac{1,936,400}{20} = 96,820 \text{ lbs.} = 48.41 \text{ ``}$$

$$\text{`` `` `` } 3 ; - \frac{2,658,490}{20} = 132,924 \text{ `` } = 66.46 \text{ ``}$$

(15).

The resultant chord stresses are found by adding groups (4) and (5) to (15), as follows :

$$\text{Upper chord } (1) = -(14.79+42.14) = -56.93 \text{ tons.}$$

$$\text{`` `` } (2) = -(23.58+64.21) = -87.79 \text{ ``}$$

$$\text{`` `` } (3) = -(26.37+67.50) = -93.87 \text{ ``}$$

$$\text{Lower chord } (1) = +(5.24+14.06) = +19.30 \text{ tons.}$$

$$\text{`` `` } (2) = +(18.03+48.41) = +66.44 \text{ ``}$$

$$\text{`` `` } (3) = +(24.72+66.46) = +91.18 \text{ ``}$$

(16).

Groups (14) and (16), therefore, give the resultant maximum stresses in all members of the truss.

Those web members, such as 1, 3 and 5, which sustain compression, are called " posts " or " struts," while those, such as 2, 4, 6 and 14, which sustain tension, are called " ties."

If the truss be divided through *CD* and either *LK* or *KH*, it is seen that more than three members must be cut; but if that number is exceeded, it is known from the first principles of statics that the stresses must become indeterminate. Hence, *when counterbraces are introduced, indetermination always results.* If provision is made for one system of legitimate stress analysis, however, the safety of the structure is assured.

Only one point more needs passing attention before the examination of the next case. It has been stated in the course of the demonstrations that the stress in certain members is tension, and compression in others. In web member 4, for example, let it be desired to determine the kind of stress. It has been seen that when the greatest main web stress exists in that member, the reaction at R is 32.058 tons, and it is evident that it is directed *upward*. At the same time the live load resting at M is 4.32 tons and is directed *down*. The difference of these forces is an *upward* shear of 27.74 tons. Hence, if the truss is divided anywhere between BM and CL, this shear will tend to move the left portion (between the line of divisions and R) *upward* and past the right portion; *i. e.*, it will tend to increase the distance between B and L, and, consequently, produce tension in web member 4. The general principle then is to determine the effect of the resultant external forces on the distance between the extremities, or any other two points in the axis of the member; if the tendency is to increase this distance, the resulting stress will be tension, and compression if the reverse is the case. In trusses with parallel chords, after a very little experience, the kind of stress in any member may readily be discovered at a glance, but in many structures with curved or polygonal outlines, resort must be made to the general principle stated above, which will be more thorougly given hereafter. This simple statement, however, is all that is needed here.

Art. 10.—Single System of Vertical and Diagonal Bracing.—Verticals in Tension.

This form of truss when built with timber compression members, has long been known as the Howe truss. The skeleton diagram of the structure to be considered is shown in Plate I, Fig. 3. The moving load will be supposed to pass along the upper chord; hence the bridge is a "deck" structure. The following are the principal dimensions and fixed load data:

Span = 98 feet.   Panel length = 14 feet.
Depth = 20 feet.   Number of panels = 7.
Upper chord fixed load = 385 lbs. per lineal foot per truss.
Lower   "   "   " = 215 "   "   "   "   "   "
Upper   "   "   " per truss panel = W' = 2.70 tons.
Lower   "   "   "   "   "   " = W = 1.50 "
$tan\ ABC = 0.7$ $sec\ ABC = 1.22.$

The moving load will consist of the two consolidation locomotives used in the preceding Art., the weights of which are
shown in Art. 77; and this load will be taken as passing
from N towards M.

The web stresses will first be determined, and the first
counterbrace needed comes first in order. As the number of
panels is seven, n in Eq. (7) of Art. 7, is equal to 7. Let it
be required to ascertain whether the compression counterbrace
LK is necessary. If the train is so placed that W_2 rests at L,
the bridge will carry the weights $(W_1 \cdots W_7) = 70.5$ tons, or,
35.25 tons on each truss. Now $7\ W_1 = 52.5$ tons, and $7\ (W_1
+ W_2) = 136.5$ tons. Since the total load on the bridge is
found to lie between these values, by the principles of Art. 7
it is placed to give the greatest *compressive* shear in KL or
tensile shear in HP. In order to find the shear by Eq. (8) of
Art. 7, the following values result from the position of the
moving load just taken; $x = 1.3$ ft., $W_n = W_7$ and $W_{n'-1} =
W_1$. Since $l = 98$ and $p = 14$, Eq. (8) of Art. 7 gives by the
aid of the tabulation on page 41;

$$S = \frac{1}{98}\left(\frac{2,414,910}{2} + \frac{141,000}{2} \times 1.3\right) - \frac{1}{14} \times \frac{121,200}{2}$$

$$= 8920 \text{ lbs.} = 4.46 \text{ tons.}$$

The fixed load shear in the same panel is $W' + W = 4.2$
tons. As the latter is less than, and opposite in kind to that
of the moving load, the counter post or strut KL must be
introduced. Since the difference in these shears is very small,
it is evident that no counterbrace between KL and O is needed,
and that conclusion may easily be verified.

By proceeding in precisely the same manner for the shears in the other panels, the following quantities are found for insertion in Eq. (8) of Art. 7, when Eq. (7) of that Article is satisfied;

For greatest shear in

$KL \ldots W_2$ at $L \ldots W_n = W_7 \ldots x = 1.3 \ldots W_{n'-1} = W_1.$
$HG \ldots W_2$ " $H \ldots W_n = W_9 \ldots x = 4.8 \ldots W_{n'-1} = W_1.$
$FE \ldots W_2$ " $F \ldots W_n = W_{11} \ldots x = 1.8 \ldots W_{n'-1} = W_1.$
$DC \ldots W_3$ " $D \ldots W_n = W_{14} \ldots x = 6.8 \ldots W_{n'-1} = W_2.$
$BA \ldots W_3$ " $B \ldots W_n = W_{17} \ldots x = 2.5 \ldots W_{n'-1} = W_2.$

That Eq. (8) then gives:

$$\text{Shear in } KL = \frac{1}{98}\left[\frac{2,414,910}{2} + \frac{141,000}{2} \times 1.3\right] - \frac{1}{14} \times \frac{121,200}{2} = 4.46 \text{ tons.}$$

$$\text{" " } HG = \frac{1}{98}\left[\frac{3,967,960}{2} + \frac{171,000}{2} \times 4.8\right] - \frac{1}{14} \times \frac{121,200}{2} = 10.05 \text{ "}$$

$$\text{" " } FE = \frac{1}{98}\left[\frac{7,009,740}{2} + \frac{210,000}{2} \times 1.8\right] - \frac{1}{14} \times \frac{121,200}{2} = 16.68 \text{ "}$$

$$\text{" " } DC = \frac{1}{98}\left[\frac{10,431,240}{2} + \frac{282,000}{2} \times 6.8\right] - \frac{1}{14} \times \frac{345,450}{2} = 25.33 \text{ "}$$

$$\text{" " } BA = \frac{1}{98}\left[\frac{15,631,350}{2} + \frac{327,000}{2} \times 2.5\right] - \frac{1}{14} \times \frac{345,450}{2} = 35.79 \text{ "}$$

The shears in the inclined web members only have been given, because each pair of braces that intersect in the chord *not* traversed by the moving load take their greatest stresses with the same position of moving load. Braces 2 and 3, 4 and 5, etc., thus go together in pairs.

The resultant web stresses, by the aid of the preceding results, will then take the following values:

Brace 1 $\ldots$ $-[3(W + W^1) + 35.79] \times 1.22 = -59.04$ tons.
" 3 $\ldots$ $-[2(W + W^1) + 25.33] \times 1.22 = -41.15$ "
" 5 $\ldots$ $-[\ W + W^1\ + 16.68] \times 1.22 = -25.48$ "
" 7 $\ldots$ $\qquad -10.05 \times 1.22 = -12.26$ "
" 8 $\ldots$ $\qquad -4.46 \times 1.22 = -5.44$ "

Brace 2 . . . $2W^1 + 3W + 25.33 = + 35.23$ tons.
"   4 . . . $W^1 + 2W + 16.68 = + 22.38$ "
"   6 . . . $W + 10.05 = + 11.55$ "

The positions of the moving load for the greatest chord stresses are found by the aid of Eq. (14) of Art. 7, and result in the following quantities:

For upper chord

  3 . . . W_7 at F . . . $W_n = W_{16}$ . . . $x = 2$ ft. . . . $W_{n'-1} = W_6$.
  2 . . . W_5 " D . . . $W_n = W_{16}$ . . . $x = 4$ " . . . $W_{n'-1} = W_4$.
  1 . . . W_3 " B

The greatest stress in upper chord 1 occurs with the maximum shear in BA, and is found by taking the product of that shear by the tangent of its inclination to a vertical line. The shear has already been determined to be 35.79 tons, and the *tangent* is 0.7. Hence;

Stress in upper chord $1 = -35.79 \times 0.7 = -25.05$ tons.

Eq. (15) of Art. 7 then gives by the aid of the tabulation on page 41, the following bending moments:

$$\text{Upper chord } 3 = \frac{3}{7}\left[\frac{13,862.310}{2} + \frac{312,000}{2} \times 2\right] - \frac{2,414.910}{2} = 1,896,754 \text{ ft. lbs.}$$

$$\text{" \quad " } \quad 2 = \frac{2}{7}\left[\frac{13,862,310}{2} + \frac{312,000}{2} \times 4\right] - \frac{1,020,450}{2} = 1,648,391 \text{ " "}$$

As the depth of the truss is 20 feet, the moving load chord stresses become:

Upper (3) $= -1,896,754 \div 20 = -94,837$ lbs. $= -47.42$ tons.
"   (2) $= -1,648,391 \div 20 = -82,420$ " $= -41.21$ "
"   (1) $=$ $= -25.05$ "

By combining these results with those due to the fixed load, the following resultant chord stresses are found:

Upper (1) $= -[25.05 + 3(W + W^1) \times 0.7] = -33.87$ tons.
"   (2) $= -[41.21 + 5(W + W') \times 0.7] = -55.91$ "
"   (3) $= -[47.42 + 6(W + W^1) \times 0.7] = -65.06$ "

It is to be observed that the upper and lower chord stresses are the same in pairs, *i. e.*, in the same oblique panel. The reason is obvious. If a panel, as *DFEC*, be divided by any line cutting *DF, DE* and *EC*, it will be evident that no horizontal forces whatever exist except the stresses in *DF* and *EC*: hence, by the first principles of statics, those stresses must be equal in amount and opposite in kind. The same result holds for any oblique panel, and, indeed, in all vertical and diagonal bracing with horizontal chords and vertical loading. The stresses in one chord can then always be written from those in the other, as is done here :

$$Lower \ (1) = + 33.87 \ tons.$$
$$`` \quad (2) = + 55.91 \quad ``$$
$$`` \quad (3) = + 65.06 \quad ``$$

In reality the moments remain the same whether the upper or the lower extremity of any vertical brace is taken for the moment origin ; hence the equality of upper and lower chord stresses in the same oblique panel.

If the truss becomes a through one (*i. e.*, with the load on the lower chord), *the same pairs of web members do not intersect in the unloaded chord*, hence a different position of the moving load must be taken for the greatest shears in half the braces. A simple inspection of the diagram will show at once that the position of the moving load for the greatest shears in the oblique or compression braces must be the same whether that load traverses the upper or lower chord. For the vertical braces, however, the moving load must be advanced in the lower chord at least one panel (more in some cases) beyond its position on the upper. Hence the vertical-brace stresses will be greater in a through truss than those found with the moving load on the upper chord, while the stresses in the oblique braces remain the same. Consequently this type of truss is better adapted to the deck than the through form.

Art. 11.—Single System of Vertical and Diagonal Bracing.—Verticals in Compression.

This type of truss is very common in American bridge

·practice. Its compression members are the shortest possible, and its details are simple in character, and both those features are conducive to economy. The skeleton diagram to which reference is to be made is given by Fig. 1, of Plate II. The bridge is supposed to be a "through" structure, hence the floor system will rest on the lower chord. The loads and stresses in this Art. will be given in pounds. The principal dimensions and fixed load data are as follows:

$$
\begin{array}{ll}
\text{Span} = 184'\ 11\tfrac{5}{8}'' & \text{Panel length} = 20'\ 6\tfrac{5}{8}'' \\
\text{Depth} = 27'\ 0'' & \text{Number of panels} = 9
\end{array}
$$

Upper chord fixed load = 230 lbs. per lin. ft. per truss.
Lower   "   " = 533   "   "   "
Upper   "   " per truss panel = $W' = $ 4726 lbs.
Lower   "   "   "   " $= W = $ 10953 "

$$\overline{\hspace{3cm}15679 \text{ lbs.}}$$

$$tan\ ABL = 0.761 \qquad sec\ ABL = 1.26$$

The moving load will consist of a train of two coupled consolidation locomotives, each with concentrated weights as shown in the diagram below, followed by a uniform train of 2,240 pounds per lineal foot. It will be taken to move from right to left.

The moving load is comparatively light, hence the fixed load is also assumed rather low.

As usual, Eq. (7) of Art. 7 will be used in determining

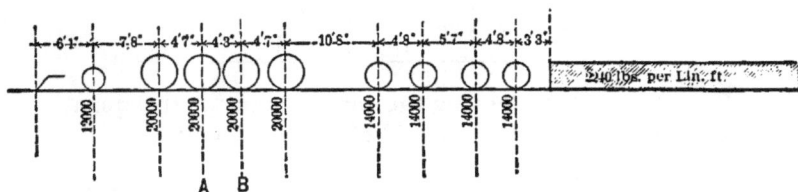

the positions of moving loads for the greatest shears, and, hence, for the greatest stresses. For a reason that will hereafter appear, the greatest shear for every panel in the truss will be found. The application of Eq. (7), Art. 7, will give the following quantities for use in Eq. (8) of the same Art.:

Max. shear in

$$OP \ldots W_1 \text{ at } P \ldots W_n = W_4 \ldots \quad x = 4.1 \text{ ft.}$$
$$MN \ldots W_2 \text{ `` } N \ldots \text{ ``} = W_9 \ldots \quad x = 2.2 \text{ ``}$$
$$FG \ldots \text{ `` `` } G \ldots \text{ ``} = W_{12} \ldots \quad x = 1.0 \text{ ``}$$
$$EH \ldots \text{ `` `` } H \ldots \text{ ``} = W_{15} \ldots \quad x = 2.0 \text{ ``}$$
$$DI \ldots W_3 \text{ `` } I \ldots \text{ ``} = W \quad\quad\quad 2x = 9.0 \text{ ``}$$
$$CJ \ldots \text{ `` `` } J \ldots \text{ ``} = W \; \Big| \; \text{Uniform } 2x = 29.5 \text{ ``}$$
$$BK \ldots \text{ `` `` } K \ldots \text{ ``} = W \; \Big/ \quad \text{load.} \quad 2x = 50.2 \text{ ``}$$
$$AB \ldots \text{ `` `` } L \ldots \text{ ``} = W \quad\quad\quad\quad 2x = 70.75 \text{ ``}$$

The tabulation to be used in connection with Eq. (8) and (15) of Art. 7, is given below and is formed precisely like that in Art. 9.

1	12,000 × 7.666	92,000	
2	32,000 × 4.583	146,666	238,666
3	52,000 × 4.25	221,000	459,666
4	72,000 × 4.583	329,976	789,642
5	92,000 × 10.666	981,332	1,770,974
6	106,000 × 4.666	459,333	2,230,307
7	120,000 × 5.583	670,000	2,900,307
8	134,000 × 4.666	625,333	3,525,640
9	148,000 × 9.333	1,381,333	4,906,973
10	160,000 × 7.666	1,226,666	6,133,639
11	180,000 × 4.583	825,000	6,958,639
12	200,000 × 4.25	850,000	7,808,639
13	220,000 × 4.583	1,008,333	8,816,972
14	240,000 × 10.666	2,560,000	11,376,972
15	254,000 × 4.666	1,185,333	12,562,305
16	268,000 × 5.583	1,496,333	14,058,638
17	282,000 × 4.666	1,316,000	15,374,638
18	296,000 × 3.25	962,000	16,336,638

Eq. (8) of Art. 7 gives by the introduction of the quantities found above, and by the aid of the tabulation:

$$\text{Shear in } OP = \frac{1}{185}\left[\frac{459,666}{2} + \frac{72,000}{2} \times 4.1\right] = 2,040 \text{ lbs.}$$

$$\text{`` `` } MN = \frac{1}{185}\left[\frac{3,525,640}{2} + \frac{148,000}{2} \times 2.2\right] - \frac{92,000}{2 \times 20.55} = 8,171 \text{ ``}$$

$$\text{`` `` } FG = \frac{1}{185}\left[\frac{6,958,639}{2} + \frac{200,000}{2} \times 1.0\right] - \text{``} = 17,110 \text{ ``}$$

$$\text{Shear in } EH = \frac{1}{185}\left[\frac{11,376,972}{2} + \frac{254,000}{2} \times 2.0\right] - \frac{92000}{2 \times 20.55} = 29,884 \text{ ''}$$

$$\text{`` `` } DI = \frac{1}{185}\left[\frac{16,336,638}{2} + \frac{612,160}{2} \times 4.5\right] - \frac{238,666}{2 \times 20.55} = 45,788 \text{ ''}$$

$$\text{`` `` } C\mathcal{F} = \frac{1}{185}\left[\text{``} + \frac{658,080}{2} \times 14.75\right] - \text{``} = 63,735 \text{ ''}$$

$$\text{`` `` } BK = \frac{1}{185}\left[\text{``} + \frac{704,448}{2} \times 25.1\right] - \text{``} = 86,132 \text{ ''}$$

$$\text{`` `` } AB = \frac{1}{185}\left[\text{``} + \frac{750,480}{2} \times 35.38\right] - \text{``} = 110,105 \text{ ''}$$

Since the fixed load shear in *HM* is 15,679 lbs. it is seen that counterbrace 10 is the first counter needed. Combining the fixed load shears with those above due to the moving load, the following resultant stresses in the inclined web members are found :

$$\text{Stress (10)} = + (\ 17,110 \qquad\qquad) \times 1.26 = +\ 21,560 \text{ } lbs.$$
$$\text{``} \qquad (9) = + (\ 29,884 \qquad\qquad ) \times \text{`` } = +\ 37,654 \text{ ``}$$
$$\text{``} \qquad (7) = + (\ 45,788 + W + W') \times \text{`` } = +\ 77,445 \text{ ``}$$
$$\text{``} \qquad (5) = + (\ 63,735 + 2W + 2W') \times \text{`` } = +\ 119,817 \text{ ``}$$
$$\text{``} \qquad (3) = + (\ 86,132 + 3W + 3W') \times \text{`` } = +\ 167,800 \text{ ``}$$
$$\text{``} \qquad (1) = - (110,105 + 4W + 4W') \times \text{`` } = -\ 217,750 \text{ ``}$$

Eq. (21) of Art. 7 shows that if W_4 be placed at the feet of the vertical brace 2, that member will take its maximum moving load stress, which, by Eq. (19) of the same Art. takes the value :

$$(2) = + \frac{4.05 \times 6,000 + (11.72 + 16.32 + 15.97)\ 10,000 + 2.92 \times 14,000}{20.55}$$
$$+ \ 10,000 = + \ 34,600 \text{ } lbs.$$

The stresses in the vertical braces become :

$$\text{Stress (8)} = - (29,884 \qquad + W') = - 34,611 \text{ } lbs.$$
$$\text{``} \qquad (6) = - (45,788 + W + 2W') = - 66,193 \text{ ``}$$
$$\text{``} \qquad (4) = - (63,735 + 2W + 3W') = - 99,819 \text{ ``}$$
$$\text{``} \qquad (2) = + (34,600 + \qquad + W) = + 45,553 \text{ ``}$$

These complete the greatest stresses in the braces.

Eq. (14) of Art. 7 gives the following values for the positions of moving load for the greatest stresses in the chords:

Upper 3 and 4 . . . $W_{n'-1} = W_{11}$ . . . $W_n = 72,800$ . . . $2x = 65.$ *ft.*

"   .2 . . .   " $= W_9$ . . . " $= 82,300$ . . . " $= 73.5$ "

"   1 . . .   " $= W_5$ . . . " $= 78,000$ . . . " $= 69.7$ "

The "$2x$" shows the distance covered by the uniform load.

Eq. (15) of Art. 7 then gives the following maximum bending moments by the aid of the tabulation already given:

For upper 3 and 4 . . . $M = \dfrac{4}{9}\left[\dfrac{16,336,638}{2} + \dfrac{737,600}{2} \times 32.5\right]$

$$- \dfrac{6,958,639}{2} = 5,478,153 \; ft. \; lbs.$$

"   "   2 . . . $M = \dfrac{3}{9}\left[\dfrac{16,336,638}{2} + \dfrac{756,640}{2} \times 36.8\right]$

$$- \dfrac{4,906,973}{2} = 4,910,013 \; ft. \; lbs.$$

"   "   1 . . . $M = \dfrac{2}{9}\left[\dfrac{16,336,638}{2} + \dfrac{748,130}{2} \times 34.9\right]$

$$- \dfrac{1,770,974}{2} = 3,830,813 \; ft. \; lbs.$$

The moving load stresses in lower chord 1 and 2 are found by taking the product of the greatest shear in brace 1 by its vertical tangent:

Lower (1) *and* (2) $= 110,105 \times 0.761 = 83,790 \; lbs.$

Since the depth is 27 ft. the greatest moving load upper chord stresses are the following:

Upper (3) *and* (4) $= -5,478,153 \div 27 = -202,900 \; lbs.$

"   (2) $= -4,910,013 \div 27 = -181,850$ "

"   (1) $= -3,830,813 \div 27 = -141,900$ "

Since $(W + W')$ *tan* $ABL = 11,932$, the resultant upper chord stresses become:

Upper   (1) $= - (141,900 + 7 \times 11,932), = -225,424$ *lbs.*

"   (2) $= - (181,850 + 9 \times 11,932) = -289,238$ "

" (3) *and* (4) $= - (202,900 + 10 \times 11,932) = -322,220$ "

In the lower chord, the resultant stresses are :

Lower (1) *and* (2) $= +83,790 + 4 \times 11,932 = +131,520$ *lbs.*

"   (3) $=$ $= +225,424$ "

"   (4) $=$ $= +289,238$ "

"   (5) $=$ $= +322,220$ "

These values complete the resultant stresses in the various members of the trusses, and they will be used hereafter in making the complete design of this bridge.

If the moving load traverses the upper chords of the trusses, the same pairs of braces as before will not take their greatest shears together. The stresses in their inclined braces will not in any way be changed, but it will be necessary to advance the moving load by at least one panel beyond the positions taken in the through bridge, in order to determine the greatest shears in the vertical braces of the deck truss. Hence, by changing the bridge from the "through" type to the "deck," the vertical braces will carry considerably increased stresses, and as they are in compression the weight of the bridge will be materially increased. Hence, this truss is best adapted to carrying the moving load along its lower chord.

Art. 12.—Two Systems of Vertical and Diagonal Bracing.—Verticals in Compression.

The style of truss shown in Fig. 2, Pl. II., next to be treated, was at one time more common than any other in American bridge practice. With increased facilities for fabricating and handling large bridge members, it has been possible to extend the use of single systems of triangulation to much longer spans than formerly. In this manner the ambiguity of the double system is avoided, and thus analytical excellence is combined with advantages of production. In long spans, however, the double system is still frequently used, and if properly designed it is not so objectionable as might at first seem to appear.

It should always be arranged with an even number of panels, as the stress ambiguity is then reduced to an unimportant matter. In order to show the extent to which ambiguity may arise, an odd number of panels has been selected in the present example. As has already been shown, the method of Art. 7 cannot be applied to a double system of triangulation. It is only possible to assume that each system acts as an independent truss, and to determine by trial the greatest possible concentrations at the head of the train, and in one system, and consider such concentrations the head of the moving load in that system.

There will be two equal concentrations represented by w' at the head of the moving load in each system, while those that follow will uniformly equal w. Ordinarily no two concentrations will be exactly equal, but the assumption is sufficiently accurate.

The truss under consideration is composed of two systems of right-angled triangulation, shown in Figs. 3 and 4 of Pl. II.

Before passing to the computations it is well to observe that although the action of the loads in one system may be considered as taking place independently of the actions of the loads in the other; at the same time equal loads symmetrically placed in reference to the centre, though resting on different systems of triangulation, may be considered counterbalanced. The web stresses due to the fixed load will be determined on the supposition that the web members shown by the dotted lines do not exist.

The data to be used are given below:

Span = 210 feet.   Depth of truss = 26 feet.
Number of panels = 15   Panel length = 14 "
W (upper) = 9100 lbs. = 4.55 tons = 650 lbs. per foot.
W' (lower) = 14000 " = 7.00 " = 1000 " " "
$w = 13$ tons. $w' = 20$ tons.

$$c = w' - w = 7 \text{ tons.}$$

Angle $QNO = \alpha$ Angle $MNO = \beta$

$$\tan \alpha = 1.077 \qquad \sec \alpha = 1.47$$
$$\tan \beta = 0.538 \qquad \sec \beta = 1.136$$

The excess c will be taken at four panel points as before; it will also be used in determining the chord stresses.

The counterbrace 16 is the first one needed. Carrying the moving load on the bridge from R, panel by panel, the greatest web stresses are found to be the following:

In brace $16 \ldots \frac{13}{18} w \sec \alpha + \frac{10}{18} e \sec \alpha = + 22.15$ tons.

"   " $15 \ldots \frac{16}{18} w \sec \alpha + \frac{13}{18} c \sec \alpha = + 28.62$ "

"   " $14 \ldots \frac{13}{18} w \qquad + \frac{10}{18} e + W = - 19.62$ "

"   " $1 \ldots (\frac{9}{18} w + \frac{14}{18} e) \sec \alpha + (W + W') \sec \alpha$
$$= + 52.06 \text{ "}$$

"   " $13 \ldots \frac{16}{18} w + \frac{13}{18} e + W = - 24.02$ "

"   " $2 \ldots (\frac{21}{18} w + \frac{16}{18} e + W + W') \sec \alpha = + 59.80$ "

"   " $3 \ldots \frac{20}{18} w + \frac{14}{18} c + 2W + W' = - 39.97$ "

"   " $4 \ldots \{\frac{30}{18} w + \frac{18}{18} e + 2(W + W')\} \sec \alpha$
$$= + 84.53 \text{ "}$$

"   " $5 \ldots \frac{25}{18} w + \frac{16}{18} e + 2W + W' = - 45.23$ "

"   " $6 \ldots \{\frac{33}{18} w + \frac{20}{18} c + 2(W + W')\} \sec \alpha$
$$= + 93.54 \text{ "}$$

"   " $7 \ldots \frac{30}{18} w + \frac{18}{18} e + 3W + 2W' = - 62.05$ "

"   " $8 \ldots \{\frac{42}{18} w + \frac{22}{18} c + 3(W + W')\} \sec \alpha$
$$= + 119.53 \text{ "}$$

"   " $9 \ldots \frac{36}{18} w + \frac{20}{18} c + 3W + 2W' = - 68.18$ "

"   " $10 \ldots \{\frac{42}{18} w + \frac{24}{18} c + 3(W + W')\} \sec \beta$
$$= + 100.33 \text{ "}$$

In brace 11..$w' + W'$ $= +$ 27.00 tons.

"   "   12..$7(w + W + W') \sec \beta + \frac{59}{13} c \sec \beta$
$$= - 221.73 \quad "$$

The stresses in each system of triangulation are found by virtually taking that system as a single truss supporting only the weights at the apices belonging to it.

The greatest chord stresses will be obtained by supposing the train to cover the entire bridge, with the four excesses c at panel points 1, 2, 3, and 4.

Greatest stress in $a = 3(w + W + W') [2 \tan \beta + (\tan \beta + \tan \alpha)] + (w + W + W') \tan \beta + c(4 \tan \beta + \tan \alpha) + \frac{1}{3} c \, 2 \tan \beta = - 236.51$ tons.

Here it should be explained that since $\frac{59}{13} c = 3\frac{1}{3} c$ is found in the reaction at R, the three c's at panel points 1, 2, and 3, and $\frac{1}{3}$ of that at 4 may be taken as passing directly to R, while $\frac{2}{3}$ of the c at 4 passes to M through 5′, 2′, 16, 14, 1, 3, etc. Counterbrace 16 thus comes into action.

Greatest stress in $b = [2(w + W + W') + \frac{1}{3} c] \tan \alpha +$
$$236.514 = - 291.91 \text{ tons.}$$

"   "   " $c = 2(w + W + W') \tan \alpha +$
$$291.907 = - 344.79 \quad "$$

"   "   " $d = [(w + W + W') - \frac{2}{3} c] \tan \alpha +$
$$344.787 = - 366.20 \quad "$$

"   "   " $e = (w + W + W') \tan \alpha +$
$$366.201 = - 392.64 \quad "$$

The panel stresses in e, f, and g will be the same; and if the loading were uniform over the whole bridge, the panels e, f, g, h, and k would all be subjected to the same stress.

Greatest stress in l and $m = [7(w + W + W') + 3\frac{1}{3} c] \tan \beta$
$$= + 105.01 \text{ tons.}$$

"   "   " $n = [3(w + W + W') + 1\frac{1}{3} c] \tan \beta +$
$$105.009 = + 149.65 \quad "$$

Greatest stress in $o = [3(w + W + W') + c]$ *tan* $\alpha +$
$$149.654 = + 236.51 \text{ tons.}$$

"   "   " $p = [2(w + W + W') + \tfrac{1}{3}c]$ *tan* $\alpha +$
$$236.513 = + 291.91 \quad \text{"}$$

"   "   " $q = 2(w + W + W')$ *tan* $\alpha +$
$$291.906 = + 344.79 \quad \text{"}$$

"   "   " $r = (w + W + W' - \tfrac{1}{3}c)$ *tan* $\alpha +$
$$344.786 = + 361.17 \quad \text{"}$$

"   "   " $s = (w + W + W')$ *tan* $\alpha +$
$$361.174 = + 387.61 \quad \text{"}$$

It is to be noticed that diagonally opposite panels in the upper and lower chords, up to the counterbrace 16, beginning with the panels a and o, are subjected to the same amounts of stress, but of opposite kinds.

If the loading were uniform over the whole bridge, this equality of the pairs would continue to the centre; also, the stresses in the panels c, f, g, h, k, and s would be equal to each other.

If the end posts were vertical, there would be obvious changes in the stresses of the panels l, m, and n (that in l would be zero). The upper chord panel stresses would not be changed.

The whole truss in Fig. 2, Pl. II., is composed of the two systems of triangulation shown in Figs. 3 and 4, and each of these is to be considered separately in checking the chord stresses by the method of moments. Denote by (R') and (R'') the reactions at the points indicated by the same letters in Figs. 3 and 4, then divide the total load supported by each system into two parts, *according to the principle of the lever*, and there will result:

$$(R') = \tfrac{16}{30}[7(w + W + W')] + \tfrac{13}{15} \cdot 2e = 103.7867 \text{ tons.}$$
$$(R'') = \tfrac{14}{30}[7(w + W + W')] + \tfrac{12}{15} \cdot 2e = 91.3966 \text{ tons.}$$

If the diagonals are in tension, according to this value of (R''), $D'L'$ should be drawn, and not $K'E'$. The latter is taken, however, for a reason that will appear presently.

The sum of (R') and (R'') is just equal to the total reaction at R in Fig. 2, as it ought to be.

Indicate by (BC), $(D'C')$, (d), etc., the stresses in the panels represented by those letters. Taking moments about H and G respectively, there result:

$$(BC) = -(HK) = [(R') \times R'H - 2(w' + W + W')(GH + \tfrac{1}{3}FG)] \div d.$$
$$(AB) = -(GH) = [(R') \times R'G - (w' + W + W')FG] \div d.$$

Also, taking moments about K' and H':

$$(C'D') = [(R'') \times R''K' - 2(w' + W + W')(H'K' + \tfrac{1}{2}G'H')] \div d.$$
$$(B'C') = -(H'K') = [(R'') \times R''H' - (w' + W + W')G'H'] \div d.$$

Similar expressions will give the chord stress in every panel of Figs. 3 and 4; and having found these, the resultant stresses in Fig. 2 are simply the sums of the proper pairs taken from Figs. 3 and 4.

Thus,
$$(d) = (BC) + (C'D')$$
$$(c) = (CD) + (C'D')$$
$$(o) = (GH) + (G'H')$$
$$\text{etc.} = \text{etc.} + \text{etc.}$$

This system of determination by moments may be applied to any truss with parallel chords, however many systems of triangulation there may be.

The method also applies to any irregular loading, for the stresses due to each panel load may be found separately, and the sum caused by all taken.

Web stresses may also be checked by the same method, since the increment of chord stress at any panel point is equal to the sum of the horizontal components of the stresses in the web members intersecting at the panel point in question. Such a check, however, is a very tedious one.

Applying the above equations to $C'D'$ in Fig. 4:

$$(C'D') = (91.397 \times 84 - 2 \times 31.55 \times 42) \div 26 = 193.35 \text{ tons.}$$

Also, to CD in Fig. 3:

$$(CD) = (103.787 \times 98 - 2 \times 31.55 \times 70 - 24.55 \times 28) \div 26 =$$
$$194.95 \text{ tons.}$$

But the sum of these two is 388.3 tons, whereas $(e) = 392.64$ tons. This discrepancy, not very great, is easily explained. The loading $(w + W + W')$ is counterbalanced in Fig. 2, but is not in Figs. 3 and 4.

In Fig. 2 all the load on the left of the centre of the span, except $\frac{2}{3}c$ at 4 or H', is assumed to pass directly to R (or R' and R''). Hence in Figs. 3 and 4, to be consistent with Fig. 2, there should be taken :

$$(R') = 4(w + W + W') + 2c = 112.2 \text{ tons.}$$
$$(R'') = 3(w + W + W') + \frac{1}{3}e = 82.983 \text{ tons.}$$

Introducing these in the general formula :

$$(e) = (C'D') + (CD) = [112.2 \times 98 + 82.983 \times 84 - 31.55 \times$$
$$84 - 31.55 \times 140 - 24.55 \times 28] \div 26 = 392.75 \text{ tons.}$$

This result agrees sufficiently well with that obtained by the trigonometrical method.

With the last value of (R''), $K'E'$ will be in tension.

It is thus seen that with an uneven number of panels a little ambiguity exists both in reference to the greatest chord stresses and the greatest web stresses, when there are two systems of triangulation. *This ambiguity always exists, whatever the number of systems, if the component systems are not symmetrical in reference to the centre line of the span, and it always disappears if they are symmetrical in reference to that line.*

With an even number of panels in the span and two systems of triangulation no ambiguity exists.

These observations in reference to ambiguity apply as well to isosceles bracing as to vertical and diagonal.

In the example taken there are only two systems of triangulation, but precisely the same method is to be followed what-

5

ever the number; in determining the web stresses, each system is supposed to carry those moving weights only which rest at its apices, and the same is true in reference to chord stresses for unsymmetrical loading, uniform loading being supposed counterbalanced for either stresses.

The slight changes to be made for an overhead bridge, or for verticals in tension and diagonals in compression, are evident from what has already been given in preceding articles.

It is seen that any two web members intersecting in the chord not traversed by the moving load receive their greatest stresses at the same time; the principle, indeed, is a general one.

When built in iron, this truss is frequently called the Linville truss.

Art. 13.—Truss with Uniform Diagonal Bracing—Two Systems of Triangulation.

This truss is shown in Pl. X., Fig. 6, and, although taken here as an ordinary pin connection bridge, precisely the same method of calculation is to be used for a "lattice" truss with riveted connections.

No locomotive excess will be taken, but a heavy moving load of uniform density will be assumed. The following are the data:

Span	= 182 feet.	Depth	= 23 feet.
Panel length =	13 "	Number of panels =	14.

Fixed load:

W (upper) = 450 pounds per foot = 2.925 tons per panel.
W' (lower) = 800 "   "   " = 5.2 "   "   "

Moving load:

w = 2800 pounds per foot = 18.2 tons per panel.

Angle $AaB = \alpha$.

$\tan \alpha$	= 0.565.	$\sec \alpha$	= 1.15.
$W \sec \alpha$	= 3.364 tons.	$W' \sec \alpha$ =	5.98 tons.
$\dfrac{w}{14} \sec \alpha$	= 1.500 "	$W \tan \alpha$ =	1.653 "
$W' \tan \alpha$	= 2.94 "	$w \tan \alpha$	= 10.28 "

The vertical members aB and tS are for tension only.

The moving load will be taken as passing from A to T, and its head will be supposed to rest at the various panel points in succession, in the determination of the web stresses.

The notation for the stresses is one which will frequently be used hereafter. The stress in any member is indicated by inclosing in a parenthesis the letters which belong to it in the figure.

Head of moving load at D.

$$(dF) = \{\tfrac{3}{2} W' + W - \tfrac{4}{14} w\} \sec \alpha = + 2.6 \times \sec \alpha.$$

Hence the stress in dF will always be tension.

Head of moving load at E.

$$(Ee) = \{\tfrac{6}{14} w - W' - \tfrac{3}{2} W\} \sec \alpha = - 1.79 \times \sec \alpha.$$

Hence the stress in Ee will always be compression.

The web stresses desired are, then, the following:

$$
\begin{aligned}
(Ff) &= -(\ \tfrac{1}{2}W' + \quad W - \tfrac{9}{14}\, w)\ \sec \alpha & &= +\quad 7.15 \text{ tons.}\\
(eG) &= (\quad W' + \tfrac{1}{2}W - \tfrac{6}{14}\, w)\quad " & &= -\quad 1.34 \quad "\\
(Gg) &= -(\qquad\quad \tfrac{1}{2}W - \tfrac{12}{14}\, w)\quad " & &= +\ 16.32 \quad "\\
(fH) &= (\ \tfrac{1}{2}W' \qquad - \tfrac{9}{14}\, w)\quad " & &= -\ 10.51 \quad "\\
(Hh) &= (\ \tfrac{1}{2}W' \qquad + \tfrac{16}{14}\, w)\quad " & &= +\ 26.99 \quad "\\
(gK) &= -(\qquad\quad \tfrac{1}{2}W + \tfrac{12}{14}\, w)\quad " & &= -\ 19.68 \quad "\\
(Kk) &= (\quad W' + \tfrac{1}{2}W + \tfrac{20}{14}\, w)\quad " & &= +\ 37.66 \quad "\\
(hL) &= -(\ \tfrac{1}{2}W' + \quad W + \tfrac{16}{14}\, w)\quad " & &= -\ 30.35 \quad "\\
(Ll) &= (1\tfrac{1}{2}W' + \quad W + \tfrac{25}{14}\, w)\quad " & &= +\ 49.83 \quad "\\
(kO) &= -(\quad W' + 1\tfrac{1}{2}W + \tfrac{20}{14}\, w)\quad " & &= -\ 41.02 \quad "\\
(Oo) &= (\ 2W' + 1\tfrac{1}{2}W + \tfrac{30}{14}\, w)\quad " & &= +\ 62.00 \quad "\\
(lP) &= -(1\tfrac{1}{2}W' + \quad 2W + \tfrac{25}{14}\, w)\quad " & &= -\ 53.20 \quad "\\
(Pp) &= (2\tfrac{1}{2}W' + \quad 2W + \tfrac{36}{14}\, w)\quad " & &= +\ 75.68 \quad "\\
(oQ) &= -(\ 2W' + 2\tfrac{1}{2}W + \tfrac{30}{14}\, w)\quad " & &= -\ 65.37 \quad "\\
(Qt) &= (\ 3W' + 2\tfrac{1}{2}W + \tfrac{42}{14}\, w)\quad " & &= +\ 89.35 \quad "\\
(pS) &= -(2\tfrac{1}{2}W' + \quad 3W + \tfrac{36}{14}\, w)\quad " & &= -\ 79.04 \quad "\\
(tT) &= -\tfrac{13}{2}(W' + \quad W + \quad w)\quad " & &= -\ 197.24 \quad "\\
(tS) &= 3\tfrac{1}{2}W' + \quad 3W + \tfrac{40}{14}\, w & &= +\ 90.68 \quad "
\end{aligned}
$$

With the moving load covering the whole bridge, the following chord stresses are found:

$$
\begin{aligned}
(ab) &= -\,(9\tfrac12 W' + 9W + 9\tfrac12 w)\ \tan\alpha && = -\,141.46 \text{ tons.}\\
(bc) &= (ab)\ -5\,(W' + W + w)\ \text{``} && = -\,215.83 \quad\text{``}\\
(cd) &= (bc)\ -4\,(\ \text{``} \quad \text{``} \quad \text{``}\ )\ \text{``} && = -\,275.33 \quad\text{``}\\
(de) &= (cd)\ -3\,(\ \text{``} \quad \text{``} \quad \text{``}\ )\ \text{``} && = -\,319.95 \quad\text{``}\\
(ef) &= (de)\ -2\,(\ \text{``} \quad \text{``} \quad \text{``}\ )\ \text{``} && = -\,349.70 \quad\text{``}\\
(fg) &= (ef)\ -\ (\ \text{``} \quad \text{``} \quad \text{``}\ )\ \text{``} && = -\,364.57 \quad\text{``}\\
(AB) &= \qquad\quad 6\tfrac12\,(\ \text{``} \quad \text{``} \quad \text{``}\ )\ \text{``} && = +\ \ 96.68 \quad\text{``}\\
(BC) &= (AB)\ +(2\tfrac12 W' + 3W + 2\tfrac12 w)\ \text{``} && = +\,134.69 \quad\text{``}\\
(CD) &= (BC)\ +5\,(W' + W + w)\ \text{``} && = +\,209.06 \quad\text{``}\\
(DE) &= (CD)\ +4\,(\ \text{``} \quad \text{``} \quad \text{``}\ )\ \text{``} && = +\,268.55 \quad\text{``}\\
(EF) &= (DE)\ +3\,(\ \text{``} \quad \text{``} \quad \text{``}\ )\ \text{``} && = +\,313.17 \quad\text{``}\\
(FG) &= (EF)\ +2\,(\ \text{``} \quad \text{``} \quad \text{``}\ )\ \text{``} && = +\,342.92 \quad\text{``}\\
(GH) &= (FG)\ +\ (\ \text{``} \quad \text{``} \quad \text{``}\ )\ \text{``} && = +\,357.80 \quad\text{``}
\end{aligned}
$$

The following operations constitute a check on the accuracy of the chord stresses.

The horizontal forces exerted at the joints g and H, respectively, are:

$$
\begin{aligned}
(fg)\ &-\tfrac12 W \tan\alpha && = -\,365.40 \text{ tons.}\\
\text{and}\qquad (GH)\ &+\tfrac12\,(W' + w)\ \tan\alpha && = +\,364.41 \quad\text{``}
\end{aligned}
$$

The horizontal force exerted at either one of these joints, as found by the moment method, is:

$$
\frac{7\,(W' + W + w) \times 0.25 \times 182}{23} = 364.5 \text{ tons.}
$$

The agreement is close.

It is to be observed that (Ff) is the greatest tensile stress in hL, also; and, on the other hand, that (hL) is the greatest compression stress Ff. Similar observations apply to the pairs of members eG, kK; Gg, Kg; fH, Hh.

These, consequently, are the only web members which need to be counterbraced.

Precisely the same methods of calculation apply, whatever

may be the number of systems of triangulation or the character of the load, or whether the truss be a through or deck one.

If Fig. 6 represented a deck truss, however, the compressive web stresses would be increased and the tensile ones diminished, while the chord stresses would remain the same. Since the increase of compression in any web member would numerically exceed the decrease of tension in the adjacent one, the truss is better adapted to a through load than a deck load.

This truss, particularly with only one system of triangulation, is frequently called the "triangular" truss.

Art. 14.—Compound Triangular Truss.

A very economical style of truss, in point of quantity of material, is that shown in Fig. 1 of Pl. III. The truss is of the ordinary isosceles bracing, and formed of two systems of triangulation, but a half of the floor system and moving load is carried by verticals directly to the intersections E, F, etc.

Half the weight of the trusses is supported at the apices of the main systems, as H and M, in the upper chord, and half at the apices, as P and R, in the lower chord. The truss chosen is a deck or overhead truss; consequently half the floor system and moving load will be supported by the verticals in compression. The weight of the floor system will be taken at 300 pounds per foot, and the moving load taken will be a uniform one made up of a load of heavy engines weighing 2700 pounds per foot. In such a case there is no excess e.

The following are the data:

Length of span $= 200$ ft. Depth of truss $= 27.75$ ft.
Upper-panel length $= 12.5$ " $tan\ CDL = tan\ \alpha = 0.9$
Lower-panel length $= 25.0$ " $sec\ CDL = sec\ \alpha = 1.345$

$$W\ (\text{upper}) = 25 \times 500 + 12\tfrac{1}{2} \times 300 = 8.125 \text{ tons.}$$
$$W_1\ (\text{middle}) = 12\tfrac{1}{2} \times 300 \qquad\qquad = 1.875 \text{ "}$$
$$W'\ (\text{lower}) = 25 \times 500 \qquad\qquad = 6.25 \text{ "}$$
$$w = w' \qquad = 12\tfrac{1}{2} \times 2700 \qquad\qquad = 16.875 \text{ "}$$

The middle loads W_1 or w are applied to the trussing as follows. The adjoining figure represents a portion of the truss in question as indicated by the same letters *HMPR* (see

figure in plate). Any weight resting at *K* is carried down to the intersection, or two apices *A* and *B*, and the proper portion of each load is hung at each apex. In the truss in question, *AC*, in the adjoining figure, will represent $1\frac{1}{16}$ of the weight at *K*, and *BD* $\frac{5}{16}$ of the

same weight. The moving load is supposed to pass on the bridge from *A*. By examination it is seen that *o* and *s* are the first members which need counterbracing. The head of the train must be at the panel point between *f* and *e* for greatest moving-load stress in *s*, and at the panel point between *f* and *g* for that in *o*, and at corresponding positions for other web members.

Fixed-load stress in $\quad s.. = \frac{1}{2}(W + W_1)\ sec\ \alpha = -\quad 6.73$ tons.

Moving-load stress in $s.. = (1 + 3 + 8 + 5)\dfrac{w}{32}\ sec\ \alpha$

$$= +\quad 12.05\ ``$$

Fixed-load stress in $\quad o.. = \frac{1}{2}(W' + W_1)\ sec\ \alpha = +\quad 5.46\ ``$

Moving-load stress in $o.. = (1 + 4 + 3 + 5 + 12)\dfrac{w}{32}\ sec\ \alpha$

$$= -\quad 17.73\ ``$$

Fixed-load stress in $\quad p.. = \frac{1}{2}\ W\ sec\ \alpha \qquad = -\quad 5.46\ ``$

Moving-load stress in $p.. = (1 + 3 + 8 + 5 + 7)\dfrac{w}{32}\ sec\ \alpha$

$$= +\quad 17.02\ ``$$

Fixed-load stress in $\quad r.. = \frac{1}{2}W'\ sec\ \alpha \qquad = +\quad 4.20\ ``$

Moving-load stress in $r.. = (1 + 4 + 3 + 5 + 12 + 7)\dfrac{w}{32} \sec \alpha$

$$= - \ 22.70 \text{ tons.}$$

Greatest stress in $\quad 1.. = (\tfrac{4.9}{3.2} w + \tfrac{1}{2} W) \sec \alpha$

$$= - \ 33.82 \text{ ``}$$

"   "   " $2.. = (\tfrac{4.1}{3.2} w + \tfrac{1}{2} W' + \tfrac{1}{2} W_1) \sec \alpha$

$$= + \ 34.53 \text{ ``}$$

"   "   " $3.. = (\tfrac{3.2}{3.2} w + \tfrac{1}{2} W') \sec \alpha$

$$= + \ 26.9 \quad \text{``}$$

"   "   " $4.. = (\tfrac{4.3}{3.2} w + \tfrac{1}{2} W + \tfrac{1}{2} W_1) \sec \alpha$

$$= - \ 41.47 \text{ ``}$$

"   "   " $5.. = (\tfrac{8.1}{3.2} w + \tfrac{1}{2}(W' + W_1) + W) \sec \alpha$

$$= - \ 59.64 \text{ tons.}$$

"   "   " $8.. = (\tfrac{7.2}{3.2} w + \tfrac{1}{2}(W' + 2 W_1) + W) \sec \alpha$

$$= - \ 68.72 \text{ tons.}$$

"   "   " $6.. = (\tfrac{4.3}{3.2} w + \tfrac{1}{2} W + W' + \tfrac{1}{2} W_1) \sec \alpha$

$$= + \ 49.87 \text{ tons.}$$

"   "   " $7.. = (\tfrac{6.9}{3.2} w + \tfrac{1}{2} W + W' + W_1) \sec \alpha$

$$= + \ 58.93 \text{ tons.}$$

"   "   " $9.. = (\tfrac{8.4}{3.2} w + \tfrac{3}{2} W + W' + W_1) \sec \alpha$

$$= - \ 86.88 \text{ tons.}$$

"   "   " $12.. = (\tfrac{9.1}{3.2} w + \tfrac{3}{2} W + W' + \tfrac{3}{2} W_1) \sec \alpha$

$$= - \ 97.38 \text{ tons.}$$

"   "   " $11.. = (\tfrac{7.2}{3.2} w + \tfrac{3}{2} W' + W_1 + W) \sec \alpha$

$$= + \ 77.13 \text{ tons.}$$

"   "   " $10.. = (\tfrac{8.5}{3.2} w + \tfrac{3}{2} W' + \tfrac{3}{2} W_1 + W) \sec \alpha$

$$= + \ 87.59 \text{ tons.}$$

"   "   " $13.. = (\tfrac{11.3}{3.2} w + \tfrac{3}{2} W' + \tfrac{3}{2} W_1 + 2 W) \sec \alpha$

$$= - 118.37 \text{ tons.}$$

Greatest stress in $16.. = \left(\frac{128}{32} w + \frac{3}{2} W' + 2W_1 + 2W\right) \sec \alpha$
$$= -130.30 \text{ tons.}$$

"   "   " $15.. = \left(\frac{97}{32} w + \frac{3}{2} W + 2W' + \frac{3}{2} W_1\right) \sec \alpha$
$$= +105.79 \text{ tons.}$$

"   "   " $14.. = \left(\frac{112}{32} w + \frac{3}{2} W + 2W' + 2W_1\right) \sec \alpha$
$$= +117.66 \text{ tons.}$$

"   "   " $17.. = \frac{128}{32} w + 2W + 2W' + 2W_1$
$$= -100.00 \quad\text{"}$$

"   "   " $18.. = W + w$ $= -25.00 \quad\text{"}$

The stress in 17 added to the vertical component of the stress in 16 is equal to $\left(8w + 4W + 4W_1 + 3\frac{1}{2}W'\right)$ the weight of the truss and its load, as it should. This constitutes a check in the work if, as was done, each web stress is found by adding a proper increment to a preceding one.

For the greatest chord stresses the load will cover the whole bridge.

Stress in a or $b.. = \left(3\frac{1}{2} w + \frac{3}{2} W + 2W' + 2W_1\right) \tan \alpha$
$$= -78.75 \text{ tons.}$$

"   " c or $d.. = \left(6 w + 3 W' + 3 W_1 + 3 W\right) \tan \alpha +$
$$78.75 = -213.75 \quad\text{"}$$

"   " e or $f.. = 2\left(2 w + W + W_1 + W'\right) \tan \alpha +$
$$213.75 = -303.75 \quad\text{"}$$

"   " g or $h.. = \left(2w + W + W_1 + W'\right) \tan \alpha +$
$$303.75 = -348.75 \quad\text{"}$$

"   " n $.. = \left(4w + 2W + 2W_1 + \frac{3}{2}W'\right) \tan \alpha$
$$= +87.19 \quad\text{"}$$

"   " m $.. = 3\left(2w + W + W_1 + W'\right) \tan \alpha +$
$$87.19 = +222.19 \quad\text{"}$$

"   " l $.. = 2\left(2w + W + W_1 + W'\right) \tan \alpha +$
$$222.19 = +312.19 \quad\text{"}$$

Stress in $k\ldots = (2w + W + W_1 + W')\ tan\ \alpha +$
$$312.19 = +\ 357.19\text{ tons.}$$

In determining these values, it is to be remembered that the increment of chord stress at any panel point is equal to the sum of the horizontal components of the stresses in the web members intersecting at that point.

The results for g or h or k may be easily verified by the method of moments. Let l be the span in feet, and d the depth of a flanged beam, in feet also ; then if w is the load per foot, the flange stress at the centre, as is well known, will be $\dfrac{wl^2}{8d}$. To apply this to the present case, $(w + W + W_1 + W')$ must be written for w, and l and d have the values respectively of 200.00 and 27.75. Hence $\dfrac{wl^2}{8d} = 360.4$ tons.

Now since the resultant stress at either of the centre joints is horizontal in direction for a uniform load from end to end of the truss, the value corresponding to the above will be found by adding to the stress in h the horizontal component of the stress in brace 1, for the supposed uniform load ; or by adding to that in panel k the horizontal component of that in brace 3.

Horizontal component in $1 = \left(\dfrac{w + W}{2}\right) tan\ \alpha = 11.25$ tons,

and $\qquad\qquad 348.75 + 11.25 = 360.00$ tons.

Horizontal component in $3 = \dfrac{W'}{2}\ tan\ \alpha = 2.8125$ tons,

and $\qquad\qquad 357.19 + 2.81 = 360.00$ tons.

Both of the above results are remarkably satisfactory verifications ; they would have agreed exactly, but 0.9 is not the exact value of *tan* α.

If the bridge were a through one, the general method of

calculation would be exactly the same ; the slight changes in the details of the operations are sufficiently obvious after what has been said before.

As a through truss there would be some saving of material, for the secondary verticals 18 would be in tension.

A much greater saving might be effected by using inclined end posts, in which case a short beam or girder would take the place of the end panels *LC*, and braces 15 would be vertical and run up to *L*, while braces 18, 14, and 17 would be omitted altogether.

Art. 15.—Methods of Obtaining Stresses—Stress Sheets.

In the preceding cases the analytical expressions for the stresses have been written in such a manner as would seem best to show in detail the principles by which they are traced.

In practice every engineer has a method best fitted for himself by either habit or taste.

The "strain sheet," or properly "stress sheet," is almost invariably made as shown in the plates. A skeleton of the truss is drawn, and along each member is written the greatest stress belonging to it.

Art. 16.—Ambiguity caused by Counterbraces.

It is important to notice, from what has preceded, that a little ambiguity always exists, both in web and chord stresses, near the middle of the truss, when *counterbraces* are used instead of *counterbracing*. This arises from the fact that even with a single system of triangulation it is impossible to divide the truss by cutting less than four members, which is equivalent, as a question of equilibrium, to having four unknown quantities and only three equations by which they are to be determined.

This ambiguity, however, has been shown to be not of a dangerous character.

CHAPTER III.

NON-CONTINUOUS TRUSSES WITH CHORDS NOT PARALLEL.

Art. 17.—General Methods.

THE determination of stresses in trusses with chords that are not parallel can usually be accomplished more conveniently by either a combination of the method of moments with the graphical method, or the graphical method alone, than in any other manner.

So long as three members, at most, are cut by any surface whatever, dividing a truss into two parts, the problem of the determination of the stresses in these members is determinate; for in such a case the problem is really one of the equilibrium of any system of three forces parallel to a given plane, for the solution of which, as is well known, there are three equations of condition.

The matter, however, requires a little attention here, in order that the particular kind of stress (tension or compression) developed in any bar may be known from the stress diagram.

Let Fig. 1 represent a portion of any truss divided into two parts by the plane (it might be any other surface) AB; let F, G, and H be the points of intersection of this plane and the three members CK, CD, and ED; and let ΣP be the resultant of all the external forces acting on that portion of the truss lying on

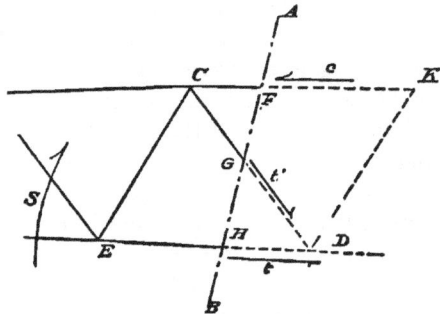

FIG. 1.

the left of AB. The external forces are known, and the stresses c, t', and t, in CK, CD, and ED respectively, are required.

Now, any one of those stresses may be determined by the method of moments, if the origin of moments be properly located. *If the origin be taken at the point of intersection of the lines of action of any two of the stresses, the moments of those stresses will be zero.*

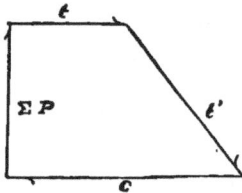

FIG. 2.

Hence, as a general principle, *in order to determine any one of those unknown stresses by the method of moments, the origin of moments is to be taken at the intersection of the lines of action of the other two;* the moment of the third unknown stress can then be placed equal to the resultant moment of the external forces, giving one equation with one unknown quantity.

Suppose in Fig. 1 that the stress in CK is to be found by moments. D is the point of intersection of CD and ED, and consequently is the origin of moments. The lever arm of that stress is of course the normal distance from D to CK. The kind of stress in CK can always be determined by the known direction of the resultant external moment. If the effect of the moment is to shorten the piece in question, the resulting stress will be compression, while tension will exist if the effect is to lengthen the piece.

If any one of the three stresses is known, either of the others may be found by moments, by taking the origin of moments *anywhere* on the line of action of the third force. This method is illustrated on page 255.

In Fig. 1 it will be assumed that the effect of the moment of the external forces on the left of AB is a tendency to turn that portion of the truss in the direction of the curved arrow, and consequently to shorten CK, thus producing compression in that member. Let c represent that compression, then, so far as the portion of the truss on the left of AB is concerned, it is equivalent to an external force acting from F toward C, as shown in the figure by the arrow c.

Now, it is known from rational mechanics that if the sides of a closed polygon represent a system of forces in equilibrium, the lines representing the forces must be laid off in the same direction around the polygon, in order that the true direction of those forces may be represented. If, for instance, the lines of the polygon in Fig. 2 represent a system of forces in equilibrium, the forces must act in the direction shown by the arrow-heads : all in the same direction around the polygon.

The portion of the truss on the left of AB, Fig. 1, is held in equilibrium by the external forces acting upon it, and the three stresses acting at F, G, and H. The amount and direction of the stress acting at F have already been found ; they are represented by c.

In Fig. 2 let ΣP represent the resultant external force acting on the left of AB, both in amount and direction, as shown by the arrow-head ; then draw c parallel to CK, representing in amount and direction of arrow-head the compression in CK; finally, complete the quadrilateral by drawing t and t' parallel to ED and CD respectively. The directions of c and ΣP are fixed already, hence t and t' must have the directions indicated by the arrow-heads in those lines. The directions and magnitudes of the forces t and t' are thus known, and H and G are their points of application. Finally, consider t and t' attached to their points of application, as indicated in Fig. 1 ; their tendencies will be to lengthen EH and CG respectively, hence ED and CD will both be in tension.

The general method, then, of determining the kind of stress existing in any member, is *to apply to the point of division of that member the force represented in magnitude and direction by the proper side of the equilibrium polygon, and observe whether the effect is to shorten or lengthen the member in question : in the first case, the stress will be compression ; in the second, tension.*

This method is perfectly general, and may always be used.

The origin of moments for the stress in CD would be the intersection of CK and ED produced, and C would be the origin for the stress in ED.

So far as the method is concerned, it is a matter of indifference which one of the three stresses is first found by moments; it is simply advisable to take that one which can be found most conveniently.

In all trusses with vertical loading, ΣP is the algebraic sum of the external forces acting on the portion of the truss in question, or the "*shearing stress.*"

Art. 18.—Curved Upper Chord—Two Systems of Vertical and Diagonal Bracing.

The preceding principles will be first applied to the truss shown in Fig. 2, Pl. III. That truss is taken first, rather than a simpler one, because the application of the method is there perfectly general, including all cases possible.

The truss taken is composed of two systems of triangulation, and it is of the greatest importance to notice that *the systems cannot be treated separately*, since the upper chord is curved, and the loads supported by one system induce stresses in the other. As an extension of this part of the matter, it is equally true that with any number of systems of triangulation, of whatever kind, any load on one system will induce stresses in all the others. The result of this is, that with a curved chord and more than one system of triangulation there is *always* ambiguity in the determination of stresses, because more than three members will be severed however the truss may be divided. Certain assumptions, however, may be made which involve no danger, and which give a determinate character to the stresses desired.

The assumption in general terms is this—*i. e.*, that at the point for which ΣP is zero the total load is divided, and each of its portions travels to the nearest abutment by the most direct path. This is a legitimate analysis, so far as simple equilibrium is concerned, but it is by no means certain that the stresses determined are those which actually exist in the truss.

In Fig. 2 of Pl. III., if a uniform load covers the entire truss it will be assumed that the counters 18', 18, 19, 20, and

22, and the corresponding ones on the other side of the centre, are not needed, and consequently not subjected to stress.

Again, suppose the left half of the truss to be loaded with the moving load, in addition to the fixed load over the entire bridge, and suppose $\Sigma P = 0$ for the panel point at the foot of diagonal 10; then it would be assumed that all the verticals are necessary, but only the diagonals 19, 20, 22, 24, 26, 28, 30, 32, 34, 36, and 37 in one part of the truss, and 10, 8, 6, 3, and 2 in the other part.

These assumptions involve no danger, because the stresses deduced by *one* legitimate method of analysis at least are provided for. Nevertheless there is some ambiguity, and when that exists there cannot be economy of material.

The greatest stresses to be determined in the case treated will be found in accordance with the preceding principles.

The moving load will consist of a train headed by four excesses e, and will be supposed brought on from left to right, first touching the truss at R.

The following are the data:

Length of span   = 200 ft. Height at centre   = 35 ft.
Length of panel   = 12.5 "   Height at ends   = 15 "
Number of panels   = 16   Radius of upper chord = 260 "

W (upper) = 500 lbs. per foot = 3.125 tons ⎱ ....Fixed load.
W' (lower) = 800 "   "   "   = 5.000 " ⎰

$w = 13.00$ tons. $w' = 17.00$ tons.

$$e = w' - w = 4.00 \text{ tons.}$$

The truss is supposed to be designed for a single-track through railroad bridge.

The stresses due to the fixed and moving loads will be found separately, and those due to the fixed load will be found first.

For any given condition of loading, the reactions at the two ends R and R' will be denoted by these letters simply.

As the notation indicates, a part of the fixed load is assigned to the upper chord, according to the principles already set forth.

For the fixed load only, $R = R' = 60.9375$ tons $= 7.5$ $(W + W')$. It was shown above that the verticals and main diagonals are the only web members which will be assumed to be stressed by the fixed load.

Fig. 1, Pl. IV., is the only diagram necessary for the determination of all the fixed-load stresses in the truss.

The truss may be divided through the members h, 16, and k without severing more than three members; hence the stresses in those three are determinate.

The stress in h is the most convenient one to find by the method of moments, hence the middle panel point of the lower chord at the intersection of k and 16 will be the origin of moments. The lever-arm of the stress in h is found by careful measurements (it might be found by calculation) on a large drawing to be 34.95 feet. Taking moments, therefore:

$$(7.5 (W + W') \times 100 - 7 (W + W') \times 50) \div 34.95 = 93.1 \text{ tons} =$$
stress in h.

Hereafter, for the sake of brevity, the stress in any member will be represented according to the notation of Art. 13.

Now let a dividing plane cut the truss in h, 16, and k; that portion on the left of it will be held in equilibrium by the external forces $\Sigma P = 7.5 (W + W') - 7 (W + W') = 4.0625$ tons and the stresses (h), (16), and (k); their directions being determined according to the preceding principles.

In Fig. 1, therefore, of Pl. IV., make LM, acting upward, equal to 4.0625 tons, and h, acting toward M, equal to 93.1 tons; then draw k and 16 parallel to the members indicated by these letters; the directions of action of the stresses are indicated by the arrow-heads, being the same around the quadrilateral in question. Stresses (16) and (k) are therefore tension.

The actual stresses were determined with a scale of ten tons to the inch; but Fig. 1 of Pl. IV. is drawn to a scale of twenty tons to the inch.

The next plane of division cuts g, 16, 15, and k, but (k) and

(16) are known, hence (15) and (k) may be determined. For this plane, $\Sigma P = 7.5\,(W + W') - 7\,W' - 6\,W = \frac{1}{2}W' + \frac{3}{2}W = 7.1875$ tons. Lay off from L downward a distance (not lettered) equal to 7.1875, and from the lower extremity of 16 draw g parallel to that panel of the upper chord until it intersects LM in O. As shown, (g) will act toward O, and the difference between LO and 7.1875 will be stress (15); it will act up, and consequently will represent tension; its value is 1.9 tons. g, of course, represents compression.

The third plane of division cuts g, 16, 14, and l. $\Sigma P = \frac{3}{2}(W + W') = 12.1875$ tons. Hence make OK equal to 12.1875, and draw 14 and l parallel to those members; their directions are indicated by the arrow-heads.

The completion of the diagram is simply a repetition of these operations; it is only necessary to use care in giving the stresses their proper directions. The stresses in the verticals will be found in the vertical line AT; the shearing stresses ΣP are also laid off on that line, acting upward.

The following are the results of the complete operation, + denoting tension, and − compression:

(h)	= − 93.1 tons.	(16)	= + 2.6 tons.
(k)	= + 91.5 "	(17)	= + 1.0 "
(g)	= − 93.25 "	(15)	= + 1.9 "
(l)	= + 89.2 "	(14)	= + 4.0 "
(f)	= − 92.1 "	(13)	= − 1.5 "
(m)	= + 83.7 "	(12)	= + 8.5 "
(e)	= − 90.4 "	(11)	= − 1.4 "
(n)	= + 78.0 "	(10)	= + 8.6 "
(d)	= − 85.8 "	(9)	= − 6.3 "
(o)	= + 66.0 "	(8)	= + 16.5 "
(c)	= − 80.5 "	(7)	= − 8.3 "
(p)	= + 49.7 "	(6)	= + 21.0 "
(b)	= − 69.6 "	(5)	= − 13.1 "
(q)	= + 20.0 "	(4)	= − 20.2 "
(a)	= − 52.7 "	(3)	= + 34.8 "
(r)	= 0	(2)	= + 31.8 "

$$(1) = -\,60.9375 - 3.125 = -\,64.0625 \text{ "}$$

6

If the diagram has been properly constructed, the sum of the vertical components of (2), (3), and (a) will be 7.5 (W + W') = 60.9375 tons. In this case it came 60.25 tons, which is sufficiently near to prove the accuracy of the work. Constant checks may also be applied in the course of the work, for the vertical component of the stress in any diagonal must be equal to the algebraic sum of the stress in the vertical which meets it at its foot and the weight which hangs from the same point.

With the exception of a few web members near the centre of the truss, the greatest web stresses will exist at the head of the train when it covers the larger segment of the span, precisely as in trusses with parallel chords. A number of the web stresses, however, which exist near the head of the train, for each of its positions, will be given for the main diagonals, since they appear in the stress diagram and may readily be scaled from it.

Bringing the moving load on from R, panel by panel, it is found by trial that 18′ is the first counter needed, the head of the train being at its foot. For this position of the moving load $R' = 6.375$ tons, hence the point for which $\Sigma P = 0$ is at the head of the train. It is then assumed that only the diagonals 6, 3, and 2 on one side, and 18′, 18, 19, 20, 22, 24, 26, 28, 30, 32, 34, 36, and 37 on the other are needed. The truss may then be divided by a plane cutting the three members d, 6, and p only, and any one of these may be found by the method of moments, but it is convenient to take d. The lever-arm of d is found to be, by a large scale, 26.7 feet. Hence,

$$(d) = (R' \times 162.5) \div 26.7 = -38.8 \text{ tons.}$$

A single four-sided stress polygon then gives:

$$(18') = +19.00 \text{ tons.}$$

The head of the train next rests at the foot of 18. The diagonals on the right of the head of the train which slope similarly to 18, and those on the left of it which slope similarly to 8, are needed ; the others are not. $R' = 10.63$, hence $\Sigma P = 0$ at the head of the train.

Hereafter the lever-arm of any upper chord panel will be denoted by l, with the proper subscript. In the present case $l_e = 29.7$ feet ; hence,

$$(e) = R' \times 150 \div 29.7 = -53.7 \text{ tons,}$$

and a single diagram gives

$$(18) = +24.8 \text{ tons.}$$

In order to save constant repetition, it may be stated as a general rule that for every position of the train all the vertical web members are needed, but *only those inclined braces which slope upward and away from that panel point for which $\Sigma P = 0$ are considered necessary.*

With the head of the train at the foot of 19, $R' = 15.68$ tons, $l_f = 32.10$ feet ; hence,

$$(f) = -67.2 \text{ tons.}$$

Diagrams give

$$(11) = +\ 4.0 \text{ tons.}$$
$$(19) = +28.8 \text{ "}$$

Only the diagrams for the next three positions of the moving load will be given, for they will sufficiently illustrate the rest ; they all embody exactly the same operations.

Figs. 2, 3, and 4, of Pl. IV., are all drawn to the same scale of 20 tons to the inch.

With the head of the train at the foot of 20, $R' = 21.56$ tons, and $l_f = 32.10$ feet ; hence,

$$(f) = (R' \times 137.5 - w'p) \div 32.10 = -85.75 \text{ tons,}$$

since $\Sigma P = 0$ at the foot of 19, and p is the panel length.

Divide the truss through f, 19, and m, then in Fig. 2, Pl. IV., make 1—6 vertical, and f parallel to that panel and equal to 85.75 tons; it acts toward 2 as shown. Make $2 - 3$ equal to $21.56 - w' = 4.56$ tons, and then draw m and 19. These latter must act in the direction shown. Now suppose g, 19, 13, and m severed. From the upper extremity of 19 draw g parallel to that panel (referring to Fig. 2) until it intersects $1 - 6$ in 1; $1 - 2$, acting *downward*, is the *tension* in 13.

It should be stated that the shear $2 - 3 = 4.56$ tons acts *downward*, as shown.

Next make $1 - 6 = 21.56$ tons, acting down, and draw l and 20, acting in the directions shown; finally, draw h and h'. $1 - 4$, which acts upward (not indicated), is the compression in 15; and, in like manner, $4 - 5$ is the compression in 17. Scaling from the diagram:

$(11) = + 5.2$ tons.	$(20) = + 16.0$ tons.
$(13) = + 3.9$ "	$(15) = - 10.1$ "
$(17) = - 9.0$ "	$(19) = + 17.6$ "

With the head of the train at the foot of 22, $R' = 28.25$ tons, and $\Sigma P = 0$ at the foot of 20; hence (g) is the stress to be found by moments. $l_g = 33.75$ feet.

$$(g) = (R' \times 125 - w'p) \div 33.75 = - 98.3 \text{ tons.}$$

Fig. 4, Pl. IV., is the diagram used for this position of the load, and is constructed in precisely the same manner as Fig. 2. $2 - 5$ is equal to $R' - w' = 11.25$ tons, the shear for g, l, and 20, and it acts downward. $1 - 7$ is the shearing stress, 28.25 tons, for h, k, 22, and 20, also acting down, as shown. $1 - 2$ is the stress in 15; $2 - 3$, that in 13; $1 - 4$, that in 17; and $4 - 6$, that in 23. Below are the numerical values:

$(15) = + 3.8$ tons.	$(13) = + 6.0$ tons.
$(22) = + 16.0$ "	$(23) = - 10.8$ "
$(20) = + 20.8$ "	$(17) = - 12.2$ "

Fig. 3 is the diagram for the head of the train at the foot of 24, and is constructed in precisely the same manner as

Figs. 1, 2, or 4. $R' = 35.75$, and $\Sigma P = 0$ for the foot of 20. $l_g = 33.75$ feet. Hence,

$$(g) = (R' \times 125 - 3w'p) \div 33.75 = -113.5 \text{ tons.}$$

$(15) = + 4.8$ tons.	$(24) = + 25.2$ tons.
$(22) = + 15.0$ "	$(23) = - 9.4$ "
$(17) = - 3.2$ "	$(25) = - 15.7$ "

With the head of the train at the foot of 26, $\Sigma P = 0$ for the foot of 22, since $R' = 35.75$ tons. $l_h = 34.65$ feet. Hence,

$$(h) = (R' \times 112.5 - 3w'p) \div 34.65 = -124.7 \text{ tons.}$$

$(22) = + 16.1$ tons.	$(26) = + 34.3$ tons.
$(17) = + 7.0$ "	$(25) = - 2.5$ "
$(24) = + 12.5$ "	$(27) = - 28.0$ "
$(23) = - 10.2$ "	

With the head of the train at the foot of 28, $R' = 53.1875$ tons, and $\Sigma P = 0$ at the foot of 22. $l_h = 34.65$ feet. Hence,

$$(h) = (R' \times 112.5 - 6w'p) \div 34.65 = -136.1 \text{ tons.}$$

$(17) = + 6.9$ tons.	$(25) = - 1.5$ tons.
$(24) = + 12.8$ "	$(28) = + 25.1$ "
$(23) = - 1.0$ "	$(27) = - 16.0$ "
$(26) = + 21.5$ "	$(29) = - 15.5$ "

With the head of the train at the foot of 30, $R' = 63.125$ tons, and $\Sigma P = 0$ for the foot of 24. $l_{h'} = 34.95$ feet.

$$\therefore (h') = (R' \times 100 - 6w'p) \div l_{h'} = -144.1 \text{ tons.}$$

$(17) = + 10.0$ tons.	$(27) = - 9.3$ tons.
$(23) = + 4.5$ "	$(30) = + 35.6$ "
$(26) = + 15.7$ "	$(29) = - 10.7$ "
$(25) = + 1.0$ "	$(31) = - 27.5$ "
$(28) = + 20.8$ "	

With the head of the train at the foot of 32, $R' = 73.875$ tons, and $\Sigma P = 0$ for the foot of 24. $l_{h'} = 34.95$ feet.

$$\therefore (h') = (R' \times 100 \times 10w'p) \div l_{h'} = -150.6 \text{ tons.}$$

(27) = − 9.9 tons.	(32) = + 24.8 tons.
(30) = + 36.0 "	(31) = − 26.8 "
(29) = − 1.0 "	(33) = − 16.7 "

With the head of the train at the foot of 34, $R' = 85.4375$ tons, and $\Sigma P = 0$ for the foot of 24. $l_{h'} = 34.95$ feet.

$$\therefore (h') = (R' \times 100 - 14w'p - wp) \div 34.95 = -154.7 \text{ tons.}$$

(17) = + 8.8 tons.	(32) = + 24.6 tons.
(25) = + 8.9 "	(31) = − 17.4 "
(23) = + 5.5 "	(34) = + 55.2 "
(30) = + 26.5 "	(33) = − 15.4 "
(29) = − 0.8 "	(35) = − 46.8 "

With the head of the train at the foot of 36, $R' = 97.8125$ tons, and $\Sigma P = 0$ for the foot of 24. $l_{h'} = 34.95$ feet. Hence,

$$(h') = (R' \times 100 - 18w'p - 3wp) \div 34.95 = -156.5 \text{ tons.}$$

(17) = + 11.0 tons.	(32) = + 12.0 tons.
(24) = − 2.0 "	(31) = − 20.3 "
(23) = + 4.5 "	(34) = + 60.0 "
(25) = + 12.0 "	(33) = − 2.8 "
(27) = − 5.0 "	(36) = + 40.0 "
(30) = + 29.8 "	(35) = − 50.8 "
(29) = + 8.4 "	

With the head of the train at the foot of 37, $R' = 111.00$ tons, and $\Sigma P = 0$ for the foot of 24. $l_{h'} = 34.95$. Hence,

$$(h') = (R' \times 100 - 22w'p - 6wp) \div 34.95 = -156.00 \text{ tons.}$$

(17) = + 10.9 tons.	(32) = + 10.5 tons.
(24) = − 2.2 "	(31) = − 16.7 "
(23) = + 4.4 "	(34) = + 54.2 "
(25) = + 11.9 "	(36) = + 37.0 "
(28) = + 1.5 "	(35) = − 43.5 "
(27) = − 5.5 "	(37) = + 78.0 "
(30) = + 26.0 "	(38) = − 111.0 "
(29) = + 8.7 "	

The chord stresses with this loading are :

(a')	$= - 87.7$ tons.	(k')	$= + 157.6$ tons.
(b')	$= - 131.2$ "	(l')	$= + 150.5$ "
(c')	$= - 137.0$ "	(m')	$= + 149.7$ "
(d')	$= - 153.5$ "	(n')	$= + 132.0$ "
(e')	$= - 152.6$ "	(o')	$= + 124.5$ "
(f')	$= - 158.7$ "	(p')	$= + 81.8$ "
(g')	$= - 156.6$ "	(q')	$= + 50.0$ "
(h')	$= - 156.0$ "	(r')	$= 0$

The same checks apply as with the fixed load. The greatest stresses are found by combining the fixed and moving load stresses. The greatest chord stresses are thus found to be the following:

(a')	$= - 140.4$ tons.	(r')	$= 0$
(b')	$= - 200.8$ "	(q')	$= + 70.0$ tons.
(c')	$= - 217.5$ "	(p')	$= + 131.5$ "
(d')	$= - 239.3$ "	(o')	$= + 190.5$ "
(e')	$= - 243.0$ "	(n')	$= + 210.0$ "
(f')	$= - 250.8$ "	(m')	$= + 233.4$ "
(g')	$= - 249.85$ "	(l')	$= + 239.7$ "
(h')	$= - 249.1$ "	(k')	$= + 249.1$ "

In the case of the web stresses the combination is effected by taking the algebraic sum. It will be seen that a number of the braces near the centre need counterbracing, *i. e.*, acting consistently with the assumptions that were made. It is uncertain to what extent the existence of the counters renders this counterbracing unnecessary; their influence is therefore neglected.

It is to be borne in mind that the greatest result of a given sign is to be selected from all the preceding moving-load stresses, and added algebraically, to the stress in the same member caused by the fixed load.

$(18') = +$ 19.0 tons.	$(17) = \begin{cases} - & 11.2 \text{ tons.} \\ + & 12.0 \text{ "} \end{cases}$
$(18) = +$ 24.8 "	
$(19) = +$ 28.8 "	$(23) = \begin{cases} - & 8.9 \text{ "} \\ + & 7.4 \text{ "} \end{cases}$
$(20) = +$ 20.8 "	
$(22) = +$ 16.1 "	$(25) = \begin{cases} - & 17.2 \text{ "} \\ + & 10.5 \text{ "} \end{cases}$
$(24) = +$ 27.8 "	
$(26) = +$ 38.3 "	$(27) = \quad - \; 29.4 \text{ "}$
$(28) = +$ 33.6 "	$(29) = \begin{cases} - & 21.8 \text{ "} \\ + & 2.4 \text{ "} \end{cases}$
$(30) = +$ 44.6 "	
$(32) = +$ 41.3 "	$(31) = \quad - \; 35.8 \text{ "}$
$(34) = +$ 81.0 "	$(33) = \quad - \; 29.8 \text{ "}$
$(36) = +$ 74.8 "	$(35) = \quad - \; 71.0 \text{ "}$
$(37) = +$ 109.8 "	$(38) = \quad - \; 175.0625 \text{ tons.}$

In actual practice it perhaps would hardly be worth while to counterbrace the web member 29.

These results, then, are the greatest values of the stresses to which the different members of the truss are subjected.

Art. 19.—General Considerations.

It is clear that in the graphical treatment of such a problem the stress diagrams should be as large as possible. The scale used for all the results obtained in Art. 18 was ten tons to the inch, and it is not usually best to use more tons to the inch than that.

Another method, but a far more tedious one, of applying precisely the same general principles is to determine the stresses produced in every member of the truss by each individual panel load, and then combine the results. The steps of the different operations in such a case are precisely the same as those gone through above.

Although the truss taken consists of but two systems of triangulation, precisely the same method is applicable to any number of any kind of systems, or to the ordinary "bow-string" truss of one system.

In the latter case there is no ambiguity unless counter-braces are used.

It is to be particularly noticed, also, that the method is perfectly independent of the character of the curve of the upper chord; it may equally well be applied to trusses with both chords curved, or to trusses with parallel chords; in fact, it is perfectly general, though usually not desirable.

Art. 20.—Position of Moving Load for Greatest Stress in any Web Member.

The position of the moving vertical load which will give the greatest stress of either kind to any web member in a truss of one system of triangulation, having one or both chords inclined either uniformly or irregularly, may easily be assigned.   Let the figure represent any such truss whatever;

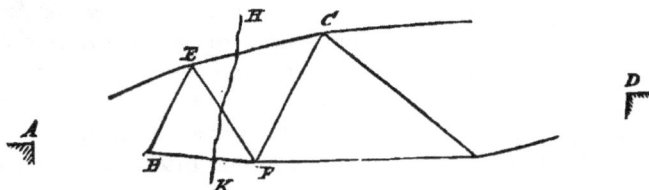

FIG. 1.

A and D being the ends of the span, or points of support. Let the truss be divided by HK, taken anywhere between EB and FC, and let the web member EF be considered.

Although the figure does not show such a condition, let EC and BF be supposed to intersect between HK and D, and let that point be taken as the origin of moments for the determination of the stress in EF. Then suppose any weight to rest between HK and D. The only force then acting on that portion of the truss between A and HK will be the corresponding reaction at A. The effect of that reaction is equivalent to a right-hand (⌢) couple, in reference to the intersection of EC and BF, and the left-hand couple required to balance it must be supplied by the stress in EF; consequently that stress must be tension. All loads on the right of HK, therefore, produce tension in EF.

Now, let a load be placed between A and HK, EC and BF

intersecting as before. The moment of the reaction at A, in reference to that point of intersection, will be greater than the moment of the load in reference to the same point, because the moments of these two forces about D are equal, and the moments in question are those moments diminished by the products of the respective forces into the horizontal distance from D to the intersection of EC and BF, while, of course, the reaction at A is less than the load itself. The resultant action, therefore, about the origin of moments is still equivalent to a right-hand couple, and the stress at EF will still be tension. Under the conditions assumed, then, every load on the truss will produce tension in EF.

The following general principle may now be stated :

In any panel of a truss, if the intersection of the chord sections is found between vertical lines passing through the points of support and above the brace located in that panel, then that brace will be subjected to tension only, and will receive its greatest stress when the whole truss is loaded ; the point of intersection being considered " *above* " the brace when it is above that point of EF, or EF prolonged, lying in the same vertical line with it.

As a limiting case, if EC and BF intersect in a vertical line passing through D, all loads on the left of HK produce no stress in EF.

In the next place, suppose the intersection of EC and BF to lie between HK and A, and let that point be taken as the origin of moments, as before. Precisely the same reasoning used above, which it is not necessary to repeat, shows that the resultant effect of every load on the truss, in reference to the origin of moments, is again equivalent to a right-hand couple, which must be balanced by a left-hand one furnished by the stress in EF; consequently that stress must be compression.

The following general principle, for this case, may then be stated :

In any panel of a truss, if the intersection of the chord sections is found between vertical lines passing through the points of support and below the brace located in that panel, then that brace will be subjected to compression only, and will receive its greatest stress when the whole truss is loaded ; the point of intersection being considered "*below*" the brace when it is below that point of *EF*, or *EF* prolonged, lying in the same vertical line with it.

If *EC* and *BF* intersect in a vertical line through *A*, all weights between *HK* and *D* will produce no stress in *EF*.

As a third case, let the intersection of *EC* and *BF* be found to the right of a vertical line through *D*. The effect of the reaction at *A*, due to any load between *HK* and *D*, will, as before, be equivalent to a right-hand couple about the point of intersection, which couple will cause tension in *EF*. If a load, however, be placed between *HK* and *A*, the result is different, for the moment of the reaction is then *less* than the moment of the weight about the point of intersection, since their moments are equal about *D*. The resultant action of all loads between *HK* and *A* is, then, equivalent to a left-hand couple, causing compression in *EF*.

Since the loads on one side of *HK* produce tension in *EF*, while those on the other produce compression, the problem resolves itself into finding that position of the moving load which will make the algebraic sum of the two stresses a maximum.

Now the general conditions of this case are in no wise affected by the location of the point of intersection of the chord sections, and if that point is located at an infinite distance, the simple case of parallel chords at once results. It follows, then, that if *EC* and *BF* intersect on the right of *D*, the conditions of maximum stress are precisely the same as for parallel chords.

The general principles for the two remaining cases may now be stated together as follows:

In any panel of a truss, if the intersections of the chord sections is found without the vertical lines passing through the points of support, and either above or below the brace located in the panel, then that brace will receive its greatest tension or compression, respectively, under the same condition of loading as for trusses with parallel chords. Those conditions are completely given in Art. 7.

These four cases cover all that exist, and the last two cases include all the trusses which are ordinarily found in the practice of the engineer.

The conditions of loading in the last two cases are exactly the same as those which hold with parallel chords.

If a web member is vertical, the position "*above*," or "*below*," of the point of intersection becomes indeterminate, which shows that the same loading may cause tension or compression, according as the adjacent web member cuts its one extremity or the other.

There is no indetermination in regard to the stresses, however, in any given truss; the conditions are simply limiting ones of those already given. Following the order taken above:

If the adjacent brace cutting the upper extremity of the vertical and the point of intersection of the chord sections are on opposite sides of that vertical brace, all loads on the truss subject it (the vertical brace) to tension; if they are on the same side, all loads subject it (the vertical brace) to compression.

For the two cases in which the point of intersection of the chord sections lies without the verticals passing through the points of support:

If the brace cutting the upper extremity of the vertical and the point of intersection are on opposite sides of the vertical brace, all loads on the segment of the truss adjacent to the point of intersection subject it (the vertical brace) to tension, while all loads on the other segment subject it to compression.

If the brace cutting the upper extremity of the vertical and the point of intersection are on the same side of the vertical brace, all loads on the segment of the truss adjacent to the point of intersection subject it (the vertical brace) to compression, while all loads on the other segment subject it to tension.

Art. 21.—Position of Moving Load for Greatest Stresses in Chords.

Since every load tends to produce the same kind of stress in any given panel, it is clear that for the greatest stress in that panel the load must cover the whole truss, as in the case of parallel chords.

Art. 22.—Horizontal Component of Greatest Stress in any Web Member —Constant Value of the Same for Vertical and Diagonal Bracing with Parabolic Chord.

Of more interest, perhaps, than real value to the engineer, is the expression for the horizontal component of the greatest stress in any web member, though it may very easily be written. It may be useful at times as a numerical check.

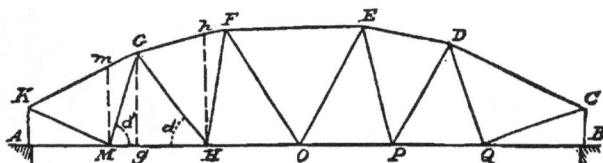

FIG. —.

Let the figure represent a truss of one system of triangulation, subjected to the action of vertical loads passing along the lower chord AB. It is desired to find the horizontal component of the greatest tensile stress in GH. Let Hh and Gg be verticals passing through H and G.

The following notation will be used:

p = panel length (uniform) in AB.

n = number of any panel from B; for PQ, n has the value 2, and 4 for OH.

rp = Mg.

d = Gg.

d' = Hh.

N = number of panels in AB.

l = $AB = Np$.

w = moving panel load.

R = reaction at B.

$n_1 p = BM$.

For the greatest tension in GH the moving load must extend from B to H.

The distance of the centre of such a load from B is $\dfrac{n_1 p}{2}$.

Hence,

$$R = \left(l - \frac{n_1 p}{2}\right)\frac{1}{l}(n_1 - 1)\,w = w\left\{(n_1 - 1) - \frac{n_1(n_1 - 1)}{2}\frac{p}{l}\right\}$$

Now let moments be taken about G.

Hence,

$$(MH) = \left(R(n_1 - r)p - (n_1 - 1)w\left\{(n_1 - r)p - \frac{n_1 p}{2}\right\}\right) \div d$$

$$\text{Or, } (MH) = \frac{wp}{2d}\frac{[l - (n_1 - r)\,p]}{l}\,n_1(n_1 - 1) \quad . \quad (1).$$

In order to obtain the horizontal component of the stress in GF, due to the assumed load, it is only necessary to take moments about H in precisely the same manner. The expression, however, can be derived immediately from Eq. (1) by putting $r = 1$, and writing d' for d.

$$\therefore \text{ Hor. Com. } (GF) = \frac{wp}{2d'}\frac{[l - (n_1 - 1)\,p]}{l}\,n_1(n_1 - 1) \quad . \quad (2).$$

The horizontal component of the greatest tensile stress in GH is the difference between the second members of Eqs. (1) and (2); let it be called H_1.

$$\therefore H_1 = \frac{wp}{2l}\left\{\frac{l - (n_1 - 1)\,p}{d'} - \frac{l - (n_1 - r)\,p}{d}\right\}n_1(n_1 - 1) \; . \; (3).$$

If α is the angle of inclination of GH to a horizontal line, then:

$$(GH) = H_1 \sec \alpha \; . \; . \; . \; . \; . \; . \; (4).$$

Eqs. (3) and (4) apply to all tensile web stresses. For compressive web stresses as typified by (GM) there would be found the Hor. Comp. (GK), instead of Hor. Comp. (GF), by taking moments about M; d' would then represent Mm. By making $r = 0$ in Eq. 1:

Hor. Comp. $(GK) = \dfrac{wp}{2d'}\dfrac{(l - n_1p)}{l} n_1(n_1 - 1)$. (5).

Hence, for the horizontal component of the greatest compressive stress in GM:

$$H_1' = \frac{wp}{2l}\left\{\frac{l - n_1 p}{d'} - \frac{l - (n_1 - r)p}{d}\right\} n_1(n_1 - 1) \ . \ (6).$$

And, $(GM) = H_1' \sec \alpha'$ . . . . (7).

By means of the Eqs. (3), (4), (6), and (7), every web stress in the truss may be determined by formula.

If GM is vertical, $r = 0$ and $d = d'$ in Eq. (6), and $H_1' = 0$, as it should.

If GH is vertical, Eq. (3) shows H_1 to be zero in the same manner.

It is to be borne in mind, in the application of these formulæ, that n is counted along the loaded segment; also that d', for tension, is taken at the head of the train, and one panel in front of it for compression.

If the moving load passes along the upper chord, exactly the same formulæ hold true, but d' taken at the head of the train will give compression, and tension when taken a panel length in front.

If the curve KFC is a parabola, with vertex at the centre of the span, if K and C coincide with A and B, respectively, and if GM and all corresponding web members are vertical, Eq. (3) becomes:

$$H_1 = \frac{wp}{2l}\left\{\frac{l - (n_1 - 1)p}{d'} - \frac{l - n_1 p}{d}\right\} n_1(n_1 - 1) \ . \ (8).$$

From the ordinary equation to the parabola:

$$y^2 = ax$$

and
$$\frac{l^2}{4} = ad_1;$$

in which d_1 is the depth of the truss at the middle of the span. Hence,

$$y^2 = \frac{l^2}{4d_1} x.$$

In this equation put $y = \frac{l}{2} - n_1 p$ and $x = d_1 - d$, then $y = \frac{l}{2} - (n_1 - 1) p$ and $x = d_1 - d'$, successively. There will result:

$$\frac{l^2}{4} - n_1\, lp + n_1^2 p^2 = \frac{l^2}{4d_1}(d_1 - d);$$

$$\frac{l^2}{4} - (n_1 - 1)\, lp + (n_1 - 1)^2 p^2 = \frac{l^2}{4d_1}(d_1 - d').$$

Remembering that $l = Np$:

$$d = d_1\, \frac{4\,(n_1\, N - n_1^2)}{N^2};$$

$$d' = d_1\, \frac{4((n_1 - 1)\, N - (n_1 - 1)^2)}{N^2}.$$

Putting these values in Eq. (8), also $Np = l$, there will result:

$$H_1 = \frac{w p^2 N^2}{8 d_1\, l} \left\{ \frac{N - (n_1 - 1)}{(n_1 - 1)\, N - (n_1 - 1)^2} - \frac{N - n_1}{n_1\, N - n_1^2} \right\} n_1\, (n_1 - 1)$$

$$\therefore H_1 = \frac{w p N}{8 d_1} = \frac{w l}{8 d_1} = \text{constant} \quad . \quad . \quad . \quad (9).$$

As this is the horizontal component of the greatest tension in any diagonal web member and constant, *that greatest stress itself is the hypothenuse, parallel to the brace in question, of a right-angled triangle of which the base is* $H_1 = \dfrac{w l}{8 d_1}.$

This furnishes a very short method of finding the stress in any inclined web member.

The similarity between H_1 and the total stress in the horizontal chord, with the truss wholly loaded, is interesting.

If the trussing is so designed that the diagonal or inclined braces sustain compression, Eq. (6) gives precisely the same general result, but with the sign changed.

In such a case there would be substituted in the parabolic equation $y = \dfrac{l}{2} - n_1 p$ and $x = d_1 - d'$ also, $y = \dfrac{l}{2} - (n_1 - 1)p$ and $x = d_1 - d$; d and d' having changed places.

If no web members are vertical, y will have for one value in the equation to the parabola, $\dfrac{l}{2} - (n_1 - r)p$ instead of $\dfrac{l}{2} - n_1 p$, the other values to be substituted remaining the same. This new value gives,

$$d = d_1 \frac{4\left[(n_1 - r)N - (n_1 - r)^2\right]}{N^2}.$$

Now making the substitutions in Eq. (3) instead of Eq. (8):

$$H_1 = \frac{wl}{8 d_1}\left(\frac{1}{n_1 - 1} - \frac{1}{n_1 - r}\right)n_1(n_1 - 1)$$

$$\therefore H_1 = \frac{wl}{8 d_1}\left(\frac{1 - r}{n_1 - r}\right)n_1.$$

Art. 23.—Bowstring Truss—Diagonal Bracing—Example.

The first form of bowstring truss to be treated is that shown in Fig. 1. All braces are inclined, and each apex in the upper chord is vertically over the centre of the panel below.

The truss is supposed to be designed for a highway bridge. There is a sidewalk on either side.

7

The greatest moving load will be assumed to be that of an advancing crowd of people, from the left end of the span, weighing eighty-five pounds per square foot.

As the span is a short one, and the roadway heavy, the whole of the fixed load will be put upon the lower chord.

The following are the data required:

Span = 72 feet.   Depth of truss at centre = 11.7 feet.

Radius of circumference of circle passing through apices in upper chord = 60 feet.

Number of panels = 6.   Panel length = 12 feet.

Width of roadway, from centre to centre of trusses = 20 feet.

Width of each sidewalk = 6 feet.

W = 900 pounds per foot = 5.4 tons per panel.

w = 32 × 85 × 12 = 16.32 tons per panel.

As usual, W and w refer to fixed and moving loads respectively.

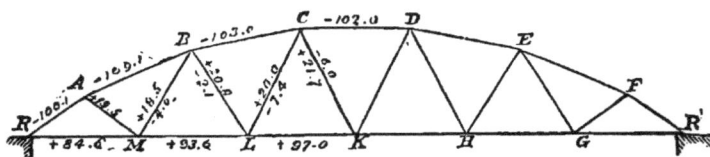

Fig. 1.

In all the diagrams that follow, the lines indicated by any two letters are parallel to the members of the truss at the extremities of which the same letters are found.

In this truss and in the two which follow, the upper and lower chord sections found in any panel intersect outside of the span RR', hence the positions of the moving load, for the greatest web stresses, are precisely the same as those which would be taken for a truss with parallel chords.

With the head of the moving load at M, the truss is first to be considered as divided through the members AB, BM, and ML; then through BC, BL, and ML.

Fig. 2 is the complete diagram for this position.

R (reaction) $= 27.1$ tons.

$(ML) = (R \times 18 - 21.72 \times 6) \div 9.3 = 38.4$ tons. 9.3 feet is the depth of truss through B.

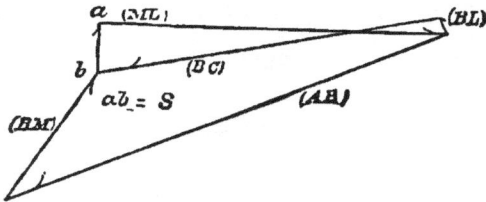

FIG. 2.

The shear is,

$$S = 27.1 - 21.72 = 5.38 \text{ tons.}$$

The diagram needs no explanation. It gives:

$(BM) = + 18.5$ tons. $(BL) = - 2.1$ tons.

With the head of the moving load at L: $R = 37.98$ tons. Fig. 3 is the complete diagram.

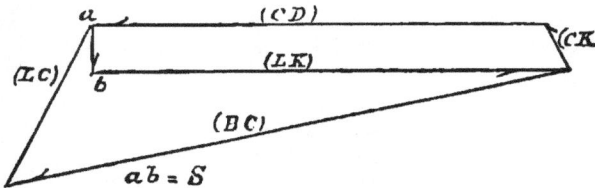

FIG. 3.

The truss is first supposed to be divided through BC, CL, and LK; then, through CD, CK, and LK.

$$(LK) = (R \times 30 - 2 \times 21.72 \times 12) \div 11.7 = 52.8 \text{ tons.}$$

The shear is,

$$S = 37.98 - 43.44 = - 5.46 \text{ tons.}$$

The diagram gives:

$$(LC) = + 20.0 \text{ tons.} \qquad (CK) = - 0.60 \text{ tons.}$$

With the head of the moving load at K: $R = 46.14$ tons. Hence,

$$(KH) = (R \times 42 - 3 \times 21.72 \times 18) \div 11.7 = 65.4 \text{ tons.}$$

Fig. 4 is the complete diagram.

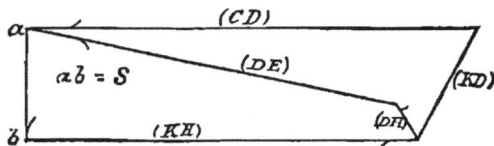

FIG. 4.

The shear is,

$$S = 46.14 - 65.16 = - 19.02 \text{ tons.}$$

The diagram gives:

$$(KD) = + 21.7 \text{ tons.} \qquad (DH) = - 7.4 \text{ tons.}$$

With the head of the moving load at H: $R' = 40.7$ tons; and $S = - (40.7 - 5.4) = - 35.3$ tons.

$$(HG) = (R' \times 18 - 5.4 \times 6) \div 9.3 = 75.3 \text{ tons.}$$

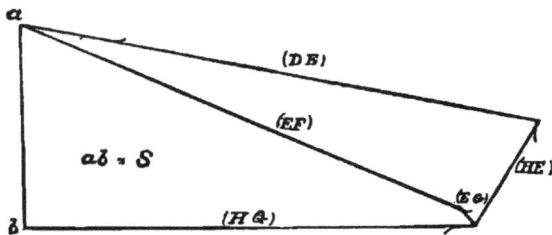

FIG. 5.

Fig. 5 is the diagram, and it gives:

$$(HE) = + 20.9 \text{ tons.} \qquad (EG) = - 4.0 \text{ tons.}$$

With the head of the moving load at *G*, or with the moving load over the whole bridge: $R = 2.5 \times (W + w) = 54.3$ tons. In this case no chord stress is found by moments, but the diagram, Fig. 6, is worked up from the end of the truss.

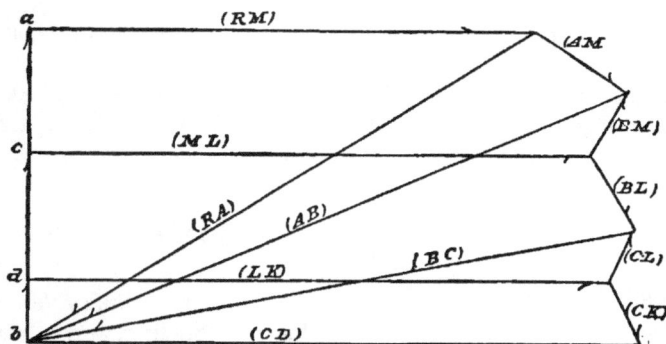

FIG. 6.

It gives:

$(RA) = -100.1$ tons. $(RM) = +84.6$ tons.
$(AB) = -109.1$ " $(ML) = +93.6$ "
$(BC) = -103.0$ " $(LK) = +97.0$ "
$(CD) = -102.0$ " $(AM) = +19.5$ "

The results of these diagrams are collected and written in Fig. 1.

Both web and chord stresses may be checked by moments as follows.

Moments about *K* give:

$(CD) = -(54.3 \times 36 - 2 \times 21.72 \times 18) \div 11.7 = -100.2$ tons.

The diagram gave 102.0 tons. The agreement is close enough for the purpose, but in an actual truss the difference ought not to be greater than one per cent. of the smallest result.

Again, *BC* and *LK* intersect in a point about 30.8 feet to the left of *R*, and the normal distance from that point to *CL* produced is about 48.5 feet. Hence, with the head of the moving load at L, and, by taking moments about the point of intersection:

$$(CL) = (2 \times 21.72 \times 48.8 - R \times 30.8) \div 48.5 = + 19.6 \text{ tons.}$$

The diagram gave $+ 20.0$ tons. The agreement is sufficiently close.

Numbers of checks like the two above should be applied.

The Fig. 1 shows that the web members MB, BL, CL, CK, DK, DH, HE, and EG must be counterbraced.

Art. 24.—Bowstring Truss—Vertical and Diagonal Bracing with Counters.

The case next to be taken is that of an ordinary "bowstring" truss with vertical and diagonal bracing, and is represented in Fig. 1. The inclined braces are not supposed capable of resistance to a compressive stress. Inasmuch as counters are almost invariably introduced in such a truss in ordinary engineering practice, they will be supposed to exist in this case.

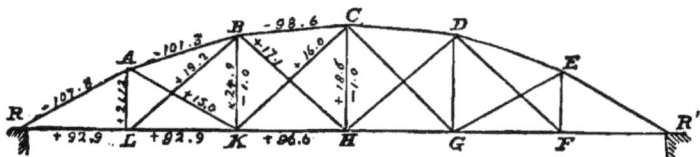

FIG. 1.

The truss is supposed to be designed for a highway bridge, furnished with sidewalk on each side.

The greatest moving load will be taken as a crowd of people weighing eighty-five pounds per square foot, covering roadway and sidewalks, advancing panel by panel from R until the truss is entirely covered.

Since the span is a short one, and the roadway very heavy, the whole of the fixed load will be taken as applied to the lower chord.

The following are the data required ; they are taken from the preceding Article :

Span $= 72$ feet. Depth of truss, 12 feet.
Radius of circumference of circle passing through upper extremities of verticals $= 60$ feet.
Number of panels $= 6$. Panel length $= 12$ feet.
Width of roadway, from centre to centre of trusses $= 20$ feet.
Width of each sidewalk $= 6$ feet.
$W = 900$ pounds per foot $= 5.4$ tons per panel.
$w = 32 \times 85 \times 12 = 32,640$ pounds $= 16.32$ tons per panel.

As in Art. 18, if the plane dividing the truss cut more than three members, some one of these members must be neglected or assumed not to exist.

For the sake of brevity, two letters inclosed by a parenthesis will denote the stress in the member indicated by these letters.

The placing the load for the greatest web stresses, is done in accordance with the general principles established in Art. 20.

When the head of the moving load is at L, the existence of AK must be ignored; for if BL be then omitted, AK will suffer compression.

With the head of the train at L, $R = 27.1$ tons; hence

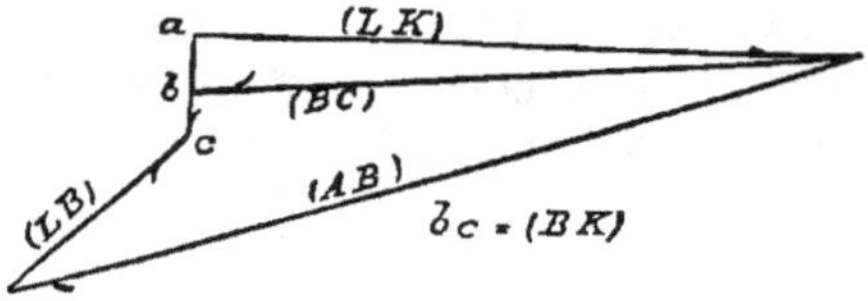

FIG. 2.

$$(LK) = (R \times 24 - 21.72 \times 12) \div (BK = 10.8) = 36.1 \text{ tons.}$$

Fig. 2 is the complete diagram for this case, and explains itself. ac is the shear $\Sigma P = 27.1 - 21.72 = 5.38$ tons. Scaling from Fig. 2 :

$$(BL) = + 19.2 \text{ tons.} \qquad (BK) = + 2.0 \text{ tons.}$$

BH is also omitted for this loading.

With the head of the moving load at K, $R = 37.98$ tons, and CG and BH are omitted. $(KH) = (R \times 36 - 2 \times 21.72 \times$

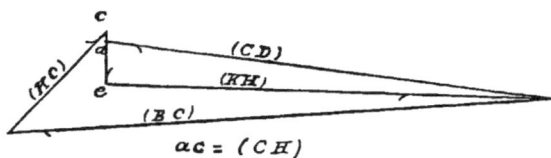

$ac = (CH)$

FIG. 3.

$18) \div (CH = 12) = 48.78$ tons.   Fig. 3 is the diagram for this case.   *ee* is the shear, acting downwards, $\Sigma P = 37.98 - 2 \times 21.72 = -5.46$ ton.

This diagram gives the results :

$$(CH) = -1.0 \text{ ton.} \qquad (KC) = +16.0 \text{ tons.}$$

With the head of the moving load at H, $R = 46.14$ tons. DF and CG will be omitted.

It is unnecessary to give the diagram for this case.   It is drawn precisely as the two preceding ones have been.

The diagram gives :

$$(DG) = -1.0 \text{ ton.} \qquad (HD) = +17.1 \text{ tons.}$$

Neither will the diagram for the head of the load at G be given, as it is constructed exactly like the others.   It gives, omitting GC and DF:

$$(GE) = +15.00 \text{ tons.}$$

For the greatest chord stresses the moving load covers the entire truss, and Fig. 4 of the next Article is the complete diagram for the case.   All the explanation which the diagram needs is there given. BL, KC, GC and FD are supposed to be omitted.

Taking moments about B, with uniform load $(w + W)$:

$$(KH) = (54.3 \times 24 - 21.72 \times 12) \div 10.8 = 96.5 \text{ tons.}$$

Others may be checked in the same way.

The most rational circumstances under which the greatest tensile stresses can be supposed to occur in the verticals, are those under which they are found in the next Article; and the results there obtained are used in this case. They are introduced, without more explanation, in Fig. 1. Their diagrams will be found in the next Article.

These results are by no means satisfactory, but nothing better can be done with such a form of truss.

Some of the web stresses should be checked by moments, by the general method.

Fig. 1 shows the greatest web stresses selected from all the preceding results.

It is far more convenient to treat the fixed and moving loads together, as has been done in this case, than to treat them separately, as, of course, may be done.

Again, the stresses caused by each panel load on all the members of the truss may be found, and their effects combined, but this also requires far more labor than the method followed.

The stress diagram for each position of the moving load might have been worked up from the end of the truss, as was that for the chord stresses, but it saves considerable labor to find one chord stress by moments, and begin the diagram with that.

Art. 25. — Bowstring Truss — Vertical and Diagonal Bracing without Counters.

It is evident, from what has preceded, that the existence of the counters causes considerable ambiguity in the web stresses,

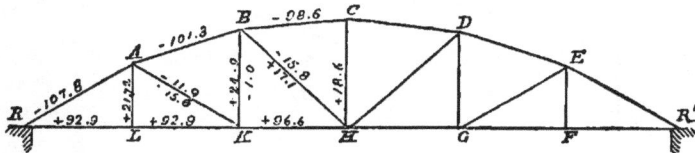

FIG. 1.

and it is much more satisfactory from a strictly technical point of view to leave them out, as shown in Fig. 1.

Fig. 1 is exactly the same as Fig. 1 of the preceding Article, with the counters omitted, and in this Article will be found the stresses existing in it with precisely the same data as were used above.

The moving load is brought on panel by panel from R, according to the general principles established in a preceding Article.

With the head of the moving load at L, $R = 27.1$ tons. As before, $(LK) = 46.9$ tons.

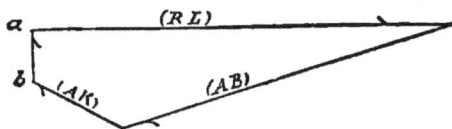

FIG. 2.

Fig. 2 is the complete diagram for this loading. $(RL) = (LK) = 46.9$ tons, and $ab = 27.1 - 21.72 = 5.38$ tons.

Hence, $(AK) = 11.9$ tons.

With the head of the moving load at K, $R = 37.98$ tons.

$(KH) = (R \times 24 - 12 \times 21.72) \div 10.8 = 60.27$ tons.

Fig. 3 is the diagram for this loading, and it explains itself. Hence,

$(BK) = + 24.9$ tons. $(BH) = - 15.8$ tons.

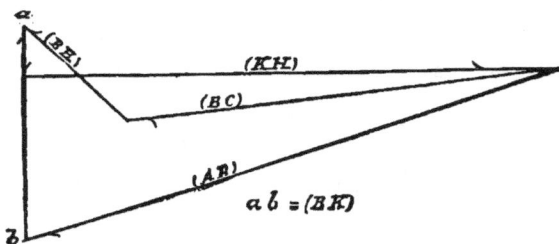

FIG. 3.

It is unnecessary to show the diagrams for the two cases of the head of the moving load at H and G. They give respectively :

$(DH) = +17.1$ tons, $(DG) = -1.0$ ton, and $(GE) = +15.0$ tons.

Fig. 4 is the diagram for the moving load over the whole bridge. *bd* is the reaction $R = R' = 54.30$ tons; *dc* $= 21.72$ tons; and *fg* is equal to $\frac{1}{2}$ (21.72) tons.

The greatest chord stresses, taken from Fig. 4, are written in Fig. 1.

The greatest web stresses are selected from all the preceding results, and also written in Fig. 1.

The stress (BC) may easily be checked by moments, as follows, by taking the origin at H. The normal distance of H from (BC) is 11.9 feet. Hence,

$$(BC) = \{2\frac{1}{2}(w + W) \times 36 - 3 \times 12 (w + W)\} \div 11.9 = 98.56 \text{ tons.}$$

Other and similar checks should also be applied.

It is seen in Fig. 1 that the diagonals need counterbracing, but there is no ambiguity, and the superiority of the design over that in the preceding Article is evident.

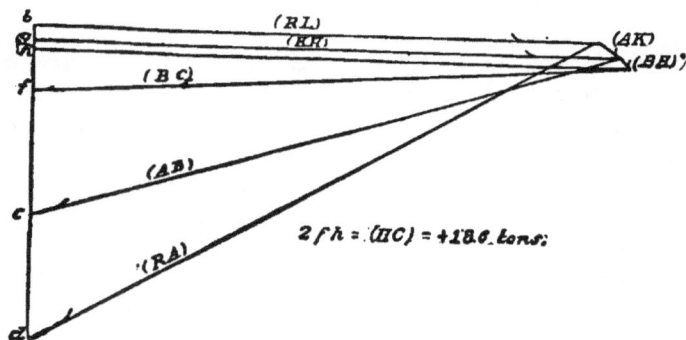

FIG. 4.

It is to be carefully borne in mind that the diagrams must always be drawn as large and as accurately as possible. Those of the present Article were constructed to a scale of ten tons to the inch. The figures do not show the scale.

Art. 26.—Deck Truss with Curved Lower Chord, Concave Downward —Example.

The truss shown in Fig. 1 is, in some respects, a peculiar one. It has one prominent characteristic which distinguishes it from the bowstring trusses which have been treated in the three preceding articles, in that the chord sections (upper and lower) in any panel, excepting those two at the centre, intersect in the upper chord within the limits of the span. All the web stresses, except those in DN and DM, will have their greatest values when the moving load covers the whole truss, as was shown in Art. 20. NM is horizontal, consequently the

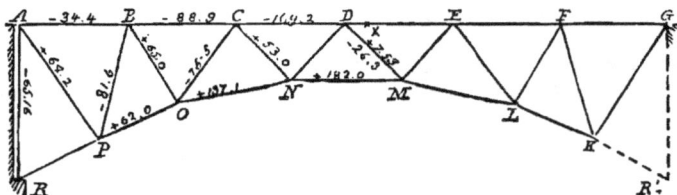

FIG. 1.

intersections of NM with CD and DE are found at an infinite distance from the truss, but ON and CD intersect between E and F (near the latter point): all other intersections are found between A and G.

The positions of moving load for the greatest stresses in DN and DM are the same, therefore, as for a truss with parallel chords.

Observations relating to the positions of the points of intersection of the chord sections in the figure apply to a truss for which the following are the data:

Radius of circumference of circle passing through the apices of the lower chord = 60 feet.
Vertical distance of centre of circle below D = 66 feet.
Depth of truss through D = 6.3 feet.
Span = AG = 72 feet.
$AR = GR'$ = 18 "

Uniform upper chord panel length = 12 feet.
NM = 12 feet.
$ON = ML = 13$ "
$OP = LK = 11.7$ "
Uniform panel fixed load = 5.4 tons = W.
" " moving load = 16.32 " = w.

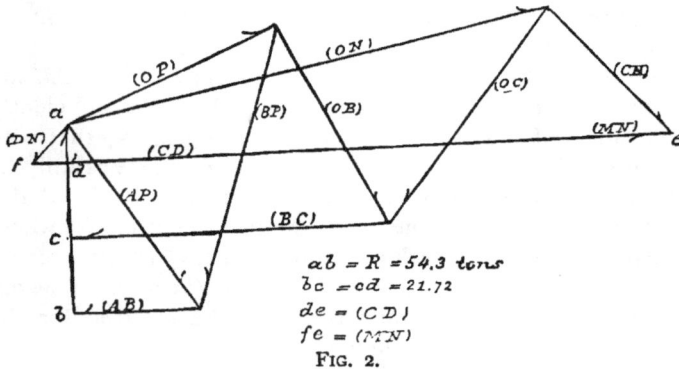

$ab = R = 54.3$ tons
$bc = cd = 21.72$
$de = (CD)$
$fe = (MN)$

FIG. 2.

The loading is the same as that used in the preceding bow-string trusses.

Let the angle which DN or DM makes with a vertical line be denoted by α. Then,

$$tan\ \alpha = 0.952; \qquad sec\ \alpha = 1.38.$$

For greatest tensile stress in DN.

With the moving load extending from A to C:

Reaction $R = 37.98$ tons.
The shear $S = 2 \times 21.72 - 37.98 = 5.46$ tons.
Hence, $(DN) = S\ sec\ \alpha = + 7.53$ tons.

For greatest compressive stress in DM.

With the moving load extending from A to D:

Reaction $R = 46.14$ tons.
The shear $S = 46.14 - 3 \times 21.72 = -19.02$ tons.
Hence, $(DM) = S\ sec\ \alpha = -26.25$ tons.

For the other web stresses the moving load must cover the whole truss, and Fig. 2 is the complete diagram for that condition of loading. With the data given below the figure, no explanation is needed. The results of the diagram will be found in Fig. 1, together with those determined above by the trigonometrical method.

One advantage inherent in this form of truss, as in all in which the chord intersections are found within the limits of the span, is the little counterbracing required.

In the truss taken, the two web members DN and DM are all that require such treatment. Indeed, with a sufficiently small radius of lower chord and centre depth, together with an odd number of upper chord panels, a truss may readily be designed which will require no counterbracing at all. A disadvantage, however, is the small depth at centre, just where a great one is needed, with the resulting heavy chord stresses.

As the Fig. 1 shows, no stress exists in PR or KR'; nevertheless those members would ordinarily be inserted for the purpose of stiffening the whole structure.

The truss may be supported directly, as at G, or there may be an end post, as AR.

The greatest stress in AR, will be the reaction R added to $\frac{1}{2}(w + W)$. Hence,

$$(AR) = -(54.3 + 10.86) = -65.16 \text{ tons.}$$

As checks, moments about D give:

$$(NM) = (54.3 \times 36 - 2 \times 21.72 \times 18) \div 6.3 = +186.0 \text{ tons.}$$

The normal distance from C to ON is 7.5 feet (nearly). Hence, by moments about C:

$$(ON) = (54.3 \times 24 - 21.72 \times 12) \div 7.5 = +139.0 \text{ tons.}$$

OP, prolonged, cuts the upper chord at a point about three feet from D toward E, and the normal distance from that point of intersection to OB, prolonged, is about 23.5 feet. Hence,

$$(OB) = (54.3 \times 39 - 21.72 \times 27) \div 23.5 = +65.1 \text{ tons.}$$

The agreement of the last result with that obtained by diagram is very close, but the other results, by the two methods, show a difference of about two per cent. This is close enough for the present purpose, but in practice the figure and diagrams should be drawn large enough to make this difference, at most, one per cent. of the smallest result.

A truss of this character, with more than one system of triangulation, gives indeterminate stresses, but the approximate method of Art. 18 may be used. Approximate determinations may also be made by treating each system, with its weights, by the methods just given, and combining the results for the chords.

Art. 27.—Crane Trusses.

A form of truss which has been used for powerful cranes,

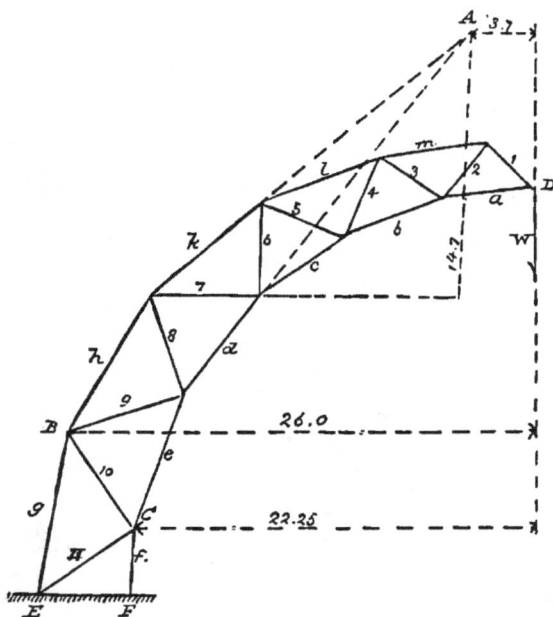

FIG. 1.

under circumstances requiring much head room, is that shown

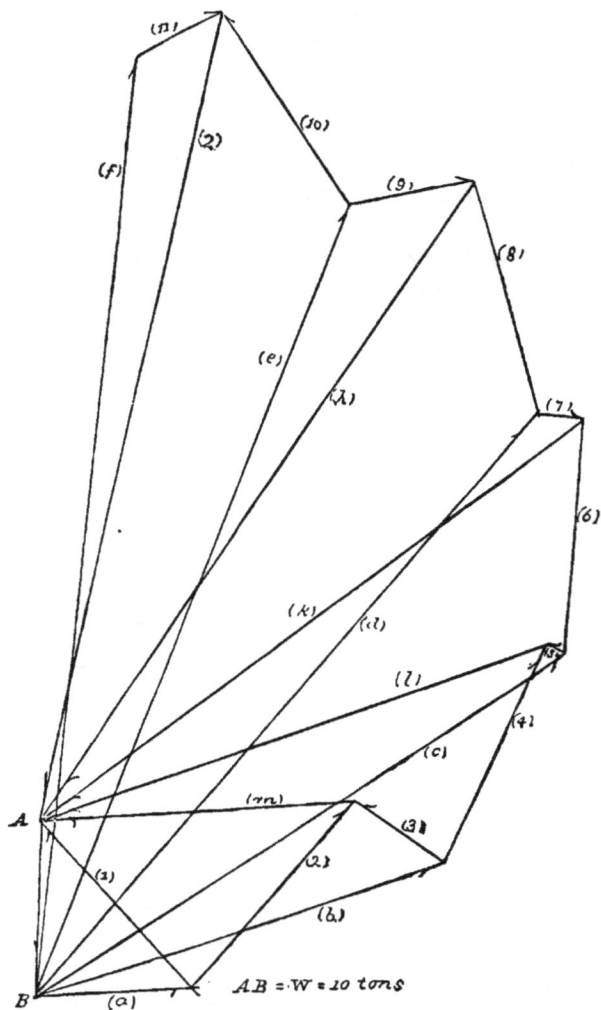

FIG. 2.

in Fig. 1.   It revolves about a vertical axis midway between
E and F.

In the example taken, the weight, *W*, hanging from *D*, the peak, is supposed to be ten tons.

Each chord of the truss *m, l, k*, etc., or *a, b, c*, etc., is made up of chords of quadrants of two circumferences of circles. The radius for the chord *mlk*, etc., is 25 feet, and that for the other chord is 22.6 feet. *EF* is 5 feet.

Denoting the chord panels by single letters:

$$a = 5 \text{ feet.}$$
$$m = b = c = 6 \text{ "}$$
$$l = d = 7 \text{ "}$$
$$k = e = 8 \text{ "}$$
$$h = 9 \text{ "}$$

Fig. 2 is the complete diagram for the stresses in the truss, supposing the *only* load to be the ten tons hanging from the peak. If it should be necessary to take into account the weight of the truss, it would be done precisely as the fixed weights of trusses have been treated in the preceding Articles.

The lines in Fig. 2, denoted by letters and figures, are parallel to lines denoted by the same letters and figures in Fig. 1.

The diagram gives the following results:

(1) = + 12.5 tons.		(2) = − 14 tons.
(3) = + 6.00 "		(4) = − 13.7 "
(5) = + 1.50 "		(6) = − 13.4 "
(7) = − 2.60 "		(8) = − 14.0 "
(9) = − 7.20 "		(10) = − 13.2 "

$$(11) = -5.8 \text{ tons.}$$

(*a*) = − 8.7 tons.		(*m*) = + 17.8 tons.
(*b*) = − 24.0 "		(*l*) = + 30.4 "
(*c*) = − 35.9 "		(*k*) = + 38.7 "
(*d*) = − 44.0 "		(*h*) = + 44.2 "
(*e*) = − 48.2 "		(*g*) = + 47.2 "

$$(f) = -54.1 \text{ tons.}$$

These results may easily be checked by moments. The

8

different lever arms, with two exceptions, to be used, are shown in Fig. 1. The normal distance from g to C is about 4.6, and from e to B about 5.3 feet. These lever arms were scaled from the drawings, and may not be *exactly* right, but near enough for the purpose. By moments about C:

$$(g) = + (10 \times 22.25) \div 4.6 = + 48.4 \text{ tons.}$$

By moments about B:

$$(e) = - (10 \times 26.0) \div 53 = - 49.0 \text{ tons.}$$

The chord sections d and k, prolonged, meet at A, and moments about that point give:

$$(7) = - (10 \times 3.7) \div 14.7 = - 2.5 \text{ tons.}$$

These results agree sufficiently well with those obtained by the diagram.

If the chain, rope, or cable pass along either chord, the tension in it will tend to produce an equal amount of compression in the panels of that chord. The resultant stress, therefore, in any panel will be the algebraic sum of this amount of compression, and the stress due to the weight W.

Fig. 3 is a skeleton diagram of the ordinary crane, which revolves about the centre line of ED as an axis. AB is the weight hung at the peak, A. BC is parallel to DA, and represents the amount of tension in that member. AC is the compression in AE due to the weight W. As before, the tension in the rope or cable tends to produce an equal amount of *compression* in any member along which it lies.

Let l denote the normal distance from DA to any point in the centre line of DE; then any section of DE will be subjected to the bending moment:

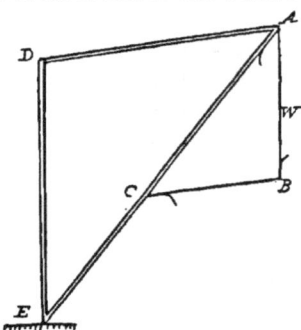

FIG. 3.

$$M = (DA) \times l.$$

DE will also be subjected to a direct stress (tension in Fig. 3) equal to the vertical component of the stress in *DA*. The greatest resultant intensity of stress in any section will be the combination of the intensities due to these two causes.

Art. 28.—Preliminary to the Treatment of Roof Trusses—Wind Pressure —Notation.

Four, only, of the principal types of roof trusses with straight rafters will be treated, since the method used for any one is precisely the same in character as that to be used for any other.

The wind will be assumed to act on one side of the roof, and its resultant action will be assumed to be normal in direction to the rafters. If such is not the case, f and v will represent empirical determinations of the horizontal and vertical components, respectively, in the Articles which follow, and the methods will remain precisely the same.

Let p be the intensity of this normal wind pressure, and let l and l_1 be the lengths of two adjacent panels of the rafter, while d is the horizontal distance between two adjacent and parallel rafters. The total normal wind pressure supposed exerted or concentrated at the point between the two panels, will then be $\dfrac{p\,d\,(l + l_1)}{2}$; or if $l = l_1$, as is usually the case, pdl.

If θ is the angle which the rafter makes with a horizontal line, the horizontal component of this normal pressure will be:

$$f = \frac{p\,d\,(l + l_1)}{2}\ sin\ \theta;\ or\ pdl\ sin\ \theta;$$

and the vertical component will be:

$$v = \frac{p\,d\,(l + l_1)}{2}\ cos\ \theta;\ or\ pdl\ cos\ \theta.$$

Finally, let the total fixed weight of the roof be supposed concentrated at the panel points of the rafters; and let the weight of such load at any panel point be represented by W_2. The total vertical load at any panel point will then be:

$$W = W_2 + v.$$

The vertical reactions due to the vertical component of the

wind pressure and the fixed load are found by the principle of the lever in the usual manner ; that at the left of the span will be called R, and that at the right, R'.

The vertical reactions due to the horizontal forces f, will, however, be called R'', and they will be found for the different cases by taking moments about any point in the horizontal line joining the feet of the rafters. If b is the vertical projection of a rafter, and $2c$ the span, or distance between the feet of two rafters meeting at the ridge, there will result :

$$R'' = bp \cdot \frac{b}{2} \cdot \frac{1}{2c} = \frac{b^2 p}{4c} .$$

At the foot of the rafter pressed by the wind, this reaction will be downward in direction ; at the foot of the other rafter it will be upward.

The total horizontal reaction at the points of support will be equal to bp, *the total horizontal force of the wind*, and its direction of action will be opposite to that of the wind. The horizontal component of the wind pressure and the horizontal reaction produce a couple, equal and opposite to that whose force is R'', and whose lever arm is $2c$.

R'' must be numerically less than R, or the wind will turn over the roof bodily.

If the foot of one rafter is supported on rollers, the horizontal reaction will be wholly exerted at the foot of the other.

If the foot of neither rafter is supported on rollers, the horizontal reaction will be assumed to be equally divided between the points of support.

The stresses for the vertical and horizontal loads are found by separate diagrams, although they might be found by one only, because the slope of the roof may, in some cases, be so small as to make it needless to consider the forces f.

If rollers are used at the foot of one rafter, the wind may press that one or the other. In treating a large roof it may, then, be necessary to take the wind first in one direction and then in the other.

These two, with the case of no rollers, make three possible cases, and an example will be taken in each one.

Art. 29.—First Example.

The truss represented in Fig. 1 is a roof truss, applicable to short spans. There is no "moving load" in such a case. The wind pressure, however, may act on one side and not on the other, and for that reason W, W_1, and W_2 are taken as differing from each other, as was explained in the preceding Article.

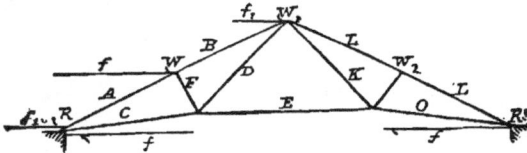

FIG. 1.

There is no essential error in this case in assuming all the load concentrated at the points indicated. The wind pressure is assumed to act on the left side of the truss, so that $W > W_1 > W_2$. Also, $W_1 = \frac{1}{2}(W + W_2)$. A and B are equal in length, and E is horizontal.

Figs. 2 and 3 are the stress diagrams for Fig. 1, and the lines in them, indicated by letters primed, are parallel to, and represent stresses in the members marked with the same letters in

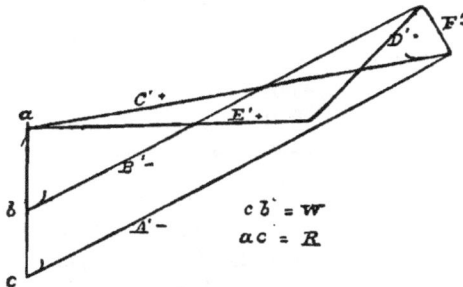

$$cb = w$$
$$ac = R$$

FIG. 2.

Fig. 1. The kinds of stresses in the different members are shown by the signs $+$ or $-$, in both figures, signifying tension or compression, respectively.

Fig. 2 is the diagram for the vertical loads, and Fig. 3 that for the horizontal loads f.

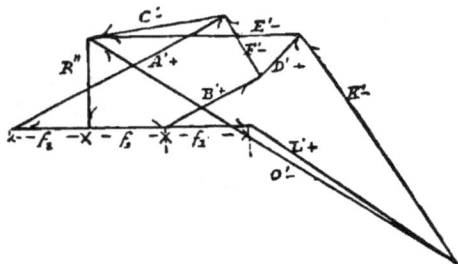

FIG. 3.

Rollers are supposed to be under the foot of neither rafter. Consequently the horizontal reaction at each end is f, as shown. f_1 is equal to $\frac{1}{2} f$.

In Fig. 1, R'' acts downward at the left of the span, and upward at the right.

The resultant stress in any member is the algebraic sum of those given by the two diagrams.

Fig. 3 does not show all its lines parallel to the members of Fig. 1. There is, of course, a diagram similar to Fig. 2, for the right half of the truss, but it is not needed.

The stresses may also be found by the method of moments, by locating the origin of moments according to the general principle stated in Art. 17.

Art. 30.—Second Example.

Fig. 1 of this Article represents a very common roof truss. As before, the total load is supposed concentrated on the rafters at the points indicated. Wind pressure is taken as acting on the left of the roof, making $W > W_1 > W_2$. W_2 is simply the panel weight of the roof, and $W_1 = \frac{1}{2}(W + W_2)$. Each rafter is divided into four equal parts, and W is taken equal to $2W_2$. Hence the reaction $R = 3W$. This does not at all affect the generality of the diagram. The lower extremity, however, of D', in Fig. 2, will not usually be found at a'.

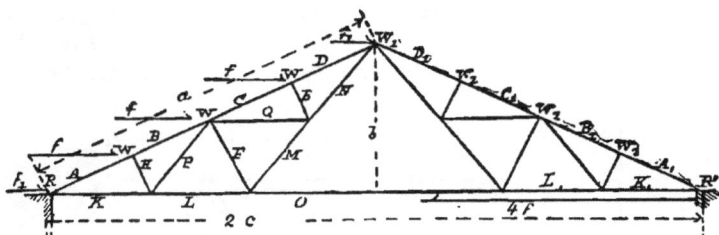

FIG. I.

Fig. 2 is the stress diagram for the vertical loading taken, and the notation has precisely the same meaning as before.

It is seen from Fig. 1, that there is some ambiguity in regard to the stresses C', Q', F', and M'. It may be assumed, however, that $E' = H'$ and $P' = Q'$, which makes $F' = 2H' = 2E'$.

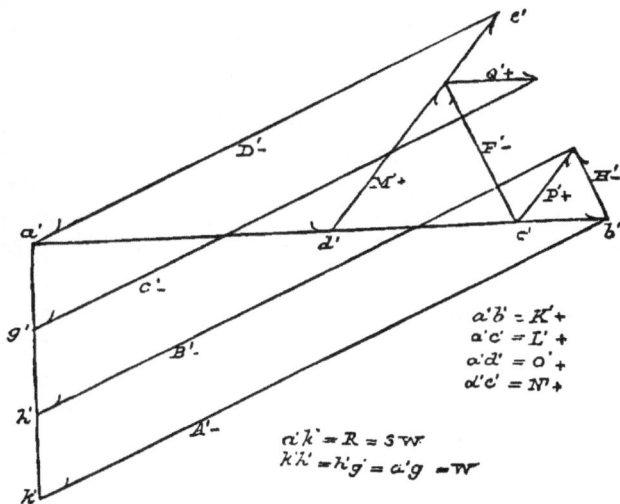

$$a'b' = K' +$$
$$a'c' = L' +$$
$$a'd' = O' +$$
$$a'e' = N' +$$

$$a'k' = R = 3w$$
$$kk' = k'g' = a'g' = w$$

FIG. 2.

The kinds of stresses are shown on the diagrams.

Fig. 3 ·is the diagram for the horizontal stresses f and $f_1 = \frac{1}{2} f$.

Rollers are supposed to be placed at the foot of A. Hence all horizontal reaction will be found at the foot of A_1. As shown, that reaction will be equal to $4f$. The vertical reac-

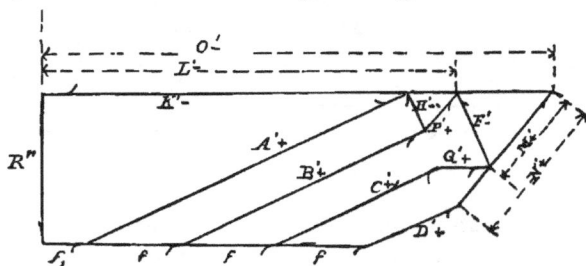

FIG. 3.

tion R'' will be directed downward at the foot of A, and upward at the foot of A_1.

Considering the horizontal forces only:

$$A_1' = B_1' = C_1' = D_1' = -A'.$$
$$K_1' = L_1' = O' = -4f - K'.$$

The resultant stress in any member is found by combining the results of the two diagrams in the usual manner.

This style of truss is so frequently used that formulæ for the stresses due to the vertical loading are given below, in which a is the length of the rafter, c the half span RR', and b the height of W_1 above O. These expressions may be readily derived from Fig. 1.

$$A' = \frac{a}{b} R, \qquad\qquad B' = A' - W\frac{b}{a},$$

$$C' = A' - 2W\frac{b}{a}, \qquad\qquad D' = A' - 3W\frac{b}{a},$$

$$H' = E' = \frac{c}{a} W, \qquad\qquad F' = 2H' = 2\frac{c}{a} W,$$

$$P' = Q' = \tfrac{1}{2} H' \frac{a}{b} = \tfrac{1}{2} \frac{c}{b} W, \qquad K' = \frac{c}{b} R,$$

$$L' = K' - P' = \frac{c}{b} (R - \tfrac{1}{2} W), \qquad M' = \tfrac{1}{2} F' \frac{a}{b} = \frac{c}{b} W,$$

$$O' = L' - M' = \frac{c}{b} (R - \tfrac{3}{2} W), \qquad N' = M' + Q' = \tfrac{3}{2} \frac{c}{b} W.$$

Art. 31.—Third Example.

Figs. 1, 2, and 3 are roof-truss and stress diagrams, respectively. The wind pressure is supposed to act on RW_1, and the total load is taken to be concentrated as shown.

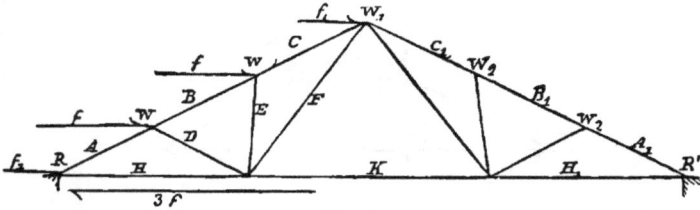

FIG. 1.

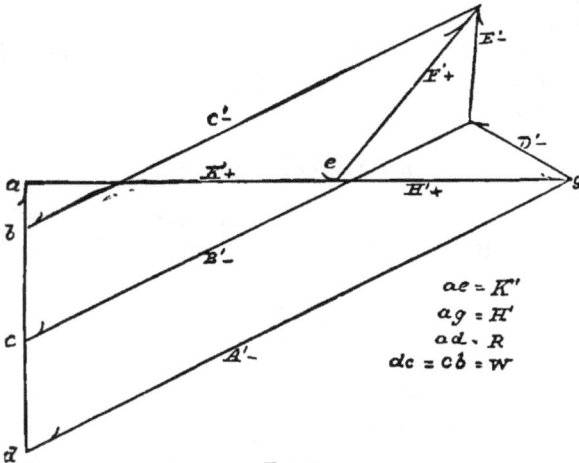

$$ae = K'$$
$$ag = H'$$
$$ad = R$$
$$dc = cb = w$$

FIG. 2.

The rafters are each divided into three equal parts.

The notation has the same signification as that used before, and it is unnecessary to explain the diagrams.

Fig. 2 is the diagram for the vertical loading, and Fig. 3 that for the horizontal forces f and $f_1 = \frac{1}{2}f$.

Rollers are supposed to be placed at the foot of A_1 in Fig. 1. Hence the total horizontal reaction will exist at the foot of A, and its value will be $3f$, as shown. As before, the direction of R'' will be upward at the foot of A_1, and downward at the foot of A.

The resultant stress in any member will be found by combining the results given in Figs. 2 and 3.

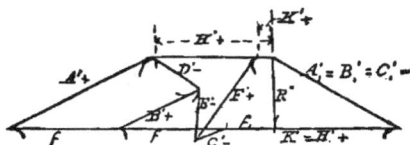

FIG. 3.

Fig. 3 does not show K' equal to H_1^1, as it is not a scale diagram.

The members FH and K are in tension, while D and E are in compression. $W > W_1 > W_2$, the last being the fixed panel weight of the roof.

Art. 32.—Fourth Example.

In this Article, as before, Fig. 1 is the truss, and Fig. 2 is its stress diagram for the vertical loading. Wind pressure is

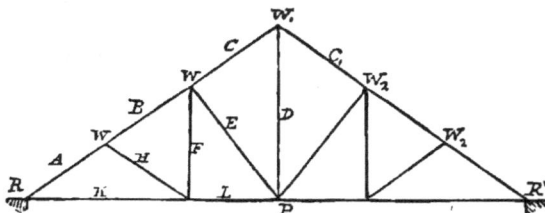

FIG. 1.

taken as acting on RW_1; the reaction R, therefore, is $\frac{1}{3}$ ($3W +$ $W_1 + W_2$) and $W > W_1 > W_2$. The rafters are each divided into three equal parts, and F and D are vertical. W_2 is the fixed panel weight of the roof, and $W_1 = (W + W_2) \div 2$.

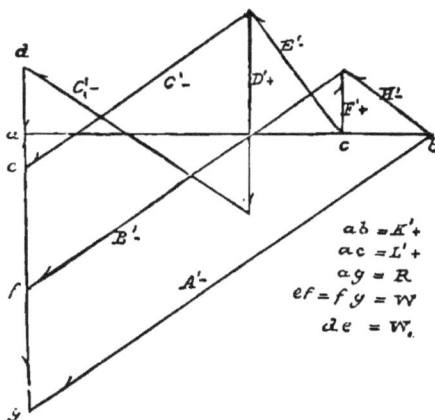

$$ab = K'+$$
$$ac = L'+$$
$$ag = R$$
$$ef = fy = w$$
$$de = w.$$

FIG. 2.

The kinds of stresses in the various members are shown in Fig. 2.

It is unnecessary to give the diagram for the horizontal components of the wind pressure. The method of drawing it will fall under some one of the three cases already given.

Art. 33.—General Considerations.

In all the preceding cases of roof trusses the stresses may be obtained by the method of moments, and therefore the graphical results may be checked by that method.

The operations are precisely similar if a part of the load is hung from points in the tie RR', in all the cases, or if the loading is even more eccentric than that assumed.

In many cases the sides of large buildings are braced to

the roof by oblique members extending from that panel point of RR' adjacent to the side, to some point in that side. In such cases the wind will cause stresses, in the different members of the roof truss, which must be determined independently of those already found and added, algebraically, to them.

CHAPTER IV.

SWING BRIDGES. ENDS SIMPLY RESTING ON SUPPORTS.

Art. 34.—General Considerations.

WITHOUT regarding the nature of the supports or attachments at the extremities of the two arms, swing bridges are divided into two classes: those with center-bearing turntables, and those with rim-bearing turn-tables. In the first class the entire reaction at the pivot pier is exerted through a central pivot, or a nest of one or more series of solid, conical rollers; usually the latter. In such a case there may be a circular drum or framework, supported on wheels running on a circular track, but they are used solely for the purpose of steadying the bridge while open.

In the case of a rim-bearing turn-table, however, the reaction at the pivot pier is exerted through the circular track on which the wheels supporting the drum or framework of the table turn. The object of a pivot, in such a case, is simply to enable the bridge to turn truly about a center.

In reference to the truss, there is evidently only one point of support, at the pivot pier, with a center-bearing turn-table, *i. e.*, at the center.

With a rim-bearing turn-table, however, there may be two or more points of support at the pivot pier, though it will be shown hereafter that, by separating the different systems of triangulation, it will never be necessary to consider more than two at once.

Again, with either turn-table there may be three different methods of supporting or securing the extremities of the two arms of the bridge. These extremities may simply rest on supports, so that the reaction will always be zero, or upward;

125

this is the only case which will receive more than a passing notice in this chapter.

Again, those extremities may be fastened down or latched to the piers when the bridge is not open. The reaction may then be nothing, upward or downward.

In these two cases the reactions at the extremities of the arms will be zero when the bridge is simply closed, and supporting no moving load.

In the first, when the moving load is on one arm, the extremity of the other may be slightly raised from its support; in the second case, however, that extremity will be held down by the latching apparatus, *i. e.*, the reaction will be downward. The object of the latching apparatus is thus seen to be the prevention of the hammering of a truss-end on its support.

Finally, the third method is to raise, by proper machinery, the truss ends, when the bridge is closed, any desired amount.

The object of this arrangement is to insure a reaction at the extremities, which will always be nothing or upward, and thus obviate the hammering before mentioned.

In this case the whole of the bridge weight does not rest at the pivot pier, as the lifting of the ends takes up a part of it; in fact, may take up the whole of it.

Recapitulating, then, the ends of a swing bridge may be:

(1.) Simply supported,

(2.) Latched down,

(3.) Lifted up.

The detailed consideration of the first case will next be taken up.

Art. 35.—General Formula for the Case of Ends Simply Supported—Two Points of Support at Pivot Pier—One Point of Support at Pivot Pier.

With two points of support at the pivot pier there usually arises the case of a continuous beam resting on four points of

Fig.1

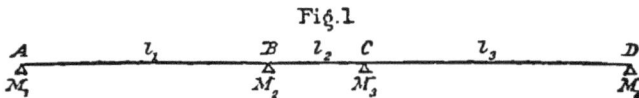

support, as shown in Fig. 1. The notations of the spans and bending moments at the different points of support are sufficiently well shown in the figure. The points of support will all be taken in the same horizontal line, as the formulæ will then also apply to any configuration belonging to a state of no stress, provided the truss may be considered straight between any two points of support (see Appendix). Any truss may be considered straight when an equivalent solid beam has a neutral surface which is plane before flexure ; a straight solid beam is " equivalent " to a straight truss when equal moments of inertia and resistance are found at the same section in the two structures.

The theorem of three moments, in the ordinary form, does not apply, then, to a continuous truss with one chord curved, and none of the following investigations apply to such a case.

Again, in the span l_2 there will be supposed no load, as such is usually the case. The load on l_2 ought always to be supported on short girders or beams resting at B and C, for there is the less complication of stresses in the trusses, and consequently less liability to uncertainties ; besides, such an arrangement is probably more economical in material.

In the present case M_1 and M_4 will each be equal to zero.

Let z denote the distance of the point of application of any force, P, from the left-hand end of the left-hand span, or right-hand end of right-hand span. In l_1, z would be measured from A, while in l_3 it would be measured from D.

The formula expressing the theorem of three moments for all supporting points in the same level becomes, by using the notation of Fig. 1 :

$$M_1 l_1 + 2M_2(l_1 + l_2) + M_3 l_2 + \frac{1}{l_1}\overset{1}{\Sigma}P(l_1^2 - z^2)z + \frac{1}{l_2}\overset{2}{\Sigma}P(l_2^2 - z^2)z = 0.$$

The symbols $\overset{1}{\Sigma}$ and $\overset{2}{\Sigma}$ indicate summations for the spans l_1 and l_2.

Applying the above equation to spans l_1 and l_2, and then to l_2 and l_3, there will result, bearing in mind the circumstances of the present case:

$$2M_2(l_1 + l_2) + M_3 l_2 + \frac{1}{l_1} \overset{1}{\Sigma}P(l_1^2 - z^2)z = 0 \quad . \text{ (1)}.$$

$$M_2 l_2 + 2M_3(l_2 + l_3) + \frac{1}{l_3} \overset{3}{\Sigma}P(l_3^2 - z^2)z = 0 \quad . \text{ (2)}.$$

If Eq. (1) be multiplied by l_2, and Eq. (2) by $2(l_1 + l_2)$; and if the results so obtained be subtracted, there will at once result:

$$M_3 = \left[\frac{l_2}{l_1} \overset{1}{\Sigma}P(l_1^2 - z^2)z - \frac{2(l_1 + l_2)}{l_3} \overset{3}{\Sigma}P(l_3^2 - z^2)z \right] \div$$
$$\{4(l_2 + l_3)(l_1 + l_2) - l_2^2\} \quad . \quad . \quad . \quad . \text{ (3)}.$$

Eq. (1) then gives:

$$M_2 = - \left\{ M_3 l_2 + \frac{1}{l_1} \overset{1}{\Sigma}P(l_1^2 - z^2)z \right\} \div 2(l_1 + l_2). \text{ (4)}.$$

Eq. (2) will evidently give another expression for M_2, but it is not necessary to write it.

Let R_1, R_2, R_3, and R_4 be the reactions at the points of support A, B, C, and D respectively, Fig. 1. Then, adapting the formulæ for reactions from the theorem of three moments (see Appendix) to the notation of the present case, there may at once be written

$$R_1 = \overset{1}{\Sigma}P\frac{l_1 - z}{l_1} + \frac{M_2}{l_1} \quad . \quad . \quad . \quad . \text{ (5)}.$$

$$R_2 = \overset{1}{\Sigma}P\frac{z}{l_1} - \frac{M_2}{l_1} - \frac{M_2 - M_3}{l_2} \quad . \quad . \quad . \text{ (6)}.$$

$$R_3 = + \frac{M_2 - M_3}{l_2} + \overset{3}{\Sigma}P\frac{z}{l_3} - \frac{M_3}{l_3} \quad . \quad . \quad . \text{ (7)}.$$

$$R_4 = \overset{3}{\Sigma}P\frac{l_3 - z}{l_3} + \frac{M_3}{l_3} \quad . \quad . \quad . \quad . \text{ (8)}.$$

As should be the case, there is found:

$$R_1 + R_2 + R_3 + R_4 = \overset{1}{\Sigma}P + \overset{3}{\Sigma}P.$$

In the different summations, Σ, of Eqs. (5), (6), (7), and (8), $\frac{1}{l_1}$ and $\frac{1}{l_3}$ are constant quantities, and may be brought outside of the signs Σ; this fact will considerably simplify the actual summations.

It may sometimes be convenient to use the following equation derived from Eq. (1):

$$M_3 = -\left\{ 2M_2(l_1 + l_2) + \frac{1}{l_1}\overset{1}{\Sigma}P(l_1^2 - z^2)z \right\} \div l_2 \quad (9).$$

These are all the equations necessary for the solution of the case of two supports at the pivot pier, frequently existing if the turn-table is rim-bearing.

If there is only one point of support at the pivot pier, the

Fig. 2

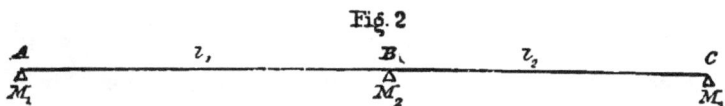

case reduces to that of a continuous beam of two spans only, as shown in Fig. 2.

As A and C are points of support only, M_1 and M_3 are each zero; hence the equation immediately preceding Eq. (1) gives:

$$M_2 = -\left\{ \frac{1}{l_1}\overset{1}{\Sigma}P(l_1^2 - z^2)z + \frac{1}{l_2}\overset{2}{\Sigma}P(l_2^2 - z^2)z \right\} \div$$
$$2(l_1 + l_2) \quad \ldots \ldots (10).$$

There will also result:

$$R_1 = \overset{1}{\Sigma}P\left(\frac{l_1 - z}{l_1}\right) + \frac{M_2}{l_1} \quad \ldots \quad (11).$$

$$R_2 = \overset{1}{\Sigma}P\frac{z}{l_1} - \frac{M_2}{l_1} + \overset{2}{\Sigma}P\frac{z}{l_2} - \frac{M_2}{l_2}. \quad \ldots \quad (12).$$

$$R_3 = \overset{2}{\Sigma}P\left(\frac{l_2 - z}{l_2}\right) + \frac{M_2}{l_2} \quad \ldots \quad (13).$$

9

In Eqs. (11), (12), and (13), also, it will be found convenient in the actual summations to bring the constant quantities $\frac{1}{l_1}$ and $\frac{1}{l_2}$ outside of the signs Σ.

There is again found:

$$R_1 + R_2 + R_3 = \overset{1}{\Sigma}P + \overset{2}{\Sigma}P.$$

These complete the general formulæ needed for the case of ends supported.

Some very important deductions are to be drawn from Eqs. (5), (6), (7), (8), (11), (12), and (13), *considering them applied to bridges with rim-bearing turn-tables.*

Those equations are so written that a positive value of R means a reaction *upward* in direction, while a negative value indicates a *downward* reaction.

In the case of Fig. 1, let the span l_3 be supposed free of any loads P, then the term involving the summation $\overset{3}{\Sigma}$ will disappear. Eq. (3) then shows that M_3 will *always* be positive; consequently Eq. (4) shows that M_2 will *always* be negative.

Using these results in connection with Eqs. (5), (6), (7), and (8), it is at once seen *that R_2 and R_4 will always be positive, while R_3 will always be negative.*

It may also be shown that Eq. (5) makes R_1 positive in such cases as always arise in an engineer's practice, although that equation apparently shows that R_1 may, under some circumstances, be negative, since M_2 is always negative in the case taken, while $(l_1 - z)$ is, of course, always positive.

By deducing the value of M_2 from Eqs. (3) and (4), and introducing it in Eq. (5), there will result:

$$R_1 = \frac{1}{l_1}\left[\overset{1}{\Sigma}P\,(l_1 - z) - \frac{1}{l_1}\left(\frac{2\,(l_2 + l_3)}{4\,(l_1 + l_2)\,(l_2 + l_3) - l_2^2} \right. \right.$$

$$\left. \left. \overset{1}{\Sigma}P\,(l_1^2 - z^2)\,z \right].$$

If l_2 is very small in comparison with l_1 or l_3, and if, at the same time, z is small, there may be written for a single "P," nearly:

$$R_1 = \frac{1}{l_1} \left[Pl_1 - \frac{1}{2l_1^2} \cdot Pl_1^2 z \right];$$

which is evidently positive.

If, on the contrary, l_2 is small and z large, there may be written for a single weight P, nearly:

$$R_1 = \frac{1}{l_1} \left[P(l_1 - z) - \frac{1}{2l_1^2} P(l_1 + z)(l_1 - z) z \right].$$

In this expression R_1 can be equal to zero only by supposing the negative quantity within the brackets to be larger, numerically, than it ought to be, *i. e.*, by making $(l_1 + z) = 2l_1$, and $z = l_1$; hence it can never be negative.

If $l_1 = l_2 = l_3 = l$ the general value of R_1 takes the form:

$$R_1 = \frac{1}{l_1} \left[P(l - z) - \frac{4}{15l^2} P(l^2 - z^2) z \right].$$

If z is small, there results nearly:

$$R_1 = \frac{1}{l_1} \left[Pl - \frac{4}{15} Pz \right].$$

If z is large, nearly:

$$R_1 = \frac{1}{l_1} \left[P(l - z) - \frac{8}{15} P(l - z) \right].$$

In neither case, therefore, can the reaction be negative.

It may, consequently, be assumed as a principle that if l_2 is small in reference to l_1 or l_3, or if it is equal to those quantities, the reaction R_1, in the case supposed, must always be positive; and within those limits must be found all cases of swing bridges.

Since any load on the span l_1 makes the reactions at A and D positive, considered by itself, so any load on l_3 will, of itself, make the reactions at D and A positive. Consequently, as there is never any load on l_2, the reactions at A and D *will always be positive, and the ends of the bridge will never tend to rise from their points of support.* No "hammering," therefore, can take place, in this case, at the ends.

The case of the two points of support, B and C, taken in connection with a center-bearing turn-table will be considered further on.

Fig. 2 represents the case of either a center or rim-bearing turn-table with only one point of support at the pivot pier; the two cases are coincident in all their circumstances.

Eq. (10) shows M_2 to be always negative. Consequently, if there is no load on l_2, R_1 and R_2 will always be positive, while R_3 will always be negative.

When span l_1 carries load, however, the span l_2 may, at the same time, support just enough load to make R_3 equal to zero; more load than that will make R_3 positive, or upward in direction.

But in the present case the point C is simply *a point of support*, consequently no negative or downward reaction can exist there. It becomes necessary, therefore, to determine just how much load on l_2, combined with the full load on l_1, will make $R_3 = 0$. For this purpose, M_2 must be taken from Eq. (10) and inserted in Eq. (13), while in the latter R_3 must be just equal to zero. For the sake of brevity, let $\frac{1}{l_1} \overset{1}{\Sigma} P (l_1^2 - z^2) z$ be represented by A.

Then from Eq. (10):

$$M_2 = - \frac{A}{2(l_1 + l_2)} - \frac{1}{2 l_2 (l_1 + l_2)} \overset{2}{\Sigma} P (l_2^2 - z^2) z . \quad (14).$$

A is a constant quantity so far as this operation is concerned.

Putting this value of M_2 in Eq. (13) after making $R_3 = 0$:

$$2 l_2 (l_1 + l_2) \overset{2}{\Sigma} P (l_2 - z) - \overset{2}{\Sigma} P (l_2^2 - z^2) z = A l_2 . \quad (15).$$

Eq. (15) indicates what disposition of the load on the span

l_2 will make $R_3 = 0$; and it will be seen to be of very easy application.

The determinations of R_1, R_2, and R_3, for all loading in excess of that indicated by Eq. (15) will require the use of Eqs. (11), (12), and (13) as they stand; for all loading less than that amount, however, $R_3 = 0$, and the reactions R_1 and R_2 are to be found by the simple principle of the lever, considering l_2 as a simple overhanging arm, or cantilever. These operations will be shown in detail hereafter.

In this case it is evident that "hammering" will take place at the ends with certain dispositions of loading.

Art. 36.—Ends Simply Supported—Four Points of Support at Center—Example.

The general formulæ of the preceding article will be applied to the truss shown by skeleton diagram in Fig. 1 of Pl. X, in which the verticals are supposed to be compression members, and the diagonals tension members. The following are the data:

$$\text{Panel length} = AB = BC = \&c. = p = 12 \text{ feet.}$$
$$NO = OO' = O'N' = 8 \text{ feet.}$$
$$AA' = 22 \times 12 + 3 \times 8 = 288 \text{ feet.}$$

$Bb = 20$ feet.	$Hh = 26$ feet.
$Cc = 21$ "	$Kk = 27$ "
$Dd = 22$ "	$Ll = 28$ "
$Ee = 23$ "	$Mm = 29$ "
$Ff = 24$ "	$Nn = 30$ "
$Gg = 25$ "	$Oo = 30$ "

Total fixed weight = 1850 pounds per foot.

Upper chord panel fixed weight $= W = 4.5$ tons.
Lower " " " " $= W' = 6.6$ "

Panel moving load $= w = 18$ tons.

No locomotive excess will be taken, but a uniform moving load of 3,000 pounds per foot is assumed.

There are four supporting points at the pivot pier, *i. e.*, *N*,

O, O', and N', yet the truss, as a whole, may be divided into two systems of triangulation, each of which is a truss in itself, possessing but two points of support at the pivot pier. One of these systems is $AcCeEgGkKmMoOo'O'$, etc.; the two points of support are O and O'. The other system evidently is $AbBdDfFhHlLnNn'N'$, etc.; the two points of support are N and N'.

Each of the systems may be considered as a truss subjected to the action of the weights resting at the apices belonging to it, since the stresses existing in one system in no way affect the other, with one unimportant exception.

This exception, for one arm, is the fact that the post Nn is subjected to stress in consequence of any stress existing in mno as a portion of that system to which Nn does not belong. If mno were straight, Nn would not be affected.

This exception apparently makes N a point of support for the system AcC, etc., but it really is not so, since the point n may be supposed held in its position by a member joining the points n and O. All stresses in the two systems may therefore be found independently of each other; but the stress in mn, considered as belonging to AcC, etc., will give to nN an amount of compression equal to the stress in mn multiplied by the sine of its inclination to a horizontal line, as will be seen hereafter.

It is seen, therefore, that all reactions may be found by applying the formulæ of the preceding article to each system of triangulation.

In the example taken there are only two systems of triangulation, but precisely the same observations would hold if there were more than two, *so long as any one system has no more than two points of support at the pivot pier.*

In determining the stresses in the different members of the truss, the simplest method of procedure is to determine all the stresses in the truss due to each weight separately, and then combine the results. But such a method, however simple, involves an amount of labor entirely unnecessary for the determination of web stresses. The results developed in the preceding article make it possible to predict what position of loading will give the greatest stresses of given kinds to the web members.

The office of all members in the arm AN sloping similarly to cE is to carry loads, so to speak, over to A, in order to make up the proper reaction at that point.

Those members and corresponding ones in the other arm may be called "*counterbraces*," or simply "*counters*."

Any reaction at A may be considered as made up of as many *whole* panel loads, fixed or moving, combined with a portion of *one*, as may be necessary ; this *one* will in general be found at that point for which the shear or ΣP is equal to zero. It will *always* be found there if the chords are horizontal.

It was shown in the previous article that any load on either arm will make both the reactions at A and A' positive. In order, therefore, to find the greatest number of counters that will ever be subjected to stress, it will be necessary to bring the moving load on the bridge from A', and then find the position of its head on the arm AN, which will make that point and the point of no shear (for which $\Sigma P = 0$, "P" including fixed as well as moving load) coincide.

As will be seen, this position may easily be found by a very few trials.

In the case taken, the moving load is uniform in intensity, but the above observations hold equally true if there is a locomotive excess.

Now, *the first counter* needed will be the one at the head of the train in the position just found ;* for in all other positions the points of no shear will either be found at this same panel point, or at others between it and the end of the truss.

The greatest stress in any other counter will be found by placing the head of the train at its foot, since the increase of reaction caused by a panel's length advance of the train will be the increase also of shearing stress in the member in question (if the chords are horizontal) consequent upon the same advance.

---

* It is possible that such a one might not be subjected to stress, in which case, of course, it might be omitted.

Thus all the greatest counter-stresses may be determined.

The greatest stress in any counter of the arm $A'N'$ will evidently exist when the train moves from A toward A'.

In connection with the greatest stress in any counter cE, there should also be found the accompanying stress in the vertical, as cC, which cuts its upper extremity, as, near the ends of the bridge, such stress may be the greatest in the members in question.

Again, confining attention to the arm AN, there is to be found the position of the train which will give the greatest stress to any web member which slopes upward and *away* from A, as Fh; such members may be called "main braces."

Since any load on $A'N'$ causes a negative reaction at either N or O, the greatest reactions will exist at the latter points when all the moving load resting on the bridge is found on the arm AN. Remembering in connection with this, that a portion of every load resting on AN passes to A, and the other portion to N or O, it follows that *the greatest shear in any of the main members under consideration will exist if the train covers the bridge from* A *to its foot.* In this manner are determined the greatest stresses in all the main web members considered.

In connection with the greatest stresses in these web members, there should be found those existing in the verticals which cut their upper extremities, as these latter will, in general, be the greatest stresses in those verticals.

The exceptions, if any, will be found near the ends.

The main web stresses in the arm $A'N'$ will evidently be found by bringing on the moving load from A'.

The same conclusions in reference to positions of loading for the greatest web stresses may be arrived at more elegantly and concisely by a consideration of the shear at either end of the arm in question.

Let S be the shear for any position of loading at either end of the arm, in conformity with the treatment of braces or counterbraces, or span, considered positive in this case. Let n and n' be the numbers of fixed and moving panel loads respectively, between the end of the arm where S exists, and any web member. Then the shear for that web member is :

$$s = S - n(W + W') - n'w.$$

The greatest stress in the web member considered occurs when s is the greatest ; but s is the greatest when $(S - n'w)$ is the greatest, since $n(W + W')$ is constant for a given section. But $n'w$ varies much more rapidly than S, although in the same direction, consequently $(S - n'w)$ will be the greatest possible when $n' = 0$.

In the present example, for the left arm of the truss and for the main web members, S would be taken at an indefinitely short distance to the left of N or O, and the results of the preceding chapter show that it will be the greatest possible when the moving load covers the whole arm, since any load on the right arm makes a positive reaction at A. But n' must, at the same time, be the least possible, or zero. Hence the moving load must extend from A to the foot of the inclined web member for the greatest stress in it, or the vertical whose upper extremity it cuts.

For the counters, S would be taken at an indefinitely short distance to the right of A. In such a case S will have its greatest possible value when the moving load covers both arms ; but n', as before, must have its least value, *i. e.*, zero. Hence the moving load must extend from the right-hand end of the right-hand arm to the foot of the counter considered, for the greatest stress in it, or, possibly, in the vertical cut by its upper extremity.

For all positions of the moving load which make $S - n(W + W') < 0$, or negative, no counters are needed.

The equation :

$$R_1 - n(W + W') = 0,$$

then gives the panel point at which the counters are to begin. R_1 is written for S because they are the same in this case.

The first counter is really found by this equation, as will presently be seen.

The preceding observations have been made as if the chords were horizontal, while in the example taken one is inclined. They hold, however, *as long as the chords are straight*, since then a *portion* of the whole shear is to be considered, which attains its maximum with the *whole*.

The central diagonals nN', $n'N$, oO', $o'O$, are subject to no stress if both arms are of equal length, and if the loading is symmetrical, for in such a case $R_2 = R_3$ and $R_1 = R_4$.

The last relation in connection with Equations (5) and (8) of the preceding Art. gives,

$$M_2 = M_3;$$

and this in Equations (5) and (6) gives :

$$R_1 + R_2 = \tfrac{1}{2}P.$$

Hence there is no shear in the span l_2. From this result it at once follows that all symmetrical loads produce no shear in the span l_2, and, consequently, that the central diagonals will sustain their greatest stresses when one arm is covered with the moving load and the other is free from it.

Since all moving loads make the end reaction positive, the bridge open is the only case in which tension can exist in the upper chords, and compression in the lower, near the ends. For this reason, and the fact that it will be advisable to find stresses due to fixed and moving loads, in the chords, separately, it will be necessary to find all the open draw chord stresses, and tabulate them.

Although positions of loading may easily be assigned for the greater portion of the greatest chord stresses due to the moving load, yet, since the same load will subject some panels to different kinds of stress, according to its position in the span, it will be best to find *all* the chord stresses in each arm due to each panel moving load, and combine the results with each other and those due to the fixed load already indicated. Such a process, as will be seen, does not involve the amount of labor which it may seem to signify, though it is surely somewhat tedious.

The preceding observations will enable all the greatest stresses in the bridge to be determined, and it is to be borne in mind that they hold when l_1 is not equal to l_3 (except in the special case where equality was assumed), when the panel loads P are not the same, and when there is, or is not, locomotive excess. It is also to be borne in mind that each system of triangulation, with its weights, is to be treated, according to the principles laid down, as if it were a truss acting by itself.

When the moving load covers, entirely or partially, one arm only, it will frequently be convenient to use the following formula :

Omitting all terms involving $\overset{3}{\Sigma}$, Equation (4) of the preceding Art. may be written, by the aid of Equation (3):

$$M_2 = - \left\{ M_3 l_2 + \frac{4(l_2 + l_3)(l_1 + l_2) - l_2^{\,2}}{l_2} M_3 \right\} \div 2\,(l_1 + l_2)\,;$$

$$\therefore\quad M_2 = - 2M_3 \left(1 + \frac{l_3}{l_2}\right) \quad . \quad . \quad . \quad . \quad . \quad . \quad (1).$$

The actual example may now be proceeded with.

In each system of triangulation $l_1 = l_3 = l$; hence the general value for M_3 from the preceding Art. may take the form :

$$M_3 = - \frac{P\left[\dfrac{l_2}{l}\overset{1}{\Sigma}(l^2 - z^2)z - 2\left(1 + \dfrac{l_2}{l}\right)\overset{3}{\Sigma}(l^2 - z^2)z\right]}{4\,(l + l_2)^2 - l_2^{\,2}}\,; \quad . \quad . \quad (2).$$

since P, the panel load, is a constant quantity.

The greatest stresses in the counters cE, bD, and bC will first be determined.

For the system of triangulation $AcCcEgG \ldots oO$ the values of z, measured from A, with quantities depending upon it, are the following :

$$z^2 = (24)^2 = 576 \quad \ldots \quad (l^2 - z^2)z = 456576$$
$$\text{``} = (48)^2 = 2304 \quad \ldots \quad \text{``} = 830208$$
$$\text{``} = (72)^2 = 5184 \quad \ldots \quad \text{``} = 1037952$$
$$\text{``} = (96)^2 = 9216 \quad \ldots \quad \text{``} = 996864$$
$$\text{``} = (120)^2 = 14400 \quad \ldots \quad \text{``} = \underline{624000}$$
$$3945600$$

The constant quantities depending on l and l_2 are:

$$l = 140 \text{ feet} \quad \therefore \quad l^2 = 19600$$
$$OO' = l_2 = 8 \text{ ``} \quad \therefore \quad l_2^2 = 64$$

$$(l + l_2)^2 = 21904 \qquad \frac{l_2}{l} = 0.05714$$

$$4(l + l_2)^2 - l_2^2 = 87552.$$

The fixed load produces no reaction at A, hence in equation (2) P will be eighteen tons, the panel moving load only. With the moving load covering both arms, or extending from C' to C, $\overset{1}{\Sigma}(l^2 - z^2)z = \overset{3}{\Sigma}(l^2 - z^2)z$;

$$\therefore \quad M_3 = -\frac{P\left(2 + \frac{l_2}{l}\right)\overset{1}{\Sigma}(l^2 - z^2)z}{4(l + l_2)^2 - l_2^2} = M_2$$

$$= -\frac{18 \times 2.05714 \times 3945600}{87552} = -1668.73 \text{ foot-tons.}$$

$\therefore$ By Eq. (5) of preceding Art., $R_1 = 31.79$ tons $\Big\}$

 " " (6) " " " $R_2 = 58.21$ " $\Big\}$ . . (3).

With moving load extending from C' to E, the summation $\overset{3}{\Sigma}$ covers the whole arm, but the summation $\overset{1}{\Sigma}$ *does not include the value of* $z = 24$. Hence Eq. (2) gives for this case:

$$M_3 = -1674.13 \text{ foot-tons.}$$

$\therefore$ By Eq. (4) of the preceding Art.;

$$M_2 = 1470.3 \text{ foot-tons;}$$

and by Eq. (5) of the same Art.,

$$R_1 = 18.3 \text{ tons.}$$

With moving load extending from C' to G, the summation $\overset{3}{\Sigma}$ remains the same as before, but values corresponding to $z = 24$ and $z = 48$ must be omitted from $\overset{1}{\Sigma}$.

By Eq. (2), and proceeding precisely as before,

$$M_3 = - 1683.83 \text{ foot-tons.}$$
$$M_2 = - 1109.4 \quad \text{``} \quad \text{``}$$
$$\therefore R_1 = 9.05$$

The fixed panel load at A evidently belongs to the system of triangulation under consideration, for it is the vertical component of the tensile stress in Ac when the bridge is open, and on account of the extra weight of the locking apparatus, etc., it will be taken at 4 tons.

Remembering that $W + W' = 11.1$ tons, also that $w = P = 18.00$ tons, and making use of the result just obtained, the number of counters needed in the system taken, and the greatest stresses in the same, can easily be determined.

With the head of the train at C, $4 + 11.1 + 18 = 33.1 > R_1 = 31.79$; $\therefore \Sigma P = 0$ at C.

With the head of the train at E, the fixed panel loads acting through A and C are less than $R_1 = 18.3$ tons by 3.2 tons; $\therefore \Sigma P = 0$ at E.

With the head of the train at G, $R_1 = 9.05$ tons $< 4 + 11.1 = 15.1$ tons; $\therefore \Sigma P = 0$ at C.

Consequently cE (if subjected to stress) is the first counter needed in approaching A from O, and it will be subjected to its greatest stress when the load extends from C' to E, in accordance with principles before determined.

With the inclined chord in this example, web stresses can be most conveniently found by a combination of the method by moments and the graphical method, as indicated in Chapter III.

Taking moments about c, with the head of the train at E, and dividing by $cC = 21$ feet, there will result :

$$\text{Stress in } CE = \frac{(18.3 - 4) \times 24}{21} = + 16.34 \text{ tons.}$$

Suppose the truss divided by any plane cutting the members cc, cE, and CE. The shear ΣP, or the algebraic sum of the external forces acting on the left of that section will be $R_1 - 15.1$ tons $= 3.2$ tons $= \Sigma P$.

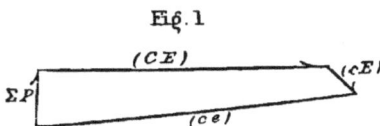

Fig. 1

The lines in Figure 1 are drawn parallel to the members indicated by the letters, and to a scale of ten tons to the inch. The Figure gives,

$$\text{Stress in } cE = + 2.4 \text{ tons.}$$

The greatest stress in bC will be found by extending the train from C' to C. Now, the panel fixed weight at b is to be taken as resting on the other system of triangulation; but since Ac cannot take compression, Ab and bc must be supposed to exist, for this system, without weight.

Taking moments about b with the head of the train at C, and dividing by $bB = 20$ feet, there will result:

$$\text{Stress in } AC = \frac{(31.79 - 4) \times 12}{20} = + 16.65 \text{ tons.}$$

Dividing the truss through AC, bC, and bc, and constructing Figure 2 in precisely the same manner as before, remembering that $\Sigma P = 27.79$ tons, the following results will be obtained :

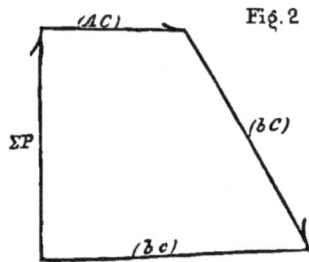

Fig. 2

Stress in $bC = + 29.00$ tons.
Stress in $bc = - 31.7$ tons.

The operations for the counters in the system $AbBdDfF \ldots nN$

are exactly similar in character to those gone through for the other system, and they will need little explanation.

The constant quantities depending upon lengths of span, are the following :

$$l_1 = l_3 = l = 132 \text{ feet.} \qquad \therefore l^2 = 17424.$$
$$l_2 = 24 \text{ `` } \qquad \therefore l_2^2 = 576.$$
$$(l + l_2)^2 = 24336 \qquad \frac{l_2}{l} = 0.182.$$
$$4 (l + l_2)^2 - l_2^2 = 96768.$$

The values of z and the quantities depending upon them, to be used in Eq. (2) of this Article, and in the equations of the preceding Article, are given in the following expressions:

$$z^2 = (12)^2 = 144 \quad \therefore \quad (l^2 - z^2)z = 207360$$
$$`` = (36)^2 = 1296 \quad \therefore \quad `` = 580608$$
$$`` = (60)^2 = 3600 \quad \therefore \quad `` = 829440$$
$$`` = (84)^2 = 7056 \quad \therefore \quad `` = 870912$$
$$`` = (108)^2 = 11664 \quad \therefore \quad `` = 622080$$

$$3110400$$

$\therefore$ By Eq. (2), for moving load on $B'B$:

$$M_3 = M_2 = - 1262.43 \text{ foot-tons.}$$
$$\begin{rcases} \therefore \quad R_1 = 39.53 \text{ tons.} \\ \therefore \quad R_2 = 50.47 \text{ tons.} \end{rcases} \quad \cdots \cdots \cdots \quad (4).$$

For moving load on $B'D$ all quantities depending on $z = 12$ must be omitted in $\overset{1}{\Sigma}$, hence:

$$M_3 = - 1269.44 \text{ foot-tons.}$$
$$M_2 = - 1171.2 \text{ `` `` }$$
$$\therefore \quad R_1 = 23.86 \text{ tons.}$$

For moving load on $B'F$ all quantities depending on $z = 12$ and $z = 36$ in $\overset{1}{\Sigma}$, must be omitted, hence:

$$M_3 = - 1289.1 \text{ foot-tons.}$$
$$M_2 = - 915.54 \text{ `` `` }$$
$$\therefore \quad R_1 = 12.7 \text{ tons.}$$

For this system there is no fixed load at A. When, there-fore, the moving load covers $B'B$ and $R_1 = 39.53$ tons, $\Sigma P = 0$ for D, since $2(11.1 + 18.00) = 58.2 > 39.53$ and $29.1 < 39.53$.

If the moving load covers $B'D$, $\Sigma P = 0$ for D, since $R_1 = 23.86 > 11.1$ and < 29.1 tons.

It is also seen that $\Sigma P = 0$ at D for the moving load on $B'F$, since $R_1 = 12.7$ tons.

The point of no shear, for which $\Sigma P = 0$, cannot, then, move to the right of D. Consequently bD is the only counter needed in this system, and it will be subjected to its greatest stress (if it receives any) when the moving load covers $B'D$.

Assuming this last condition, and taking moments about b, there results, after dividing by $bB = 20$ feet:

$$\text{Stress in } AD = \frac{23.86 \times 12}{20} = + 14.32 \text{ tons.}$$

Fig. 3

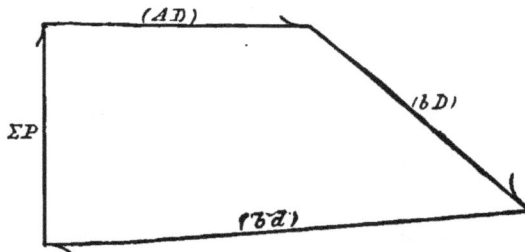

Figure 3 shows the stress diagram existing when the truss is divided through bd, bD, and BD. It is drawn to a scale of ten tons to the inch, and gives the stress in bD as 16.4 tons; $\Sigma P = 12.76$ tons. The greatest stress in bB will be simply,

$$6.6 + 18.00 = + 24.6 \text{ tons};$$

it will be found when the load covers $B'B$, as well as for some other positions.

The greatest stresses (tensile) in the counters, considering *bB* as one, are then :

$$\text{Stress in } cE = + \quad 2.4 \text{ tons.}$$
$$\text{``} \quad bD = + \ 16.4 \quad \text{``}$$
$$\text{``} \quad bC = + \ 29.00 \quad \text{``}$$
$$\text{``} \quad bB = + \ 24.6 \quad \text{``}$$

It should be stated that as all verticals are considered compression members, *bB* is only subjected to tension as a counter ; it might be taken as a tension member only, with possible ultimate economy, though more material would probably be required, but in such a case the fixed panel load at *b* would have to be taken as resting on the system *AcCcE*, etc.

The greatest main web stresses will next be found.

The main web members are the only ones subjected to stress when the bridge is open ; for this reason the fixed load must be taken into consideration in the determination of the main web stresses, although it was neglected in finding the reactions for the stresses in the counters.

Since all the fixed load is carried directly to the center supports, "*P*," in Eqs. (1) and (2), and the general expressions for R_2, R_3, and R_4, will refer to the moving load only.

The secants of the following angles will be needed :

$$sec \ AbB \ = 1.166.$$
$$\text{``} \quad AcC \ = 1.519.$$
$$\text{``} \quad oO'o' = 1.0349.$$
$$\text{``} \quad nN'n' = 1.281.$$

Also let the angle of inclination of the upper chord to a horizontal line be denoted by α.

Then :
$$tan \ \alpha = 10 \div 120 = 0.0833$$
$$sin \ \alpha = 0.0831$$
$$cos \ \alpha = 0.9965.$$

The greatest stress in *Ab* will exist when both arms of both systems are loaded with the moving load, for the reaction at *A* will then be the greatest. With the moving load on both

10

arms it has already been found, in Eqs. (3) and (4), that the greatest reaction at A will be:

$$31.79 + 39.53 = 71.32 \text{ tons.} \quad \text{Hence:}$$

Stress in Ab $= (71.32 - 4) \times sec\ AbB = -78.5$ tons.

Compressive stresses will be indicated by negative signs.

The greatest stress (tensile) in Ac will exist when the bridge is open, and it will simply be due to the fixed panel weight of 4 tons at A. Hence:

Stress in Ac $= 4 \times sec\ AcC = +6.08$ tons.

In some cases a vertical may sustain its greatest stress when the counter is subjected to its greatest stress, but in the example taken such a circumstance does not occur. It may· easily be found by trial, for instance, that Ac and cC take their greatest stresses together; not cE and cC. If Ac be taken to represent the stress (6.08 tons) in that member, and if a line be drawn parallel to bn from A until it cuts cC, the portion found between c and the point of intersection will represent the stress in cC due to that in Ac. By the aid of such a triangle there will be found:

Stress in cC $= -(3.8 + 4.5) = -8.3$ tons.

In the determination of other web stresses, each system of triangulation will be treated separately by a combination of the graphical and moment methods, as was done in finding the counter stresses. The train will be brought on from A, panel by panel.

The most convenient formula to be used for the determination of the reaction R_1 is that given in the previous article, and is the second one following Eq. (13).

Two force diagrams are required for each position of the train, but only those used for one position will be given, for they are all alike and very simple.

The system $AcCcEgG$, etc., will first be taken; and the following values of $(l^2 - z^2)\,z$, belonging to it and already used, divided by $4 (l + l_2)^2 - l_2^2 = 87552$, will be found convenient :

$$\frac{456576}{87552} = 5.22 ;$$

$$\frac{830208}{87552} = 9.48 ;$$

$$\frac{1037952}{87552} = 11.86 ;$$

$$\frac{996864}{87552} = 11.39 ;$$

$$\frac{624000}{87552} = 7.13.$$

$$\frac{2 (l_2 + l)}{l^2} = 0.0151.$$

The formula for R_1 is :

$$R_1 = \frac{\overset{1}{\Sigma} P (l - z)}{l} - \frac{2(l_2 + l)}{l^2} \cdot \frac{\overset{1}{\Sigma} P (l^2 - z^2) z}{4 (l + l_2)^2 - l_2^2}. \quad (5).$$

As usual, the general expression for the shear at any section of the system will be ΣP.

For brevity the stress in any member will be indicated by putting a parenthesis around the letters belonging to it.

In determining the shear, "P," *will include the fixed load,* *i. e.*, in general $P = 11.1 + 18 = 29.1$ tons. The fixed load will also be included in taking moments about any point.

With the head of the train at C, z has only the value 24 :

$$R_1 = 13.46 \text{ tons.}$$

Taking moments about e :

$$(C E) = - 10.62 \text{ tons.}$$

Taking a section of the system through ce, Ce, and CE, the shear will be :

$$\Sigma P = (29.1 + 4) - R_1 = 19.64 \text{ tons.}$$

The following diagram (ten tons to the inch) gives:

$$(Ce) = + \ 29.7 \text{ tons.}$$

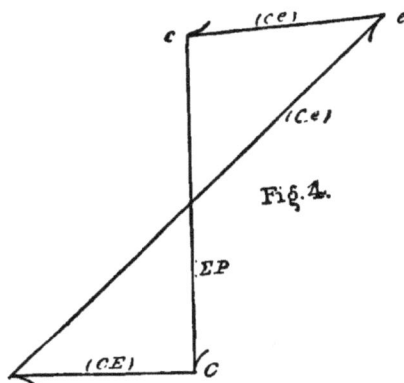

Fig. 4.

Figure 4 is exactly similar to other stress diagrams used in determining the stresses in the inclined web members, except that *e* will be usually found on the left of *cC*.

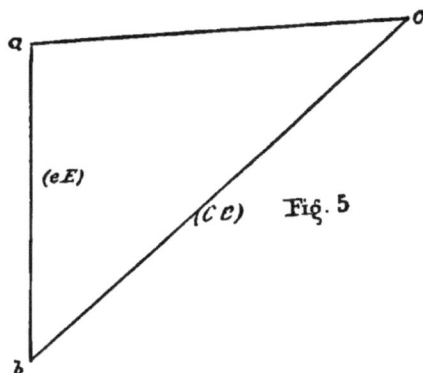

Fig. 5

Figure 5 shows a diagram, or triangle, exactly similar to all figures used in determining stresses in the verticals. *ac* is parallel with the upper chord, and *ab* is vertical, and in Fig. 5 represents the partial stress due to (*Ce*) in (*eE*) to a scale of ten tons to the inch.

$$\therefore (eE) = - \ (18.9 + 4.5) = - \ 23.4 \text{ tons.}$$

Head of train at E.

z has the values 24 and 48; $\therefore R_1 = 22.7$ tons, and $\Sigma P = 39.5$ tons. Taking moments about g:

$$(EG) = -\,29.95 \text{ tons.}$$

By taking a section through cg, Eg and EG, the diagrams give:

$$(Eg) = +\,56.0 \text{ tons.}$$
$$(gG) = -\,(37.4 + 4.5) = -\,41.9 \text{ tons.}$$

Head of train at G.

z has the values 24, 48, and 72; $\therefore R_1 = 28.27$ tons, and $\Sigma P = 63.03$ tons. Taking moments about k:

$$(GK) = -\,68.9 \text{ tons.}$$

By taking the section through gk, Gk and GK, the diagrams give:

$$(Gk) = +\,83.0 \text{ tons.}$$
$$(kK) = -\,(57.8 + 4.5) = -\,62.3 \text{ tons.}$$

Head of train at K.

z has the values 24, 48, 72, and 96; $\therefore R_1 = 30.83$ tons, and $\Sigma P = 89.57$ tons. Taking moments about m:

$$(KM) = -\,129.81 \text{ tons.}$$

By taking the section through km, Km and KM, the diagrams give:

$$(Km) = +\,109.2 \text{ tons.}$$
$$(mM) = -\,(78.8 + 4.5) = -\,83.3 \text{ tons.}$$

Head of train at M.

z has the values 24, 48, 72, 96, and 120; $\therefore R_1 = 31.46$ tons, and $\Sigma P = 118.04$ tons. In this case moments must be taken about M, and the stress in mn, consequently, will be found.

Taking such moments (the lever arm of (mn) is $29 \times cos \, \alpha = 28.9$ feet):

$$(mn) = 127.6 \text{ tons (tension).}$$

By cutting mn, Mo and MN, the diagrams give:

$$(Mo) = + 129.5 \text{ tons.}$$
$$\text{Vertical component of } (Mo) = 107.5 \text{ tons.}$$

The *apparent* compression in oO will be the reaction R_2 at O; it will be interesting to find R_3 (at O') and R_4 (at A') also. Equations (2) and (1) give, after omitting all terms involving $\overset{3}{\Sigma}$:

$$M_3 = 46.33 \text{ foot-tons.}$$
$$M_2 = - 1714.21 \text{ foot-tons.}$$

Hence, from the general formulæ of the previous article:

$$R_2 = 278.6 \text{ tons.}$$
$$R_3 = - 220.4 \text{ tons.}$$
$$R_4 = \quad 0.33 \text{ tons.}$$

It is important to notice the large negative value of R_3. The *apparent* fixed load supported at O is:

$$6 \times 11.1 + 4.0 = 70.6 \text{ tons.}$$

The word "*apparent*" has been used because the general formulæ for the reactions are really based on the supposition that a compression member extends from n to O, or, that there exists something equivalent to that arrangement, and that it belongs to this system only. As the system actually stands, however, a part of the reaction $R_2 = 278.6$ tons will be found at N; this part is the vertical component of $(mn.)$

Vert. comp. of $(mn.) = 127.6 \times sin \, \alpha = 10.6$ tons.

Consequently:

$$(oO) = - (278.6 + 59.5 + 4.5 - 10.6) = - 332.0 \text{ tons.}$$

Or indirectly, by using the vertical component of (Mo) and R_3 and R_4:

$$(oO) = - 107.5 - 4.5 - (220.4 - 0.33) = - 332.07.$$

The agreement is nearer than will frequently be found. The diagram which gave (Mo), gives also :

$$(MO) = - 199.00 \text{ tons} = (OO').$$

Or, by taking moments about o, of the fixed load of the right arm, and of R_3 and R_4:

$$(MO) = (- 220.28 \times 8 - 0.12 \times 16 - 55.5 \times 76 - 4 \times 148 + 50.46 \times 8 + 0.33 \times 148 + 9.04 \times 16) \div 30 = - 199.23 \text{ tons,}$$

an agreement sufficiently close.

The terms of this equation allow for the modified reactions at O' and N' in a manner that will be shown farther on.

There remains only to be found the stress in oO'; and evidently :

$$(oO') = - (R_3 + 0.12 + R_4) \times sec\, oO'o' = + 227.65 \text{ tons.}$$

The amount $(+ 0.12 \text{ ton})$ is the vertical component of $+ (m'n')$ caused by R_4 (at A') $= 0.33$ tons. In other words, it is that part of R_3 found at N'.

The system $AbBdDfF$, etc., will next be taken up. The values of $(l^2 - z^2)z$, divided by $4(l + l_2)^2 - l_2^2$, are the following:

$$\frac{207360}{96768} = 2.14;$$

$$\frac{580608}{96768} = 6.00;$$

$$\frac{829440}{96768} = 8.57;$$

$$\frac{870912}{96768} = 9.00;$$

$$\frac{622080}{96768} = 6.43.$$

$$\frac{2\,(l + l_2)}{l_2^2} = 0.0179.$$

Head of train at B.

z has the value 12; $\therefore$ $R_1 = 15.68$ tons, and $\Sigma P = 29.1 - R_1$ $= 13.42$ tons. Taking moments about d:

$$(BD) = -\,6.09 \text{ tons.}$$

By taking a section through bd, Bd, and BD, the diagrams give :

$$(Bd) = +\,21.3 \text{ tons.}$$
$$(dD) = -\,(12.9 + 4.5) = -\,17.4 \text{ tons.}$$

Head of train at D.

z has the values 12 and 36; $\therefore$ $R_1 = 26.83$ tons, and $\Sigma P = 31.37$ tons. Taking moments about f:

$$(DF) = -\,20.22 \text{ tons.}$$

By taking a section through df, Df, and DF, the diagrams give :

$$(Df) = +\,46.00 \text{ tons.}$$
$$(fF) = -\,(30.00 + 4.5) = -\,34.5 \text{ tons.}$$

Head of train at F.

z has the values 12, 36, and 60; $\therefore$ $R_1 = 33.89$ tons, and $\Sigma P = 53.41$ tons. Taking moments about h:

$$(FH) = -\,57.68 \text{ tons.}$$

By taking a section through fh, Fh, and FH, the diagrams give :

$$(Fh) = +\,71.3 \text{ tons.}$$
$$(hH) = -\,(48.1 + 4.5) = -\,52.6 \text{ tons.}$$

Head of train at H.

z has the values 12, 36, 60, and 84 ; $\therefore$ $R_1 = 37.54$ tons, and $\Sigma P = 78.86$ tons.   Taking moments about l:

$$(HL) = -104.63 \text{ tons.}$$

By taking a section through hl, Hl, and HL, the diagrams give:

$$(Hl) = +99.6 \text{ tons.}$$
$$(lL) = -(70 + 4.5) = -74.5 \text{ tons.}$$

Head of train at L.

z has the values 12, 36, 60, 84, and 108 ; $\therefore$ $R_1 = 38.75$ tons, and $\Sigma P = 106.75$ tons.   Taking moments about n ·

$$(LN) = -178.7 \text{ tons.}$$

By taking a section through ln, Ln, and LN, the diagrams give:

$$(Ln) = +124.5 \text{ tons.}$$
$$(ln) = +102.5 \text{ tons.}$$
$$\text{Vert. comp. of } (Ln) = 98.00 \text{ tons.}$$

That portion of the stress in nN due to the stress in ln is:

$$102.5 \times sin\ \alpha = 8.52 \text{ tons.}$$

The compression in nN will be the reaction R_2, added to the fixed load of one arm of this system and the vertical component of (mn), 10.6 tons, determined for the other system. The moments M_2 and M_3 must therefore be determined for this system.   Eqs. (2) and (1) give:

$$M_3 = 105.3 \text{ foot-tons.}$$
$$M_2 = -1368.9 \text{ foot-tons.}$$

The general formulæ of the preceding article then give:

$$R_2 = 112.7 \text{ tons.}$$
$$R_3 = -62.23 \text{ tons.}$$
$$R_4 = 0.8 \text{ tons.}$$
$$\therefore (nN) = -(98.00 + 8.52 - (R_3 + R_4) + 10.6 + 4.5) = -183.05 \text{ tons.}$$

As a check, with this position of load, the vertical com-
ponent of (Ln) added to $(ln) \times \sin \alpha$ and R_1, should equal the
total load on one arm, *i. e.*, $5 \times (18 + 11.1)$ tons.

The results obtained give:

$$98.00 + 8.52 + 38.75 = 145.27 \text{ tons.}$$
But, $$5 \times (18 + 11.1) = 145.50 \text{ "}$$

The agreement is close.

The shear, or vertical component, in (nN') is

$$- (R_3 + R_4) = 61.43 \text{ tons}; \quad \therefore$$
$$(nN') = + 61.43 \times \sec nN'n' = + 78.69 \text{ tons.}$$

Evidently: $(LN) = (NN')$.

This completes the determination of the greatest web
stresses; those in the chords remain to be found.

As has been seen, the chord stresses at the centre, with the
moving load on one arm, were found in connection with
the center web stresses, because the same diagrams gave
both.

The reactions which are in general necessary to the determination of the chord stresses, arising from the application of the moving load, are R_1 and R_4. Since the two arms
in the present case are perfectly symmetrical, it will only be
necessary to suppose a single weight, P, applied successively
to the different panel points of one arm. The weight $P = 18$
tons will be taken as applied to the arm l_1; consequently
there will be no load on l_3.

For a single weight Eq. (5) gives:

$$R_1 = P\left(\frac{l-z}{l}\right) - P\frac{2(l+l_2)}{l^2} \cdot \frac{(l^2 - z^2)z}{4(l + l_2)^2 - l_2^2}. \quad \cdots \quad (6).$$

Eq. (2), in connection with Eq. (8) of the preceding article,
gives:

$$R_4 = \frac{M_3}{l} = P\frac{l_2}{l^2} \cdot \frac{(l^2 - z^2)z}{4(l + l_2)^2 - l_2^2}. \quad \cdots \quad (7).$$

The numerical values of $\dfrac{(l^2 - z^2)z}{4(l + l_2)^2 - l_2^2}$ already determined,
can be most conveniently used in the following operations.
The system $AcCcE$, etc., will first be taken.

$$2\left(\frac{l + l_2}{l^2}\right) = 0.0151\,; \qquad \frac{l_2}{l^2} = 0.00041.$$

$$P = 18 \text{ tons.}$$

$$P \times 2\left(\frac{l + l_2}{l^2}\right) = 0.2718\,; \qquad P\frac{l_2}{l^2} = 0.00738.$$

	P at panel point C.	
$z = 24$ ft.	$R_1 = 13.50$ tons.	$R_4 = 0.039$ ton.
	P at panel point E.	
$z = 48$ ft.	$R_1 = 9.25$ tons.	$R_4 = 0.07$ ton.
	P at panel point G.	
$z = 72$ ft.	$R_1 = 5.52$ tons.	$R_4 = 0.088$ ton.
	P at panel point K.	
$z = 96$ ft.	$R_1 = 2.55$ tons.	$R_4 = 0.084$ ton.
	P at panel point M.	
$z = 120$ ft.	$R_1 = 0.645$ ton.	$R_4 = 0.053$ ton.

Only two decimals in all the values of R_1, except the last,
have been retained; however, if three had been retained, the
sum of those values would give:

$$\Sigma R_1 = 31.468 \text{ tons.}$$
Also $\qquad\qquad \Sigma R_4 = 0.334 \text{ ``}$

These agree well with the values already found by taking
all the loads together.

For the system $AbBdD$, etc., the following values are found:

$$2\left(\frac{l_2 + l}{l^2}\right) = 0.0179. \qquad \frac{l_2}{l^2} = 0.00137.$$

$$P \times 2\left(\frac{l_2 + l}{l^2}\right) = 0.3222. \qquad P \times \frac{l_2}{l^2} = 0.02466.$$

	P at panel point B.	
$z = 12$ ft.	$R_1 = 15.67$ tons.	$R_4 = 0.053$ ton.
	P at panel point D.	
$z = 36$ ft.	$R_1 = 11.153$ tons.	$R_4 = 0.148$ ton.
	P at panel point F.	
$z = 60$ ft.	$R_1 = 7.05$ tons.	$R_4 = 0.212$ ton.
	P at panel point H.	
$z = 84$ ft.	$R_1 = 3.652$ tons.	$R_4 = 0.222$ ton.
	P at panel point L.	
$z = 108$ ft.	$R_1 = 1.204$ tons.	$R_4 = 0.159$ ton.

Taking the sums of the reactions:

$$\Sigma R_1 = 38.729 \text{ tons.}$$
$$\Sigma R_4 = 0.794 \text{ "}$$

These last also agree closely with the values found by taking the train on BL.

These operations constitute good checks on the numerical work.

The stresses for the loads on the system $AcCcE$, etc., will first be found.

It will be assumed that the counters come into action when it is possible for them to do so. The counter bC will then be under stress for all moving loads, on this system, in the left span, while both bC and cE will be under stress for moving loads at and on the right of E.

Such an assumption, or some other one no better, must be made in order to avoid ambiguity. Such defects always accompany the existence of counters.

The chord stresses for each panel load, P, will be found by the method of moments, and will be shown in detail for one panel load only, as the operations are exactly alike for all.

Let the panel moving load be taken at E.

$$R_1 = 9.25 \text{ tons.} \qquad \therefore \qquad R_1 \times 12 = 111.00.$$

$$(bc) = \frac{-2R_1 \times 12}{Cc \times \cos \alpha} = -10.61 \text{ tons.}$$

$$(cg) = \frac{-4R_1 \times 12}{Ee \times \cos \alpha} = -19.37 \text{ tons.}$$

$$(gk) = \frac{-6R_1 \times 12 + 2 \times 18 \times 12}{Gg \times \cos \alpha} = -9.4 \text{ tons.}$$

$$(km) = \frac{-8R_1 \times 12 + 4 \times 18 \times 12}{Kk \times \cos \alpha} = -0.9 \text{ tons.}$$

$$(mn) = \frac{-10R_1 \times 12 + 6 \times 18 \times 12}{Mm \times \cos \alpha} = +6.44 \text{ tons.}$$

$$(nN) = -(mn) \sin \alpha = -0.535 \text{ tons.}$$

$$(no) = (mn) \cos \alpha = +6.42 \text{ tons.}$$

$$(AC) = \frac{R_1 \times 12}{Bb} = +5.55 \text{ tons.}$$

$$(CE) = \frac{2 R_1 \times 12}{Cc} = +10.6 \text{ tons.}$$

$$(EG) = \frac{6R_1 \times 12 - 2 \times 18 \times 12}{Gg} = +9.36 \text{ tons.}$$

$$(GK) = \frac{8R_1 \times 12 - 4 \times 18 \times 12}{Kk} = +0.9 \text{ tons.}$$

$$(KM) = \frac{10R_1 \times 12 - 6 \times 18 \times 12}{Mm} = -6.41 \text{ tons.}$$

$$(MN) = (NO) = (OO') = \frac{R_1 \times 140 + (Nn) \times 8 - 18 \times 92}{Oo} =$$

$$- 11.9 \text{ tons.}$$

By exactly the same method the stresses in the following tables are found :

	(bc)	(cg)	(bc)	(cg)	(gk)	(km)	(mn)
P at C	-15.5	-9.43			-4.34	0.00	$+3.74$
P at G			-6.33	-11.55	-16.00	-3.64	$+7.00$
P at K			-2.92	-5.34	-7.37	-9.10	$+4.36$
P at M			-0.74	-1.35	-1.87	-2.3	-2.68

	(AC)	(CE)	(EG)	(GK)	(KM)	(Nn)	(no)	(MO')
P at C	$+8.1$	$+9.4$	$+4.32$	0.00	-3.72	-0.31	$+3.73$	-6.52
P at G	$+3.31$	$+6.31$	$+15.9$	$+3.62$	-6.95	-0.58	$+6.98$	-14.86
P at K	$+1.53$	$+2.92$	$+7.34$	$+9.06$	-4.34	-0.36	$+4.34$	-14.4
P at M	$+0.39$	$+0.74$	$+1.86$	$+2.29$	$+2.67$	$+0.22$	-2.67	-8.93

Since all the loads on the left arm produce the same kinds of stresses in the panels of the right, all of those loads may be taken as acting together in finding the stresses in the system $A'c'C'e'E'$, etc. The reaction (upward) at A' will then be $\Sigma R_4 = 0.334$ tons.

By taking moments precisely as before, there will result:

$(c'b') = -0.38$ tons. $(n'o') = -1.4$ tons.
$(c'g') = -0.70$ " $(A'C') = +0.2$ "
$(g'k') = -0.96$ " $(C'E') = +0.38$ "

$(k'm') = -1.19$ tons. $(E'G') = +0.96$ tons.
$(m'n') = -1.39$ " $(G'K') = +1.19$ "
$(N'n') = +0.116$ " $(K'M') = +1.38$ "

$$(M'N') = (N'O) = -(oo') = +1.53 \text{ tons.}$$

The stresses in the system $AbBdD$, etc., are next to be found, and although the operations are precisely the same as those shown in the other system, the expressions for the stresses caused by one panel load will be given in detail.

As before, it will be assumed that the counter comes into action whenever possible.

Let the panel load at B be taken. The following stresses will then be found :

$$(bd) = \frac{-R_1 \times 12}{Bb \times \cos\alpha} = -9.43 \text{ tons.}$$

$$(df) = \frac{-3R_1 \times 12 + 2 \times 18 \times 12}{Dd \times \cos\alpha} = -6.03 \text{ tons.}$$

$$(fh) = \frac{-5R_1 \times 12 + 4 \times 18 \times 12}{fF \times \cos\alpha} = -3.19 \text{ tons.}$$

$$(hl) = \frac{-7R_1 \times 12 + 6 \times 18 \times 12}{Hh \times \cos\alpha} = -0.78 \text{ tons.}$$

$$(ln) = \frac{-9R_1 \times 12 + 8 \times 18 \times 12}{Ll \times \cos\alpha} = +1.28 \text{ tons.}$$

$$(AB) = \frac{R_1 \times 12}{Bb} = +9.4 \text{ tons.}$$

$$(BD) = \frac{3R_1 \times 12 - 2 \times 18 \times 12}{Dd} = +6.01 \text{ tons.}$$

$$(DF) = \frac{5R_1 \times 12 - 4 \times 18 \times 12}{Ff} = +3.18 \text{ tons.}$$

$$(FH) = \frac{7R_1 \times 12 - 6 \times 18 \times 12}{Hh} = +0.78 \text{ tons.}$$

$$(HL) = \frac{9R_1 \times 12 - 8 \times 18 \times 12}{Ll} = -\ 1.27 \text{ tons.}$$

$$(LN) = (NN') = \frac{11R_1 \times 12 - 10 \times 18 \times 12}{Nn} = -\ 3.05 \text{ tons.}$$

The same method gives the following tables :

	(bf)	(fh)	(hl)	(ln)
P at D	-18.31	-9.99	-2.8	$+3.29$
P at F	-11.58	-17.69	-6.18	$+3.68$
P at H	-6.00	-9.16	-11.83	$+1.36$
P at L	-1.9	-3.01	-3.88	-4.64

	(AD)	(DF)	(FH)	(HL)	(LN')
P at D	$+6.69$	$+9.88$	$+2.79$	-3.28	-8.54
P at F	$+4.26$	$+17.63$	$+6.16$	-3.67	-12.18
P at H	$+2.19$	$+9.13$	$+11.8$	-1.35	-12.74
P at L	$+0.72$	$+3.00$	$+3.87$	$+4.63$	-9.12

For the stresses in the same system in the right arm, the
panel moving loads at B, D, F, H, and L will be taken as
acting together. The reaction at A' will be $\Sigma R_4 = 0.794$
ton, and precisely the same method by moments gives the
following results :

$(b'f') = -\ 1.3$ tons.	$(A'D') = +\ 0.48$ tons.
$(f'h') = -\ 1.99$ "	$(D'F') = +\ 1.99$ "
$(h'l') = -\ 2.57$ "	$(F'H') = +\ 2.56$ "
$(l'n') = -\ 3.07$ "	$(H'L') = +\ 3.06$ "

$$(L'N') = -\ (nn') = +\ 3.5 \text{ tons.}$$

The chord stresses caused by the fixed weights alone still

remain to be determined. These stresses exist when the bridge is open, also when it is closed and unloaded. Consequently the counters do not sustain stress.

The same method by moments will be used for the lower chord, but all the weights will be taken together.

The system $AcCeE$, etc., will be taken first.

Taking moments about c, e, g, k, and m, respectively, there will result, by taking $w_1 = W + W' = 11.1$ tons:

$$(AC) = -\frac{4 \times 24}{21} = -4.57 \text{ tons.}$$

$$(CE) = -\frac{w_1 \times 24 + 4 \times 48}{23} = -19.90 \text{ tons.}$$

$$(EG) = -\frac{2w_1 \times 36 + 4 \times 72}{25} = -43.50 \text{ tons.}$$

$$(GK) = -\frac{3w_1 \times 48 + 4 \times 96}{27} = -73.42 \text{ tons.}$$

$$(KM) = -(no) = -\frac{4w_1 \times 60 + 4 \times 120}{29} = -108.4 \text{ tons.}$$

The upper chord stresses are determined by simply dividing the lower by $cos\,\alpha$, or multiplying them by $sec\,\alpha$. This method might have been followed in finding all the upper chord stresses. Hence:

$$(ce) = \frac{(AC)}{cos\,\alpha} = +4.57 \text{ tons.}$$

$$(eg) = \frac{(CE)}{cos\,\alpha} = +20.00 \text{ tons.}$$

$$(gk) = \frac{(EG)}{cos\,\alpha} = +43.70 \text{ tons.}$$

$$(km) = \frac{(GK)}{cos\,\alpha} = +73.7 \text{ tons.}$$

$$(mn) = \frac{(KM)}{\cos \alpha} = + 108.8 \text{ tons.}$$

Also, $- (mn) \sin \alpha = (Nn) = -9.04$ tons.

Hence, by taking moments about O:

$$(MM') = -(oo') = -\frac{5w_1 \times 68 + 4 \times 140 - 9.04 \times 8}{Oo = 30} =$$
$$-142.07 \text{ tons.}$$

The expression for (MO) given on page 49 will now be easily understood by taking into consideration the values given above for (Nn) and $(N'n')$.

The chord stresses induced by the fixed load in the system $AbBdD$, etc., are found in exactly the same manner; consequently the expressions for the moments will not be given. The following are the stresses:

$(BD) = -12.1$ tons.	$(df) = +12.1$ tons.
$(DF) = -33.3$ "	$(fh) = +33.4$ "
$(FH) = -61.5$ "	$(hl) = +61.7$ "
$(HL) = -95.14$ "	$(ln) = +95.5$ "

$$(LL') = -(nn') = -133.2 \text{ tons.}$$

Out of all these results the greatest chord stresses are to be found. The operation is a very simple one; it consists simply in inspecting the previous results and taking the algebraic sum, for any one panel, of the fixed load stress and all the stresses of the same or opposite kind caused by the moving load. It should be remembered that no calculations for moving panel loads on the right arm are needed, since the truss is symmetrical in reference to the center. If the truss were unsymmetrical, such calculations would be necessary. The greatest tensile stress in mn, for example, is found with moving panel loads at C, E, G, K, B, D, F, and H, and its value is:

$$6.44 + 3.74 + 7.00 + 4.36 + 1.28 + 3.29 + 3.68 + 1.36 + 108.8$$
$$+ 95.5 = + 235.45 \text{ tons.}$$

The preceding results show that the only tendency to compressive stress in *mn* is caused by the moving loads at L and M, and those on the right arm. The combined effect of such loads is:

$$(mn) + (ln) + (m'n') + (l'n') = -(2.68 + 4.64 + 1.39 + 3.07) = -11.78 \text{ tons.}$$

As this is less than $108.8 + 95.5 = 204.3$ tons, the fixed load tensile stress (mn), can never be changed to compression.

Again, the fixed load only causes tensile stress in *cf*, and its amount is:

$$20.00 + 12.1 = 32.1 \text{ tons.}$$

The tendency to compressive stress by the moving load is:

$$-(9.43 + 19.37 + 11.55 + 5.34 + 1.35 + 6.03 + 18.31 + 11.58 + 6.00 + 1.9 + 0.70 + 1.3) = -92.86 \text{ tons.}$$

Hence, $3.21 - 92.86 = -60.76$ tons

is the greatest compressive stress in *cf*.

These operations are sufficient to show the simple methods of combination in all cases. Such methods give the following greatest chord stresses:

$$
\begin{aligned}
(bc) &= &&- 85.00 \text{ tons.} \\
(cd) &= + &&4.57 \text{ tons}; &&- 97.76 \text{ "} \\
(de) &= + &&16.67 \text{ "}; &&- 82.26 \text{ "} \\
(ef) &= + &&32.10 \text{ "}; &&- 60.76 \text{ "} \\
(fg) &= + &&53.40 \text{ "}; &&- 39.37 \text{ "} \\
(gh) &= + &&77.10 \text{ "}; &&- 7.87 \text{ "} \\
(hk) &= + &&105.40 \text{ "} \\
(kl) &= + &&135.40 \text{ "} \\
(lm) &= + &&178.81 \text{ "} \\
(mn) &= + &&235.45 \text{ "} \\
(no) &= + &&263.07 \text{ "} \\
(oo') &= + &&275.27 \text{ "}
\end{aligned}
$$

$$
\begin{aligned}
(AB) &= -\quad 4.57 \text{ tons}; \ + 38.25 \text{ tons.} \\
(BC) &= -\quad 16.67 \text{ " }; \ + 22.76 \text{ "} \\
(CD) &= -\quad 32.00 \text{ " }; \ + 18.70 \text{ "} \\
(DE) &= -\quad 53.20 \text{ " }; \ + 21.96 \text{ "} \\
(EF) &= -\quad 76.80 \text{ " }; \ + 7.75 \text{ "} \\
(FG) &= -\ 105.00 \text{ "} \\
(GH) &= -\ 134.92 \text{ "} \\
(HK) &= -\ 178.13 \text{ "} \\
(KL) &= -\ 234.53 \text{ "} \\
(LM) &= -\ 308.65 \text{ "} \\
(MN) &= -\ 377.51 \text{ "} \\
(NO) &= -\ 423.17 \text{ "} \\
(OO') &= -\ 479.81 \text{ "}
\end{aligned}
$$

The portions ch of the upper, and AF of the lower chord must be counterbraced; also corresponding portions in the right span.

These chord stresses are obtained, as has been seen, by treating each panel weight in each system separately; there is consequently involved the condition that the members Nn', $N'n$, Oo', and oO' are subjected to stress when the moving load covers the whole bridge from end to end. Such stresses may or may not exist, in reality, since the shear will be zero at the center in either case, consequently no little ambiguity arises in regard to the centre chord stresses (no), (oo'), (NO), and (OO'). This will be at once apparent if the center diagonals, named above, be omitted, and if the chord stresses be found on that supposition, the moving load being taken over the whole bridge.

For the system $AcCcE$, etc., $R_1 = 31.468 + 0.334 = 31.8$ tons; $(Nn) = -1.785 + 0.34 = -1.445$ tons.

Taking moments about m:

$$
-(KM) = (no) = \frac{-R_1 \times 120 + 4 \times w \times 60}{m.M} = +17.4 \text{ tons.}
$$

Taking moments about O:

$$
-(MM') = (oo') = \frac{5 \times w \times 68 - R_1 \times 140 - 1.445 \times 8}{Oo} =
$$

$$
+ 55.21 \text{ tons.}
$$

Again, for the system $AbBdD$, etc., $R_1 = 38.729 + 0.794 = 39.52$ tons. Taking moments about N:

$$-(LL') = (nn') = \frac{5 \times w \times 72 - R_1 \times 132}{Nn} = + 42.10 \text{ tons.}$$

These are moving load stresses, and by combining them with the fixed load stresses, as already given, there will result:

$$(no) = 133.2 + 108.4 + 42.10 + 17.4 = + 301.10 \text{ tons.}$$
$$(oo') = 133.2 + 142.07 + 42.10 + 55.21 = + 372.58 \text{ „}$$
$$(MM') = - 372.58 \text{ tons.}$$

Now, if the moving load be taken *over the whole of both arms*, the preceding results, by single weights, give:

$$(no) = + 252.00 \text{ tons}; \ (oo') = + 265.21 \text{ tons}; \ (NO) = - 421.64$$
$$\text{tons}; \text{ and } (OO') = - 479.81 \text{ tons.}$$

Thus the differences between the results of the two methods are seen to be very great.

The amounts of these differences are the horizontal components of the greatest stresses in the center diagonals.

This ambiguity also exists for any even number of equal weights symmetrically located in reference to the center.

Omitting the center diagonals for a symmetrical load over both arms is simply equivalent to omitting the negative reactions for that load, and it certainly seems reasonable to do so.

Some of the greatest chord stresses are caused by positions of loading which are not likely to occur often, but they are possible, and should be guarded against. The following examples illustrate this matter.

For $(HK) = - 178.13$ tons, moving loads are found only at B, D, F, and H; and the same is true for $(lm) = + 178.81$ tons.

For $(NO) = - 423.17$ tons, the moving load must cover the whole of the left arm, and be found also at the points B', D', F', H', L' in the right; while for (no) moving loads are taken at $C, E, G,$ and K only.

The advisability of finding the chord stresses by single weights may now be appreciated, though it could have been anticipated that the moving load on the whole bridge would cause the greatest compression in the upper chord and tension in the lower, near the extremities.

Collecting and arranging the greatest web stresses:

(cE)	= +	2.4	tons.	(bD)	= +	16.4 tons.
(bC)	= +	29.00	"	(bB)	= +	24.6 "
(Ab)	= −	78.50	"	(bB)	= −	4.5 "
(Ac)	= +	6.08	"	(cC)	= −	8.3 "
(Bd)	= +	21.30	"	(dD)	= −	17.4 "
(Cc)	= +	29.70	"	(cE)	= −	23.4 "
(Df)	= +	46.00	"	(fF)	= −	34.5 "
(Eg)	= +	56.00	"	(gG)	= −	41.9 "
(Fh)	= +	71.3	"	(hH)	= −	52.6 "
(Gk)	= +	83.0	"	(kK)	= −	62.3 "
(Hl)	= +	99.6	"	(lL)	= −	74.5 "
(Km)	= +	109.2	"	(mM)	= −	83.3 "
(Ln)	= +	124.5	"	(nN)	= −	183.1 "
(Mo)	= +	129.5	"	(oO)	= −	332.0 "
(oO')	= +	227.7	"	(nN')	= +	78.69 "

The vertical bB is the only web member which must be counterbraced.

Reviewing the whole work, perhaps the most striking points are the large negative reactions at the center when the moving load covers one arm only.

It has been seen that a small distance between the points of support at the center, for any system, causes very large negative reactions. Hence, to make those reactions as small as possible, that distance should be as great as possible.

These negative reactions may be so great as to require special weights to be used at the center to *hold down* the points of support. At O' or O, for instance, the negative reaction may be $220.4 - 0.12 = 220.28$ tons, and this will exist with no moving load on one arm, consequently it must be balanced by the fixed weight of the system $AcCcE$, etc., and a portion of the weight at the turn-table.

Deducting that portion of the fixed weight which passes down Nn, or $N'n'$, the downward pressure at O or O', due to the fixed load, is:

$$6 \times 11.1 + 4.00 - 9.04 = 61.56 \text{ tons.}$$

An extra weight, therefore, of $220.28 - 61.56 = 158.72$ tons will be required at the turn-table at each of the points O and O'.

This shows in a marked manner at least one of the disadvantages of such a system of construction.

The other defect is the great ambiguity shown to exist in the chord stresses at the center.

On the other hand, the bridge needs no latching down or lifting up at the ends, for they can never rise.

Precisely the same principles and conditions of loading hold true, in this case, with only one system, or more than two systems of triangulation.

Art. 37.—Ends Simply Supported—Four Points of Support at Center— Partial Continuity—Example.

A number of very satisfactory swing bridges have been built with what may be called "partial continuity." These trusses have two points of support at the center for each system of triangulation, *but the main central diagonals nN', $N'n$, oO', and Oo' are omitted*. Light diagonals are, however, put in the place of those members, for the sole purpose of steadying the bridge while open; they are not supposed to affect the continuity or non-continuity of the truss.

The effect of this arrangement is assumed to make either arm of the bridge a simple truss supported at each end, for all moving loads on that arm, *so long as the other arm carries no moving load.*

As before, it is supposed that the span l_2, included between the central points of support, sustains no load.

If the moving load partially covers both arms, or the whole of one arm and a part of the other, it is to be divided into three parts. Two of these parts, together with the reactions

R_1 and R_4 due to them, produce equal and opposite moments at the center. For these two parts, consequently, the truss is one of perfect continuity, with the main diagonals at the center omitted. This last condition is admissible, because for these two parts the shear at the center will be zero.

For convenience these two parts will be called "*balanced*," while the third part will be called "*unbalanced*."

For the unbalanced part, the arm or span in which it is found will be a simple truss supported at each end.

If the panel loads are uniform in amount, balanced loads will be symmetrically placed in reference to the center.

If the panel loads are not uniform in amount, the balanced portions would be determined by equating M_2 to M_3, with the aid of Eqs. (3) and (4) of Art. 35.

Coupling these statements with the principles deduced in the preceding articles, it will at once be seen that the greatest reaction R_1 at A, Fig. 1, Pl. X, will exist with the moving load over the whole of the left arm, for any moving load on the right arm will balance a part of that on the left, and relieve, to some extent, the reaction at A.

Resuming, for the counters, the general formula of the previous article:

$$s = R_1 - n(W + W') - n'w,$$

it is now seen *that any counter will sustain its greatest stress when the moving load extends from the center to its foot;* for R_1 has its greatest possible value with the load over the whole of the arm, but n' must be zero.

The panel point at which the counters are to begin is found, as before, by the equation:

$$R_1 - n(W + W') = 0.$$

In this case R_1 *is found by the simple principle of the lever.*

It results, from what has already been stated, that the greatest possible shear, S, immediately adjacent to, and outside of, the middle supports, will occur when both arms are loaded over their entire lengths with balanced moving loads. Hence, in the equation

$$s = S - n\,(W + W') - n'w,$$

$(S - n'w)$ will have its greatest value for any main web member when $n' = 0$, *i. e., when the moving load extends from the extremity of the arm in which it is found to the web member in question, at the same time being balanced by moving load on the other arm.*

It is to be remembered that verticals in compression will sustain their greatest stresses at the same time with the diagonal tension members which cut their upper extremities, if the moving load traverses the lower chord.

As one arm of the bridge is a simple truss supported at each end, for the unbalanced loads on it, it is evident *that the greatest compression in the upper chord and tension in the lower will exist, near the ends, for the moving load over the whole of one arm.*

Since moving loads on both arms at the same time balance each other, it results *that the greatest tension in the upper chord and compression in the lower, at the centre, will exist with the moving load over the whole of both arms.*

This is true for the *centre* only. For other panels adjacent to the centre, *it will be necessary to take single-balanced panel moving weights, and find for each, all panels in which the stress is of the same kind as that caused by the fixed load alone, and the amount of that stress in those panels.*

This operation is precisely the same as that used in the preceding Article, and the results there obtained will be used in this, since, as will be seen, the data are to be exactly the same.

Having obtained the results for each pair of balanced weights, they are to be combined in the manner already shown.

The greatest compression in the lower chord and tension in the upper, *near the ends,* however, *will exist with the bridge open or closed and subjected to its own weight only.*

From these results the greatest chord stresses are to be found.

The example may now be proceeded with.

The truss and its loading will be the same as were taken in the preceding Article, and the data there used are here reproduced.

Panel length $= AB = BC =$, etc., $= p = 12$ feet.

$$NO = OO' = O'N' = 8 \text{ feet.}$$
$$AA' = 22 \times 12 + 3 \times 8 = 288 \text{ feet.}$$

Bb	$= 20$ feet.		Hh	$= 26$ feet.
Cc	$= 21$ "		Kk	$= 27$ "
Dd	$= 22$ "		Ll	$= 28$ "
Ee	$= 23$ "		Mm	$= 29$ "
Ff	$= 24$ "		Nn	$= 30$ "
Gg	$= 25$ "		Oo	$= 30$ "

Total fixed weight $= 1850$ pounds per panel.

Upper chord panel fixed weight $= W = 4.5$ tons.
Lower " " " " $= W' = 6.6$ "
Uniform panel moving load $= w = 18$ "

The verticals are compression members, and the diagonals tension members, except the end posts Ab and $A'b'$. The moving load traverses the lower chord.

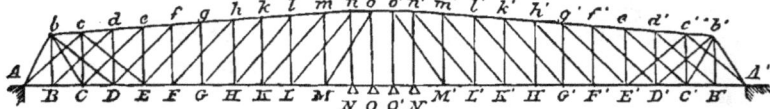

With the main central diagonals omitted, the truss will present the appearance shown in the figure above.

The stresses in the counter braces will first be found.

As before, a fixed panel weight of 4 tons will be taken at A.

By way of variety all the inclined web stresses will be found by the method of moments.

Each system of triangulation may be so divided through any inclined web member that it and one panel in each chord will be severed. By the principles of Art. 17, therefore, the origin of moments for any inclined web stress will be at the intersection of the chords prolonged.

Let this point of intersection be called i; then will

$$iB = \frac{Bb}{Nn - Bb} \times BN = \frac{20 \times 120}{10} = 240 \text{ feet.}$$

The following values may at once be written :

$$
\begin{aligned}
iA &= 228 \text{ feet.} \\
iB &= 240 \text{ "} \\
iC &= 252 \text{ "} \\
iD &= 264 \text{ "} \\
iE &= 276 \text{ "} \\
iF &= 288 \text{ "} \\
iG &= 300 \text{ "} \\
iH &= 312 \text{ "} \\
iK &= 324 \text{ "} \\
iL &= 336 \text{ "} \\
iM &= 348 \text{ "} \\
iN &= 360 \text{ "}
\end{aligned}
$$

It will presently be shown that the same three counters, bC, bD, cE, which were found necessary in the preceding Article, are the only ones needed in this truss, or rather in the left arm of it.

The lever arms of the inclined web stresses will be the normal distances from i to the web members prolonged.

Let l_1 be the normal distance from i to cE prolonged.
" l_2 "   "   "   " bD "
" l_3 "   "   "   " bC "
" l_4 "   "   "   " Ac "

Let l_5 be the normal distance from i to Bd prolonged.

" l_6 "   "   " Ce "

·   ·   ·   ·   ·   ·   ·

" l_{14} "   "   " Mo "

Then there will result the following values for the lever arms :

$$l_1 = iE \times \sin CEc = 182 \text{ feet.}$$
$$l_2 = iD \times \sin BDb = 169.0 \text{ "}$$
$$l_3 = iC \times \sin BCb = 216.0 \text{ "}$$
$$l_4 = iA \times \sin cCA = 150.3 \text{ "}$$
$$l_5 = iB \times \sin dBD = 162.2 \text{ "}$$
$$l_6 = iC \times \sin cCE = 174.4 \text{ "}$$
$$l_7 = iD \times \sin fDF = 186.6 \text{ "}$$
$$l_8 = iE \times \sin gEG = 199.0 \text{ "}$$
$$l_9 = iF \times \sin hFH = 211.7 \text{ "}$$
$$l_{10} = iG \times \sin kGK = 224.2 \text{ "}$$
$$l_{11} = iH \times \sin lHL = 237.0 \text{ "}$$
$$l_{12} = iK \times \sin mKM = 249.5 \text{ "}$$
$$l_{13} = iL \times \sin nLN = 262.4 \text{ "}$$
$$l_{14} = iM \times \sin oMO = 289.5 \text{ "}$$

The stresses in the counters may now be determined, and for that purpose, as in the preceding Article, z will denote the distance from A to any panel point.

The same signification will be attached to z in the determination of all the web stresses, but in the right span it will be measured from A'.

The general expression for the reaction at the end of either arm, for unbalanced loads in that arm, is, by the law of the lever :

$$R_1 = \Sigma P\left(\frac{l-z}{l}\right) = \frac{P}{l}\Sigma(l-z);$$

in which P is the *panel moving load.*

Let the counters in the system $AcCeE$, etc., be first considered.

With the head of the train at G, z will have the values 72, 96, and 120 feet. Hence,

$$R_1 = 16.98 \text{ tons.}$$

Since the fixed load at the points A, C, and E is $4 + 11.1 + 11.1 = 26.2$ tons, $s = 0$ for the point E.

With the head of the train at E, z will have the values 48, 72, 96, and 120 feet. Hence,

$$R_1 = 28.81 \text{ tons.}$$

The fixed load at the points A and C is $4 + 11.1 = 15.1$ tons, consequently $s = R_1 - n(W + W') = 0$ at E, and cE is the first counter needed.

Dividing the system under consideration through the members cc, cE, CE, and taking moments about i, the point of intersection of the chords:

$$(cE) = \frac{(R_1 - 4) \times 228 - 11.1 \times 252}{l_1} = + 15.7 \text{ tons.}$$

Laying off, therefore, from c on cE, by any convenient scale, 15.7 tons, and drawing from the point thus found a line parallel to ce until it cuts cC, it will be found that the distance from the point of intersection to c represents by the same scale 11.5 tons. Hence,

$$(cC) = -(11.5 + 4.5) = -16.0 \text{ tons.}$$

With the head of the train at C, z has the values 24, 48, 72, 96, and 120 feet. Hence,

$$R_1 = 43.72 \text{ tons.}$$

Dividing bc, bC, and BC, and taking moments about i:

$$(bC) = \frac{(R_1 - 4) \times 228}{l_3} = + 41.9 \text{ tons.}$$

Proceeding in precisely the same manner for the system $AbBdD$, etc., it is found that bD is the first counter required. With the head of the train, therefore, at D, z will have the values 36, 60, 84, and 108 feet. Hence:

$$R_1 = 32.74 \text{ tons.}$$

Cutting bd, bD, and BD, and taking moments about i, there will result:

$$(bD) = \frac{R_1 \times 228 - 11.1 \times 240}{l_2} = + 28.4 \text{ tons.}$$

bB is the only vertical which will be subjected to tension, and it will receive its greatest stress when the head of the train is at B. The value of that stress is:

$$(bB) = 18.00 + 6.6 = W' + w = + 24.6 \text{ tons.}$$

In the determination of the main web stresses the moving panel loads on one arm will be balanced by equal loads symmetrically placed on the other; the reactions required, therefore, may be taken from those already determined in the preceding Article on pages 135 and 136.

In the system $AcCcE$, etc., the member Ac will only be subjected to stress when the bridge is open, and the stress will be caused by the fixed load only. From the preceding Article there may then be taken, both for Ac and cC:

$$(Ac) = + 6.08 \text{ tons.}$$
$$(cC) = - 8.3 \text{ "}$$

It is seen therefore that cC will sustain its greatest stress in connection with the counter cE.

Main web member Ce.

It has been shown under what condition of loading any main web member receives its greatest stress. The condition for Ce is a balanced moving load at C, or, in other words, equal moving loads at C and C'. The preceding Article, page 135, then gives:

$$R_1 = 13.50 + 0.039 = 13.539 \text{ tons.}$$

Taking moments about i:

$$(Ce) = \frac{-(R_1 - 4) \times 228 + 29.1 \times iC}{l_6} = +29.6 \text{ tons.}$$

The method of finding the stresses in the vertical members will be shown in connection with the other system.

Main web member Eg.

This requires balanced moving loads at C and E, or equal moving loads at C, E, C', and E'. The preceding Article, page 135, then gives:

$$R_1 = 13.50 + 9.25 + 0.039 + 0.07 = 22.859 \text{ tons.}$$

Taking moments about i:

$$(Eg) = \frac{-(R_1 - 4) \times 228 + 2 \times 29.1 \times iD}{l_8} = +55.6 \text{ tons.}$$

Main web member Km.

This member requires balanced loads at C, E, G, and K. In the same manner as before:

$$R_1 = 22.859 + 5.52 + 2.55 + 0.088 + 0.084 = 31.101 \text{ tons.}$$

Taking moments about i:

$$(Km) = \frac{-(R_1 - 4) \times 228 + 4 \times 29.1 \times iF}{l_{12}} = +109.6 \text{ tons.}$$

It is unnecessary to find the other stresses in this system, as the method is exactly the same for all.

In the system *AbBdD*, etc., the member *bB* is compressed by the fixed weight at its upper extremity only, when the bridge is open or simply closed.

Hence :

$$(bB) = - 4.5 \text{ tons.}$$

Main web member Bd.

A balanced moving load at *B* is required, or equal moving loads at *B* and *B'*.

Page 136 of the preceding Article gives :

$$R_1 = 15.67 + 0.053 = 15.723 \text{ tons.}$$

Taking moments about *i* :

$$(Bd) = \frac{- R_1 \times 228 + 29.1 \times iB}{l_5} = + 20.96 \text{ tons.}$$

Main web member Df.

Balanced moving loads at *B* and *D* give, by page 136 of the preceding Article :

$$R_1 = 15.723 + 11.153 + 0.148 = 27.024 \text{ tons.}$$

Taking moments about *i* :

$$(Df) = \frac{- R_1 \times 228 + 2 \times 29.1 \times iC}{l_7} = + 45.6 \text{ tons.}$$

It is unnecessary to find the stresses in any other inclined web member. The process is exactly the same for all. The reaction R_1 is first found, and then moments are taken about *i*.

The stresses in the verticals are readily determined when those in the diagonals are known. The method is that used in the preceding Article, and also for the stress in *cC* in connection with that in *cE*.

The figure below is the one used in finding the vertical stresses of the system *AbBdD*, etc.

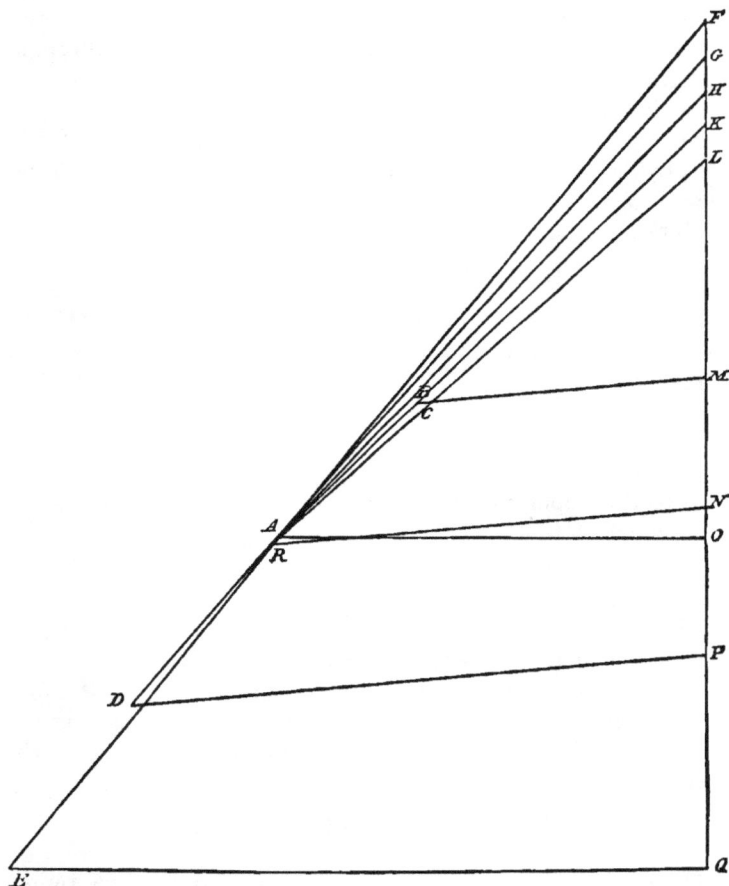

AO is drawn horizontally, and by scale, to represent 24 feet or two panel lengths.

FQ is vertical, and passes through *O*. *OL* represents in length *Dd*; *OK*, *Ff*; *OH*, *Hh*; *OG*, *Ll*, and *OF*, *Nn*, or 30 feet. The lines drawn from *A* through *L*, *K*, *H*, *G*, *F* are

then parallel to the inclined web members, the greatest
stresses in which are already known. These known stresses
are next laid off on the radial lines through A, from L, K, H,
G, and F, and through their extremities are drawn lines
parallel to the upper chord until they cut the vertical through
O. On this last line are found the stresses desired.

As examples :

$LC = (Bd) = $ 20.96 tons. CM is parallel to the upper chord.
LM, by the same scale (ten tons per inch in this case), represents 12.8 tons.

Hence :

$$(dD) = - (12.8 + 4.5) = - 17.3 \text{ tons.}$$

KB (twenty tons per inch) $= (Df) = $ 45.6 tons. BM is
parallel to the upper chord, and KM represents 29.6 tons.

Hence :

$$(fF) = - (29.6 + 4.5) = - 34.1 \text{ tons.}$$

$HR = (Fh) = $ 71.9 tons (twenty tons per inch). RN is parallel to the upper chord, and HN represents 48.6 tons.

Hence :

$$(hH) = - (48.6 + 4.5) = - 53.1 \text{ tons.}$$

The figure gives the others in the same manner. EQ is
perpendicular to QF, because FE is the stress in Ln.

A portion of the stress in nN is indirect, and due to the
stress in mn, as has been noticed before. It will presently be
shown that the stress in mn, with the moving load over both
arms, is :

$$(mn) = + 223.54 \text{ tons.}$$

Denoting the inclination of mn to a horizontal line by α,
the indirect stress in nN is :

$$(mn) \times \sin \alpha = 18.58 \text{ tons.}$$

The diagram above gives the vertical component of (Ln)
as 98.1 tons.

Hence:

$$(nN) = -(98.1 + 4.5 + 18.58) = -121.18 \text{ tons.}$$

The greatest stress in Ab occurs when the reaction at A is the greatest; and that reaction will always be the greatest when the moving load covers one arm only.

With this condition of the loading:

$$R_1 = 43.72 + 49.1 = 92.82 \text{ tons.}$$

Hence:

$$(Ab) = -(R_1 - 4) \times scc \, AbB = -103.62 \text{ tons.}$$

Finally, collecting and arranging :

$(cE) = +$ 15.7 tons.	$(bD) = +$ 28.4 tons.
$(bC) = +$ 41.9 "	$(bB) = +$ 24.6 "
$(Ab) = -$ 103.62 "	$(bB) = -$ 4.5 "
$(Ac) = +$ 6.08 "	$(cC) = -$ 16.0 "
$(Bd) = +$ 20.96 "	$(dD) = -$ 17.3 "
$(Ce) = +$ 29.6 "	$(eE) = -$ 23.3 "
$(Df) = +$ 45.6 "	$(fF) = -$ 34.1 "
$(Eg) = +$ 55.6 "	$(gG) = -$ 42.0 "
$(Fh) = +$ 71.9 "	$(hH) = -$ 53.1 "
$(Gk) = +$ 82.6 "	$(kK) = -$ 61.0 "
$(Hl) = +$ 98.8 "	$(lL) = -$ 74.0 "
$(Km) = +$ 109.6 "	$(mM) = -$ 83.5 "
$(Ln) = +$ 125.35 "	$(nN) = -$ 121.18 "
$(Mo) = +$ 128.9 "	$(oO) = -$ 112.1 "

As has been observed, the detailed expressions for all these stresses are not given.

Of all the web members the vertical bB alone must be counterbraced.

The greatest stresses in Nn and Oo, added to the total corresponding reaction at A, ought to equal the total fixed and moving load on the left arm, or indeed on the right.

Total $R_1 + (Nn) + (Oo) = 39.52 + 31.8 + 112.1 + 121.18$
$$= 304.6 \text{ tons.}$$

The fixed and moving load for one arm is:

$$10 \times 29.1 + 4 + 2 \times 4.5 = 304.0 \text{ tons.}$$

The agreement is close.

The chord stresses due to the fixed load are precisely the same as those found in the preceding Article, and the values there given will be taken without recalculation.

Denoting the inclination of the upper chord to a horizontal line by α, there will be found:

$$sec\ \alpha = 1.0035.$$

The compression in the upper, and tension in the lower chord, near the end of the arm will first be found. It will then be necessary to cover the whole of the arm with the moving load.

System AcCcE, etc.

$$R_1 = 43.72 \text{ tons};\qquad R_1 - 4 = 39.72 \text{ tons.}$$
$$(R_1 - 4) \times 12 = 476.64.$$
$$(W + W' + w) \times 12 \times 2 = 29.1 \times 12 \times 2 = 698.4.$$

The counter cE comes into action.
Taking moments about b:

$$(AC) = \frac{(R_1 - 4) \times 12}{20} = + 23.83 \text{ tons.}$$

Taking moments about c:

$$(CE) = \frac{(R_1 - 4) \times 12 \times 2}{21} = + 45.39 \text{ tons.}$$

$$(bc) = - (CE) \times sec\ \alpha = - 45.55 \text{ tons.}$$

Taking moments about E:

$$(cg) = -\frac{(R_1 - 4) \times 12 \times 4 - 29.1 \times 12 \times 2}{23} \times sec\ \alpha =$$

$$-52.71 \text{ tons.}$$

Taking moments about g:

$$(EG) = \frac{(R_1 - 4) \times 12 \times 6 - 2 \times 29.1 \times 12 \times 3}{25} =$$

$$+ 30.59 \text{ tons.}$$

$$(gk) = -(EG) \times sec\ \alpha = -30.7 \text{ tons.}$$

Taking moments about k:

$$(GK) = \frac{(R_1 - 4) \times 12 \times 8 - 3 \times 29.1 \times 12 \times 4}{27}.$$

The numerator of this expression is negative, showing that GK is subjected to compression. No other chord stresses in this system are therefore needed.

System AbBdD, etc.

$$R_1 = 49.1 \text{ tons}; \qquad R_1 \times 12 = 589.2.$$
$$(W + W' + w) \times 12 \times 2 = 29.1 \times 12 \times 2 = 698.4.$$

The counter bD comes into action.

Taking moments about b:

$$(AD) = \frac{R_1 \times 12}{20} = + 29.46 \text{ tons.}$$

Taking moments about D:

$$(bf) = -\frac{3 \times R_1 \times 12 - 29.1 \times 12 \times 2}{22} \times sec\ \alpha = -48.77 \text{ tons.}$$

Taking moments about f:

$$(DF) = \frac{R_1 \times 5 \times 12 - 2 \times 29.1 \times 12 \times 3}{24} = + 35.45 \text{ tons.}$$

$$(fh) = -(DF) \times sec\ \alpha = -35.57 \text{ tons.}$$

Taking moments about h :

$$(FH) = \frac{R_1 \times 7 \times 12 - 3 \times 29.1 \times 12 \times 4}{26} = -2.54 \text{ tons.}$$

$$(hl) = -(FH) \times \sec \alpha = +2.54 \text{ tons.}$$

This shows that (FH) is compression, consequently no other chord stresses in this system are needed.

That the counters come into action in this case is shown by the reactions at A.

There remain only to be found the tension in the upper chord and compression in the lower, at and near the center. In this case the moving load is to cover the whole bridge, for the stresses (oo'), (MN), (NO), and (OO') only, and these will first be found.

The reaction for all balanced moving loads may be taken directly from the preceding Article, remembering that in this case the reaction R, at A, will be :

$$R = R_1 + R_4.$$

System AcCcE, etc., with moving load over the whole of both arms.

$$R = \Sigma R_1 + \Sigma R_4 = 31.8 \text{ tons (see page 135).}$$

$$R - 4 = 27.8 \text{ tons;} \qquad (R - 4) \times 12 = 333.6.$$

$$(W + W' + w) \times 12 \times 2 = 29.1 \times 12 \times 2 = 698.4.$$

On account of the indirect stress in Nn it will first be necessary to find (mn). Taking moments about M :

$$(mn) = -\frac{(R - 4) \times 10 \times 12 - 4 \times 29.1 \times 12 \times 5}{29} \times \sec \alpha =$$
$$+ 126.24 \text{ tons.}$$

$$(mn) \times \sin \alpha = \text{indirect stress in } Nn = 10.49 \text{ tons.}$$

Finally, taking moments about o :

$$(MO) = (OO') = - (oo') =$$
$$\frac{(R-4) \times 140 + 10.49 \times 8 - 5 \times 29.1 \times 68}{30} = -197.27 \text{ tons.}$$

System AbBdD, etc., *with moving load over the whole of both arms.*

$$R = \Sigma R_1 + \Sigma R_4 = 39.52 \text{ tons (see page 136).}$$

$$R_1 \times 12 = 474.24.$$

Taking moments about n :

$$(LN) = (NN') = (-nn') =$$
$$\frac{R \times 12 \times 11 - 5 \times 29.1 \times 12 \times 6}{30} = -175.31 \text{ tons.}$$

Hence :

Resultant $(oo') = 197.27 + 175.31 = +372.58$ tons.

The value $+175.31$ tons of (nn') will be used in finding (no), since it is the *centre* stress for the system *AbBdD*, etc.

As was to be expected, the results of the preceding Article show that a balanced weight at M causes compression in mn and no; all other balanced weights in the same system, how-ever, cause tension in those parts. With moving panel weights at C, E, G, and K :

$$\Sigma R_4 = 0.039 + 0.07 + 0.088 + 0.084 = 0.281.$$

Hence, for that loading :

$$n'o' = -\frac{281}{334} \times 1.4 = -1.18 \text{ tons.}$$

Finally, the resultant stress :

$$(no) = 3.73 + 6.42 + 6.98 + 4.34 - 1.18 + 175.31 + 108.4 =$$
$$+ 304.00 \text{ tons.}$$

The results of the preceding Article show that only the balanced moving loads extending from B to K, inclusive, produce tension in mn. With the balanced moving loads at B, D, F, and H:

$$\Sigma R_1 = 0.053 + 0.148 + 0.212 + 0.222 = 0.635 \text{ tons.}$$

Hence, for that loading:

$$(l'n') = -\frac{635}{794} \times 3.07 = -2.46 \text{ tons.}$$

The resultant stress is then :

$$(mn) = 3.73 + 6.44 + 6.98 + 4.34 + 1.28 + 3.29 + 3.68 + 1.36$$
$$-1.18 - 2.46 + 108.8 + 95.5 = +231.76 \text{ tons.}$$

Balanced moving loads at B, D, F, and H, are the only ones causing tension in lm. Again, for that loading:

$$(l'n') = -\frac{635}{794} \times 3.07 = -2.46 \text{ tons.}$$

Hence, the resultant stress:

$$(lm) = 1.28 + 3.29 + 3.68 + 1.36 - 2.46 + 73.7 + 95.5 =$$
$$+176.35 \text{ tons.}$$

The moving loads cause no farther tension in the upper chord, and the remaining upper chord tensions are taken from those caused by the fixed load alone.

The lower chord stresses are found by precisely the same method, with the same conditions of loading; or by multiplying the corresponding upper chord stresses by $cos\ \alpha$. Thus :

$$(KL) = -(mn) \times cos\ \alpha = -230.95 \text{ tons;}$$
$$(HK) = -(lm) \times cos\ \alpha = -175.73 \text{ tons;}$$

$cos\ \alpha$ having the value 0.9965.

By collecting and arranging, the following greatest chord stresses may be written:

(*bc*)	= +	0.00 tons ;	— 94.32 tons.
(*cd*)	= +	4.57 " ;	— 101.48 "
(*de*)	= +	16.67 " ;	— 101.48 "
(*ef*)	= +	32.10 " ;	— 101.48 "
(*fg*)	= +	53.40 " ;	— 88.28 "
(*gh*)	= +	77.10 " ;	— 66.27 "
(*hk*)	= +	105.40 " ;	— 28.16 "
(*kl*)	= +	135.40 "	
(*lm*)	= +	176.35 "	
(*mn*)	= +	231.76 "	
(*no*)	= +	304.00 "	
(*oo'*)	= +	372.58 "	

(*AB*)	= —	4.57 tons ;	+ 53.29 tons.
(*BC*)	= —	16.67 " ;	+ 53.29 "
(*CD*)	= —	32.00 " ;	+ 74.85 "
(*DE*)	= —	53.20 " ;	+ 80.84 "
(*EF*)	= —	76.80 " ;	+ 66.04 "
(*FG*)	= —	105.00 " ;	+ 28.05 "
(*GH*)	= —	134.92 "	
(*HK*)	= —	175.73 "	
(*KL*)	= —	230.95 "	
(*LM*)	= —	304.00 "	
(*MN*)	= —	372.58 "	
(*NO*)	= —	372.58 "	
(*OO'*)	= —	372.58 "	

It is thus seen that the portions *bk* of the upper, and *AG* of the lower chord must be counterbraced ; also corresponding portions of the other arm.

Interesting and important comparisons may now be made between the greatest stresses in this case and the preceding one.

Although the same counters are needed in the two trusses, yet, as might have been anticipated, the stresses are by far greater in the present case.

The main web stresses,* with the exception of those at the centre, are nearly the same in the two cases. This result, also, was to be expected, since the values of R_4 in the preceding Article were so small. If R_4 were large, the difference between the results of the two cases would be large also; for then the reactions at A, with the same amount of moving load on the left arm, would be very different, and the moments by which the web stresses are determined would be correspondingly different.

The end post Ab sustains the greatest stress with partial continuity. The most marked difference, however, is to be found with the stresses (oO) and (nN). With perfect continuity, the stress (oO) is nearly three times as great as with partial, while (nN) is about once and a half as great.

The chord stresses at the centre are alike or different in the two cases, according to the method by which they are determined, in that of perfect continuity. This results from the ambiguity in the latter case, already pointed out.

As was to be expected, more counterbracing in the chords, and to a greater degree, is found with partial continuity than with perfect. Many of the remaining chord stresses are the same in the two trusses, while none of them present great differences.

With partial continuity, therefore, great stresses (chord and web), ambiguity and negative reactions are avoided at the centre; while, on the other hand, somewhat greater stresses are found near the ends.

At the same time there arises the question, does the assumption made at the beginning of the Article, in reference to unbalanced loads, hold strictly true? In other words, can an arm of the bridge be strictly considered a simple truss supported at each end, for unbalanced moving loads, while other, but balanced moving loads, rest on the bridge?

Such an assumption is probably not exactly true, for any

---

* In comparing results, in the two trusses, irregularities in some of the main web stresses, to the extent of a few tenths of a ton, will be noticed; these are due to small errors in the diagrams used in the preceding Article.

unbalanced load will change the inclination of the truss at the centre, and cause, consequently, somewhat of a disturbance in the "balance" of the other loads. This disturbance can probably never be great with the ordinary proportions between the centre and end spans of draw-bridges. It can only be said with certainty, however, thus far, that a number of large and very satisfactory swing bridges have been built on this system.

No locomotive excess has been taken, but precisely the same conditions of loading for the greatest stresses must be assumed if the train is headed with such an excess. When the train covers one arm, however, the excess must be taken first at one end and then at the other; the greatest corresponding chord stresses are then to be selected from the two sets of results.

Formulæ for the web and chord stresses, in either this or the preceding case, may easily be written, but the number of trigonometrical functions required makes their use more tedious than the methods employed, and hence undesirable.

If the chords are parallel, the work of finding the greatest stresses is very much lessened by the use of trigonometric functions.

It is worthy of notice that, either in this case or the preceding, if the moving load covers first one arm, then both, and finally be taken from both, and the stresses found in each condition, the results will not be far from the greatest stresses; in fact, may be near enough for a preliminary estimate.

Precisely the same principles and conditions of loading hold true, in this case, with only one system, or more than two systems of triangulation.

Art. 38.—Ends Simply Resting on Supports—One Support at Centre— Example.

The general principles fundamentally involved in this case are not different from those of the two preceding ones, except in the number of points of support at the first pier. All the fixed load of the bridge is carried to the central point of support, whether the bridge is open or closed; the end supports furnish reactions for the moving load only.

The truss to be taken as an example is the one shown in the accompanying figure, in which the arms are of equal length.

The general formulæ to be used for the reactions at A, B, and C, and for the bending moment at the centre, are equations (11), (12), (13), and (10) respectively of Art. 35. These equations may be written as follows, remembering that $l_1 = l_2 = l$, and $M_2 = M$:

$$M = -\frac{1}{4l^2}\left\{ \overset{1}{\Sigma}P(l^2 - z^2)z + \overset{2}{\Sigma}P(l^2 - z^2)z \right\} \quad . \quad . \quad (1).$$

$$R_1 = \frac{1}{l}\left\{ \overset{1}{\Sigma}P(l - z) + M \right\} \quad . \quad . \quad . \quad . \quad . \quad . \quad . \quad (2).$$

$$R_2 = \frac{1}{l}\left\{ \overset{1}{\Sigma}Pz + \overset{2}{\Sigma}Pz - 2M \right\} \quad . \quad . \quad . \quad . \quad . \quad . \quad (3).$$

$$R_3 = \frac{1}{l}\left\{ \overset{2}{\Sigma}P(l - z) + M \right\} \quad . \quad . \quad . \quad . \quad . \quad . \quad (4).$$

It is to be remembered that z is measured from A or C, according as the left or right arm is considered.

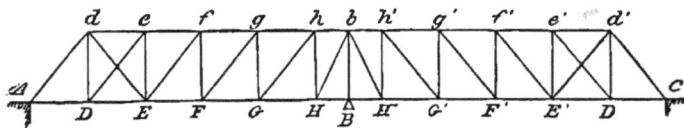

The following are the data to be used:

     Total length $= AC = 2l = 2AB = 2BC = 144$ feet.
     Uniform depth $= dD = bB = 16$ feet.
     Panel length $= AD = DE = $ etc. $= 13$ feet.
            $BH = BH' = 7$ feet.

Total fixed weight per foot   = 1,200 pounds (nearly).
Upper chord panel fixed weight = W = 2.73 tons.
Lower   "   "   "   " = W' = 5.00 "
Uniform panel moving load   = w = 19.50 "

The moving load traverses the lower chord, and the weight of the floor system is taken at nearly 350 pounds per foot.

On account of the extra weight of the locking apparatus, the fixed weight at A, or C, will be taken at 3 *tons*, and will be denoted by w_1.

As is clear from the figure, all inclined web members, except the end posts, are for tension only, while the verticals are compression members.

As the ends A and C are neither latched down nor lifted up, either arm is a single truss simply supported at each end, for all moving loads which rest upon it, *so long as there are no moving loads on the other arm.*

For exactly the same reasons, therefore, as those given in the preceding Article, *any counter, as* dE, *will sustain its greatest stress when the moving load extends from its foot to the centre, if no other moving load rests on the bridge.*

It must still be borne in mind that in connection with any counter stress, the stress in the vertical which cuts its upper extremity is to be found, for such a one may be the greatest stress in the vertical.

Again, resume the general expression for the shear in any web member:

$$s = S - n(W + W') - n'w;$$

in which S is the shear at one extremity of the *arm*, and n and n' the numbers of fixed and moving panel weights respectively between the same end of the arm and the web member in question. In considering the main web members, S will be taken adjacent to the centre, and, in the present example, at an indefinitely short distance from B in the arm AB.

For a given condition of loading in AB, it is evident that the smaller is R_1 the greater will be S. But Eq. (1) shows that M is always negative. Hence so long as $\overset{1}{\Sigma}P(l-z)$ remains the same, Eq. (2) shows that R_1 decreases as M increases (numerically).   •

Again, Eq. (1) shows that M will have its greatest numerical value, other things remaining the same, when $\overset{2}{\Sigma}P(l^2-z^2)z$ has the greatest value possible; *i. e.*, when the moving load covers the whole of the arm BC. With a given value, therefore, for $\overset{1}{\Sigma}P(l-z)$, R_1 will be the least possible when the moving load covers the whole of the other arm, or the whole of BC; consequently S will be the greatest under the same conditions. Now having found under what circumstances S is the greatest, precisely the same reasoning used in the preceding Articles shows that s will be the greatest, under the same circumstances, when n' is zero.

Any inclined main web member, then, will sustain its greatest tensile stress when the moving load extends from its foot to the free extremity of the arm in which it is found, and covers at the same time the whole of the other arm.

A few main web stresses in all trusses of this case are a little singular in character, but are no exceptions to this rule. Those for the example taken will be noticed in the proper place.

Any vertical web member, unless acting as a counter, will sustain its greatest compression in connection with the greatest tension in the inclined main web member which cuts its upper extremity.

In seeking the greatest main web stresses, it may happen that the reaction R_1 becomes zero; this, however, changes nothing in the method.

Since either arm is a simple truss for all moving loads resting on it (supposing none on the other), every such load tends to cause the same kind of stress throughout the same chord. Consequently, as in the previous Article, *the greatest tension in the lower chord, and compression in the upper, will exist when the moving load covers one arm only.* These stresses will be found in that portion of the arm adjacent to its free extremity.

The greatest chord stresses of the same kind as those

caused by the fixed load, can be found with the least labor by *first determining all the stresses due to the fixed load alone,* and tabulating them.

The stresses caused by the moving load alone are then to be determined by the aid of the following considerations.

Let that moment be considered negative which causes tension in the upper chord and compression in the lower. All moments, then, caused by the fixed or moving loads are negative, and all those produced by the upward reactions R_1 are positive. Now the compression in any lower chord panel may be found by taking moments (such will be negative if compression exists) about the panel point vertically over that extremity nearest the centre. The general expression for such compression will be:

$$\frac{R_1 np - n'wt}{d};$$

in which n is the number of the panel from the free extremity of the arm, n' the number of the moving panel loads on the arm, t the distance of the centre of gravity of the moving loads $n'w$ from the origin of moments, p the panel length, and d the depth of the truss.

The numerator of this expression must be negative, and it is desired to find what value of n' will give it its greatest negative value.

Now since every panel moving load on the arm AB increases R_1 (as a positive quantity), it appears from the figure that n' must not be greater than $(n - 1)$, and, farther, it must belong to loads between the panel considered and the free extremity of the arm. Since, however, t varies with n' the above expression may have its greatest negative value when n' is less than $(n - 1)$.

These considerations are independent of the general character of R_1; it has already been seen, however, that, with a given loading on AB, R_1 will be the least when the moving load covers the whole of BC also. Hence in order to find the greatest tension in the upper chord and compression in the lower due to the moving load, the method of procedure is as follows :

Throughout the whole operation the moving load is to entirely cover one arm, as BC. The moving load is then to cover the other arm from the free extremity to any panel point, and the stresses in the panels situated between the end of the train and the centre are to be computed. This operation is to be repeated for every panel in that arm not wholly covered by the moving load. From these results the greatest stresses may be selected and then added to the fixed load stresses.

The character of these operations and the reasons for them will be much more evident after the example is treated.

The following values will be needed, and depend only on the data already given.

$$z = 13 \quad . \quad . \quad l - z = 59 \quad . \quad . \quad (l^2 - z^2)z = \;\;65,195.00$$
$$z = 26 \quad . \quad . \quad l - z = 46 \quad . \quad . \quad (l^2 - z^2)z = 117,208.00$$
$$z = 39 \quad . \quad . \quad l - z = 33 \quad . \quad . \quad (l^2 - z^2)z = 142,857.00$$
$$z = 52 \quad . \quad . \quad l - z = 20 \quad . \quad . \quad (l^2 - z^2)z = 128,960.00$$
$$z = 65 \quad . \quad . \quad l - z = \;\;7 \quad . \quad . \quad (l^2 - z^2)z = \;\;62,335.00$$

Angle $AdD = \alpha$.
" $HbB = \beta$.

$$\tan \alpha = 0.8125. \qquad\qquad \sec \alpha = 1.29.$$
$$\tan \beta = 0.4375. \qquad\qquad \sec \beta = 1.09.$$

$$\frac{1}{4l^2} = \frac{1}{20736} = 0.00004823$$

$$P = 19.5 \text{ tons} = w; \qquad\qquad \frac{P}{7} = 0.271.$$

The following values of M are found by substituting the proper numerical values in Eq. (1). The moving load is taken to cover the whole of BC and so much of AC as is indicated by the values of z.

In arm AB; $z=$ 13 . . . $M = -547.00$

" " " ; $z = \begin{cases} 13 \\ 26 \end{cases}$ . . . $M = -657.00$

" " " ; $z = \begin{cases} 13 \\ 26 \\ 39 \end{cases}$ . . . $M = -792.00$

" " " ; $z = \begin{cases} 13 \\ 26 \\ 39 \\ 52 \end{cases}$ . . . $M = -913.00$

" " " ; $z = \begin{cases} 13 \\ 26 \\ 39 \\ 52 \\ 65 \end{cases}$ . . . $M = -972.00$

The following values may now be written. Those of R_1 are found by simply substituting the proper numerical quantities in Eq. (2); the moving load, as the values of M show, is taken to cover the whole of BC.

$z =$ 13 . . $\dfrac{P}{l} \Sigma \tfrac{1}{2} (l - z) = 15.99$ tons . . $R_1 =$ 8.39 tons.

$z = \begin{cases} 13 \\ 26 \end{cases}$ . . " $= 28.46$ " . . $R_1 = 19.33$ "

$z = \begin{cases} 13 \\ 26 \\ 39 \end{cases}$ . . " $= 37.40$ " . . $R_1 = 26.40$ "

$z = \begin{cases} 13 \\ 26 \\ 39 \\ 52 \end{cases}$ . . " $= 42.82$ " . . $R_1 = 30.14$ "

$z = \begin{cases} 13 \\ 26 \\ 39 \\ 52 \\ 65 \end{cases}$ . . " $= 44.72$ " . . $R_1 = 31.22$ "

13

The stresses in the counters will first be sought, *i.e.*, those in the arm AB.

As a trial let the moving load cover the points F, G, and H, and, as before, let R_1 be the general expression for the reaction at A. Hence:

$$R_1 = 3 \times 19.5 \times \frac{13 + 7}{72} = 16.25 \text{ tons.}$$

Since $3 + 2 \times 7.73 > R_1$, $\Sigma P = 0$ at the panel point E, and no counter is needed between e and F.

Moving load over EH.

$$R_1 = 4 \times 19.5 \times \frac{26.5}{72} = 28.71 \text{ tons.}$$

As $3 + 2 \times 7.73 + 19.5 > R_1$, $\Sigma P = 0$ at E, and dE is the first and only counter needed.

The vertical component of the stress in dE is

$$s = R_1 - 3 + 7.73 = 17.98 \text{ tons.}$$

Hence, $(dE) = 17.98 \times sec\ \alpha = + 23.19 \text{ tons.}$

The greatest compression in the end post Ad, and tension in the vertical $d'D$ will exist when the moving load covers the whole of the arm AB, the other carrying none, and with such loading :

$$R_1 = 5 \times 19.5 \times \frac{33}{72} = 44.69 \text{ tons.}$$

Hence, $(Ad) = - (R_1 - 3) \times sec\ \alpha = - 53.78 \text{ tons.}$

At the same time :

$$(d'D) = + (19.5 + 5.00) = + 24.5 \text{ tons.}$$

The stresses in the main web members are next to be determined.

Those in Ad and $d'D$ will occur under circumstances to be indicated hereafter.

The following operations are in accordance with the principles already shown.

Moving load on BC and at D.

$R_1 = 8.39$ tons; hence, for the shear in De:

$$s = -R_1 + (3 + 7.73 + 19.5) = 21.84 \text{ tons.}$$

Hence, $(De) = s \times sec\, \alpha = + 28.17$ tons.

Also, $(eE) = -(s + 2.73) = -24.57$ tons.

Moving load on BC and DE.

$R_1 = 19.33$ tons; hence, for the shear in Ef:

$$s = -R_1 + (3 + 2 \times 7.73 + 2 \times 19.5) = 38.13 \text{ tons.}$$

Hence, $(Ef) = s \times sec\, \alpha = + 49.19$ tons.

Also, $(fF) = -(s + 2.73) = -40.86$ tons.

Moving load on BC and DF.

$R_1 = 26.40$ tons; hence, for the shear in Fg:

$$s = -R_1 + (3 + 3 \times 7.73 + 3 \times 19.5) = 58.29 \text{ tons.}$$

Hence, $(Fg) = s \times sec\, \alpha = + 75.19$ tons.

Also, $(gG) = -(s + 2.73) = -61.02$ tons.

Moving load on BC and DG.

$R_1 = 30.14$ tons; hence, for the shear in Gh:

$$s = -R_1 + (3 + 4 \times 7.73 + 4 \times 19.5) = 81.78 \text{ tons.}$$

Hence, $(Gh) = s \times sec\, \alpha = + 105.5$ tons.

Also, $(hH) = -(s + 2.73) = -84.51$ tons.

Moving load over BC and AB.

$R_1 = 31.22$ tons; hence, for the shear in Hb:

$$s = -R_1 + (3 + 5 \times 7.73 + 5 \times 19.5) = 107.93 \text{ tons.}$$

Hence, $(Hb) = s \times sec\,\beta = + 117.64$ tons.

The same panel weights have been taken for H and H' as for D, E, F, etc., though, strictly speaking, they would be a little smaller. At b, however, the fixed weight will be taken as $2.73 \times 7 \div 13 = 1.47$ tons.

Hence, $(bB) = - (2 \times 107.93 + 1.47) = - 217.33$ tons.

Thus the web stresses, with the exceptions noticed, are completed.

It has been shown that the greatest compression in the upper chord and tension in the lower will exist when the moving load covers the whole of one arm, as AB, for which condition of loading, as has already been seen:

$$R_1 = 44.69 \text{ tons.}$$

Now, $R_1 - (3 + 7.73 + 19.5) = 14.46$ tons is that part of the total panel load (fixed and moving) at E, which may be considered as passing directly to A; while $(19.5 + 7.73) - 14.46 = 12.77$ tons is the remainder, which may be taken as passing directly to B.

The following values will now be needed:

$$14.46 \times tan\,\alpha = 11.75 \text{ tons.}$$
$$12.77 \times tan\,\alpha = 10.37 \text{ ``}$$
$$(7.73 + 19.5) \times tan\,\alpha = 22.12 \text{ ``}$$

The chord stresses then follow:

$$(AD) = (DE) = (R_1 - 3) \times tan\,\alpha = + 33.87 \text{ tons.}$$

$$(de) = (cf) = - (2 \times 14.46 + 27.23) \times tan\,\alpha = - 45.62 \text{ tons.}$$

$$(fg) = (ef) + 12.77 \times tan\,\alpha = - 35.25 \text{ tons.}$$

$$(EF) = - (fg) = + 35.25 \text{ tons.}$$

$$(gh) = (fg) + 27.23 \times tan\,\alpha + 12.77 \times tan\,\alpha = - 2.76 \text{ tons.}$$

$$(FG) = - (gh) = + 2.76 \text{ tons.}$$

(hb) will evidently be tension, and (GH) compression; no other stresses, therefore, are needed.

The following stresses, by moments, serve as checks:

$$(dc) = (cf) = - \frac{(R_1 - 3) \times 26 - 27.23 \times 13}{16} = - 45.62 \text{ tons.}$$

$$(gh) = - \frac{(R_1 - 3) \times 52 - 4 \times 27.23 \times 19.5}{16} = - 2.75 \text{ tons.}$$

The chord stresses due to the fixed load alone are the following:

$$(AD) = - (dc) = - 3 \times tan\, \alpha \qquad\qquad = - \;\; 2.44 \text{ tons.}$$
$$(DE) = - (cf) = (AD) - (3 + 7.73) \;\; \times tan\, \alpha = - 11.16 \text{ "}$$
$$(EF) = - (fg) = (DE) - (3 + 2 \times 7.73) \times tan\, \alpha = - 26.16 \text{ "}$$
$$(FG) = - (gh) = (EF) - (3 + 3 \times 7.73) \times tan\, \alpha = - 47.44 \text{ "}$$
$$(GH) = - (bh) = (FG) - (3 + 4 \times 7.73) \times tan\, \alpha = - 75.00 \text{ "}$$
$$(BH) = (GH) - (3 + 5 \times 7.73) \times tan\, \beta \qquad = - 93.22 \text{ "}$$

As a check:

$$(BH) = - \frac{3 \times 72 + 5 \times 7.73 \times 33}{16} = - 93.21 \text{ tons.}$$

The tension in the upper chord and compression in the lower, due to the moving load only, still remain to be found.

Moving load over AB and BC.

$R_1 = 31.22$ tons. Moments about b give:

$$(BH)' = \frac{R_1 \times 72 - 5 \times 19.5 \times 33}{16} = - 60.6 \text{ tons.}$$

Moving load over BC and AG.

$R_1 = 30.14$ tons. Moments about h give:

$$(GH)' = - (hb)' = \frac{R_1 \times 65 - 4 \times 19.5 \times 32.5}{16} = - 35.99 \text{ tons.}$$

Moving load over BC and AF.

$R_1 = 26.40$ tons.   Moments about g give:

$$(FG)' = -(gh)' = \frac{R_1 \times 52 - 3 \times 19.5 \times 26}{16} = -9.26 \text{ tons.}$$

Moving load over BC and AE.

$R_1 = 19.33$ tons.   Moments about f give:

$$(EF)' = -(fg)' = \frac{R_1 \times 39 - 2 \times 19.5 \times 19.5}{16} = -0.4 \text{ tons.}$$

Moving load over BC and at D.

$R_1 = 8.39$ tons.   Moments about e give:

$$(DE)' = -(ef)' = \frac{R_1 \times 26 - 19.5 \times 13}{16} = -2.21 \text{ tons.}$$

$(EF)' = -(fg)' = -2.21 - (19.5 - 8.39) \times tan\ \alpha = -11.24$ tons.
$(FG)' = -(gh)' = -11.24 - 9.03 = -20.27$ tons.
$(GH)' = -(hb)' = -20.27 - 9.03 = -29.30$ tons.

Other chord stresses, with the different conditions of loading taken, are not indicated, as they were found to be less, for the same panels, than those that are given. They might be needed, however, in some cases.

A very important result occurs, which has not before been noticed, when the moving load covers BC and one panel load rests at A.

In such a case Eq. (1) gives:

$$M = -\frac{P}{4l^2} \overset{2}{\Sigma} (l^2 - z^2)\ z = -486.00.$$

And Eq. (2):

$$R_1 = \frac{M}{l} = -6.75 \text{ tons.}$$

Under the circumstances just named, therefore, the condition of things at A is equivalent to *hanging* a weight of 6.75 tons at that point, as the end of the overhanging arm AB.

With such a weight, the following stresses result:

$$6.75 \times tan\,\alpha = 5.48 \text{ tons.}$$
$$(AD)'' = - (de)'' = - \quad 5.48 \text{ tons.}$$
$$(DE)'' = - (ef)'' = - \quad 10.97 \quad ''$$
$$(EF)'' = - (fg)'' = - \quad 16.45 \quad ''$$
$$(FG)'' = - (gh)'' = - \quad 21.94 \quad ''$$
$$(GH)'' = - (bh)'' = - \quad 27.42 \quad ''$$
$$(BH)'' = - (27.42 + 6.75\ tan\,\beta) = - 30.37 \text{ tons.}$$

From these results are to be selected the greatest chord stresses.

For examples:

$$\text{Resultant } (GH) = - (35.99 + 75.00) = - 110.99 \text{ tons.}$$
$$\text{'' } \qquad (FG) = - (21.94 + 47.44) = - \quad 69.38 \quad ''$$

In short, precisely as the operation has been done before.

The resultant web stresses caused by this negative reaction, are:

$$(Ad) = (3 + 6.75) \times sec\,\alpha = + 12.58 \text{ tons.}$$
$$(dD) = - (3 + 2.73 + 6.75) = - 12.48 \text{ tons.}$$

These are the " singular " stresses already mentioned.

Collecting and arranging the results, the following resultant stresses are obtained:

$$(dE) = + \quad 23.19 \text{ tons.}$$

$(Ad) = + \quad 12.58$ ''	$(Ad) = - \quad 53.78$ tons.	
$(dD) = + \quad 24.50$ ''	$(dD) = - \quad 12.48$ ''	
$(De) = + \quad 28.17$ ''	$(eE) = - \quad 24.57$ ''	
$(Ef) = + \quad 49.19$ ''	$(fF) = - \quad 40.86$ ''	
$(Fg) = + \quad 75.19$ ''	$(gG) = - \quad 61.02$ ''	
$(Gh) = + \quad 105.50$ ''	$(hH) = - \quad 84.51$ ''	
$(Hb) = + \quad 117.64$ ''	$(bB) = - 217.33$ ''	

$$(AD) = - \quad 7.92 \text{ tons}; \; + 33.87 \text{ tons.}$$
$$(DE) = - \quad 22.13 \quad '' \; ; \; + 33.87 \quad ''$$

$$(EF) = -\ 42.61 \text{ tons}; \ +\ 35.25 \text{ tons.}$$
$$(FG) = -\ 69.38 \ \text{``} \ \ ;\ +\ 2.76 \ \text{``}$$
$$(GH) = -\ 110.99 \ \text{``}$$
$$(BH) = -\ 153.82 \ \text{``}$$
$$(dc)\ \ = +\ 7.92 \ \text{``} \ \ ;\ -\ 45.62 \text{ tons.}$$
$$(ef)\ \ = +\ 22.13 \ \text{``} \ \ ;\ -\ 45.62 \ \text{``}$$
$$(fg)\ \ = +\ 42.61 \ \text{``} \ \ ;\ -\ 35.25 \ \text{``}$$
$$(gh)\ \ = +\ 69.38 \ \text{``} \ \ ;\ -\ 2.76 \ \text{``}$$
$$(hb)\ \ = +\ 110.99 \ \text{``}$$

The same stresses exist, of course, for corresponding members in the arm BC.

It is thus seen that the portions dh, $d'h'$, AG, and CG', of the chords must be counterbraced. Ad, Cd', dD, and $d'D'$, only, of the web members need the same treatment.

It may happen that, with a moving panel load at D, the reaction at A will be negative, in the search for main web stresses. In such a case the method of operation is simply an extension of that used in the example. If the numerical value of this negative reaction is equal to, or less than, a moving panel load (which may rest at A), a weight equal to this reaction is to be taken as hung from A, and the panel load (moving) at D is to be taken as hung from that point, while the arm AB is to be considered as an overhanging one. If the reaction, however, is greater than a moving panel load, then two such loads are to be taken as hanging from A and D with the overhanging condition of the arm.

A whole panel moving load is taken at A for prudential reasons. If the load were of uniform density, then a half panel moving load would be taken at A.

Negative reactions by the formula for any number of moving panel loads near the end are to be treated in exactly the same way : for it is to be remembered that negative reactions in an actual truss, in this case, cannot exist.

It may be urged that the case of partial continuity, taken in the preceding Article, should be treated according to the principles developed in this, by taking the middle span equal to zero.

Making such an assumption, however, would be a depart-ure from the real state of the truss. The *safe* way would be to determine the greatest stresses by both methods, and select the greatest of the two sets of results.

Differences would be found only in the upper chord ten-sion, lower chord compression, and main web stresses.

In the present case, if there are two or more systems of triangulation, each is to be treated precisely as the example has been.

This case really includes that of a centre-bearing turn-table with two points of support at the centre, as shown in the

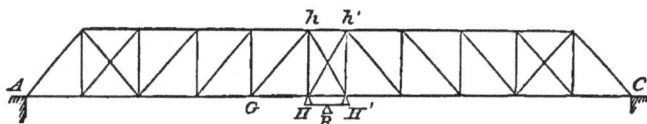

figure. HH' is free to "rock" on the central point B, and as the motion is always very small, BH (horizontal distance) is essentially equal to BH' (also horizontal) at all times. From this it results that the reaction at H will always be equal to that at H', consequently the diagonals Hh' and $H'h$ must be introduced.

Now as HH' is really a part of the truss, attached to and moving with it, the whole bridge, AC, is simply a continuous truss of two spans supported on the fixed point B. All the conclusions and formulæ, therefore, of this Article, apply to it directly. R_2 will be the reaction at B, and M will be the moment over the same point.

According to the principles established, Hh' will receive its greatest stress when AB, only, carries moving load. Since the pressure on H is always equal to that on H' also to a half of the reaction at B, there results :

$$(Hh') = \frac{R'_2 \times \sec hHh'}{2} ;$$

in which R'_2 is the reaction at B due to the moving load on

AB only, considered as a simple truss. The greatest stress in $H'h$ is, of course, equal to (Hh').

The greatest stress in Hh (equal to $(H'h)$) is found, as before, by putting the moving load on BC and AG.

No locomotive excess has been taken, but precisely the same conditions of loading hold whether such excess is taken or not.

It will only be necessary to remember that the locomotive may be at either end of the train, and that the greatest results arising from the two positions are to be selected.

CHAPTER V.

SWING BRIDGES. ENDS LATCHED TO SUPPORTS.

Art. 39.—General Considerations.

IT has already been stated that the object of fitting the ends of a swing bridge with a latching apparatus is to enable those ends to resist a negative reaction, or in other words, to prevent their rising from the points of support. All "hammering" of the ends will thus be prevented.

It has further been shown in the preceding Chapter that if there are *always* two points of support at the center, for each system of triangulation, the ends will never tend to rise. It was also observed in the preceding Article that with a pivot, or centre-bearing turn-table, the bridge always presents the case of continuity with two spans only, whatever may be the number of *apparent* points of support at the centre.

In this chapter, then, it will only be necessary to consider the one case of continuity with a single point of support between the extremities of the bridge.

Art. 40.—Ends Latched Down—One Point of Support Between Extremities of Bridge—Example.

The general formulæ required in this case are Eqs. (1), (2), (3), and (4), of Article 38, and they are here reproduced.

$$M = -\frac{1}{4l^2}\left\{\overset{1}{\Sigma}P(l^2 - z^2)z + \overset{2}{\Sigma}P(l^2 - z^2)z\right\} \quad . \quad . \quad . \quad (1).$$

$$R_1 = \frac{1}{l}\left\{\overset{1}{\Sigma}P(l - z) + M\right\} \quad . \quad . \quad . \quad . \quad . \quad . \quad (2).$$

$$R_2 = \frac{1}{l} \left\{ \overset{1}{\Sigma} P z + \overset{2}{\Sigma} P z - 2M \right\} \quad \cdots \quad \cdots \quad (3).$$

$$R_3 = \frac{1}{l} \left\{ \overset{2}{\Sigma} P(l - z) + M \right\} \quad \cdots \quad \cdots \quad (4).$$

These involve the condition $l_1 = l_2 = l$, which will appear in the example.

If this condition does not exist in any case, the formulæ to be used are Eqs. (10), (11), (12), and (13) of Article 35, but they are to be used in precisely the same manner as will be Eqs. (1), (2), (3), and (4).

This case was essentially treated in the preceding Article, insomuch that with ordinary moving loads precisely the same conditions of loading, for the greatest stresses, are required in the two cases. The results themselves, however, will be different for the upper chord compression, lower chord tension, and counter stresses.

It will probably be as expeditious and labor saving, nevertheless to find the reactions and chord stresses due to each moving panel load, and then combine the results thus found with those due to the fixed load alone, in the usual manner. Such is the method to be used in the example.

Since Eq. (1) shows that M is always negative, Eq. (2) shows that with a given value of $\overset{1}{\Sigma} P (l - z)$, R_1 will have its greatest positive value when no moving load is upon the span l_2. The expression for the shear in any counter:

$$s = R_1 - n'w - n \left(W + W' \right);$$

(in which n' is the number of moving loads between R_1 and the counter, and n the number of fixed loads similarly located), will have its greatest value for $n' = 0$. Hence, *for the greatest stress in any inclined counter, the moving load must extend from the centre to the foot of the counter in question.* This is precisely the condition used previously.

As usual, the stress in the vertical which cuts the upper

extremity of the counter must be found, for it may be the greatest in that member.

Precisely the same reasoning used previously shows that *the greatest stress in any main web member (inclined) exists when the moving load covers the whole of one span, and that portion of the other included between the free end and the foot of the member considered.*

The stress in the vertical which cuts the upper extremity of the inclined web member is to be found with the same condition of loading; it will usually be the greatest possible.

The truss to be taken for an example, and all the data, are exactly the same as those used in Article 38. The figure and the data are reproduced below:

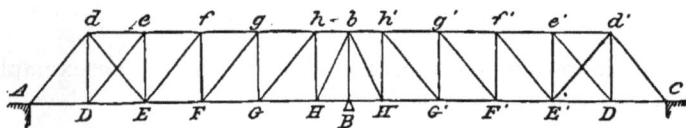

Total length $= AC = 2l = 2AB = 2BC = 144$ feet.
Uniform depth $= dD = bB = 16$ feet.
Panel length $= AD = DE = $ etc. $= 13$ feet.
$$BH = BH' = 7 \text{ feet.}$$
Total fixed weight per foot $= 1200$ pounds (nearly).
Upper chord panel fixed weight $= W = 2.73$ tons.
Lower chord panel fixed weight $= W' = 5.00$ tons.
Uniform panel moving load $= w = 19.50$ tons.

The inclined web members, except the end posts, are for tension, and the verticals for compression.

The moving load traverses the lower chord. The weight of the floor system is taken at about 350 pounds per foot of track. The fixed weight at A will be taken at three (3) tons, and that at b at 1.47 tons. Full panel loads of both kinds will be taken at H and H'.

For a single moving panel load on the arm AB:

$$M = -\frac{w}{4l^2}(l^2 - z^2)z \quad . \quad . \quad . \quad . \quad . \quad (5).$$

$$R_1 = \frac{1}{l} \left\{ w\,(l - z) + M \right\} \quad \cdots \quad (6).$$

$$R_3 = \frac{M}{l} \quad \cdots \cdots \cdots \quad (7).$$

The distance z is to be measured from A or C according as the arm AB or CB is considered.

The trigonometrical quantities used in this example are the same as those employed in the preceding Article. They are the following:

$$tan\ AdD = tan\ \alpha = 0.8125$$
$$sec\ \text{``}\ \ = sec\ \alpha = 1.29$$
$$tan\ HbB = tan\ \beta = 0.4375$$
$$sec\ \text{``}\ \ = sec\ \beta = 1.09$$

The following quantities are also taken from the example in the preceding article:

$$z = 13 \ \ . \ . \ \ l - z = 59 \ \ . \ . \ \ (l^2 - z^2)z = \ 65195.00$$
$$z = 26 \ \ . \ . \ \ l - z = 46 \ \ . \ . \ \ (l^2 - z^2)z = 117208.00$$
$$z = 39 \ \ . \ . \ \ l - z = 33 \ \ . \ . \ \ (l^2 - z^2)z = 142857.00$$
$$z = 52 \ \ . \ . \ \ l - z = 20 \ \ . \ . \ \ (l^2 - z^2)z = 128960.00$$
$$z = 65 \ \ . \ . \ \ l - z = \ 7 \ \ . \ . \ \ (l^2 - z^2)z = \ 62335.00$$

$$\frac{1}{4l^2} = \frac{1}{20736} = 0.00004823.$$

$$0.00004823 \times 19.5 = 0.00094.$$

$$\frac{w}{l} = \frac{19.5}{72} = 0.271.$$

By using these quantities in Eqs. (5), (6), and (7):

w at D . .	$R_1 = + 15.14$ tons . .	$R_3 = - 0.851$ tons.
w " E . .	$R_1 = + 10.94$ " . .	$R_3 = - 1.53$ "
w " F . .	$R_1 = + 7.07$ " . .	$R_3 = - 1.87$ "
w " G . .	$R_1 = + 3.74$ " . .	$R_3 = - 1.68$ "
w " H . .	$R_1 = + 1.09$ " . .	$R_3 = - 0.81$ "

$$\Sigma R_1 = + 37.98 \qquad\qquad \Sigma R_3 = - 6.741 \quad \text{``}$$

The greatest negative reaction at the extremity of one arm will exist when the whole of the other is covered by the moving load, and its value is seen to be − 6.741 tons. The resistance of the latching apparatus must be sufficient to oppose this with a proper safety factor.

Under the same circumstances, with ends not latched down, it was found that the reaction at A was 44.69 tons (see Article 38); but 37.98 + 67.41 = 44.721 tons, which is essentially equal to 44.69 tons, as it should be.

Counter Stresses.

dE is the only counter needed, since with moving loads at E, F, G, H, the reaction at A is:

$$R_1 = 10.94 + 7.07 + 3.74 + 1.09 = + 22.84 \text{ tons };$$

consequently $\Sigma P = 0$ at E.

The shear, or vertical component of the stress, in dE is:

$$s = 22.84 − (3 + 2.73 + 5.00) = 12.11 \text{ tons.}$$

Hence, $\quad (dE) = s \times \sec \alpha = + 15.62 \text{ tons.}$

With the moving load covering DH, dD acting as a counter will sustain a tensile stress equal to

$$(dD) = + (19.5 + 5.00) = + 24.5 \text{ tons.}$$

With the same condition of loading, Ad receives its greatest compressive stress :

$$(Ad) = −R_1 \times \sec \alpha = − 37.98 \sec \alpha = − 48.99 \text{ tons.}$$

Main Web Stresses.

The main web stresses are found precisely as in Article 38, and there is no need of repeating the operation here.

The values of the stresses will be reproduced in the proper place.

Chord Stresses.

The chord stresses due to the fixed load alone are the same as those determined on page 195 of Article 38. They will not be reproduced, but references will be made to them as they are.

Those caused by the moving load alone will be determined by placing a panel load at each panel point successively, and finding all the chord stress in both arms due to it, then tabulating the results, and, finally, combining them in the manner already shown in several instances.

The counter dE will be supposed to come into action for the weights E, F, G, and H.

The panel loads at F, G, and H will cause *apparent* compression in some, or all, of the inclined members Ef, Fg, and Gh. The resultant action of fixed and moving loads in those members, however, will in all cases be tension.

The detailed expressions for the chord stresses due to one moving panel load only will be given, as all the others are like it. For this purpose take w at F.

$$R_1 = + 7.07 \text{ tons}; \qquad R_3 = - 1.87 \text{ tons}.$$
$$w - R_1 = 19.5 - 7.07 = 12.43 \text{ tons}.$$

$$(AD) = \quad (DE) = + R_1 \tan \alpha = + 5.74 \text{ tons}.$$
$$(dc) = \quad (cf) = - 2 \times R_1 \times \tan \alpha = - 11.49 \text{ tons}.$$
$$(EF) = - (fg) = + 11.49 + 5.74 = + 17.23 \text{ tons}.$$
$$(FG) = - (gh) = 17.23 - 12.43 \times \tan \alpha = + 7.13 \text{ tons}.$$
$$(GH) = - (hb) = 7.13 - 12.43 \times \tan \alpha = - 2.97 \text{ tons}.$$
$$(HB) = - 2.97 - 12.43 \times \tan \beta = - 8.41 \text{ tons}.$$

$$(CD') = - (d'c') = \qquad R_3 \times \tan \alpha = - 1.52 \text{ tons}.$$
$$(D'E') = - (c'f') = 2 \times \text{``} \qquad \text{``} \quad = - 3.04 \quad \text{``}$$
$$(E'F') = - (f'g') = 3 \times \text{``} \qquad \text{``} \quad = - 4.56 \quad \text{``}$$
$$(F'G') = - (g'h') = 4 \times \text{``} \qquad \text{``} \quad = - 6.08 \quad \text{``}$$
$$(G'H') = - (h'b) = 5 \times \text{``} \qquad \text{``} \quad = - 7.60 \quad \text{``}$$
$$(H'B) = - 7.60 - R_3 \times \tan \beta \qquad = - 8.42 \quad \text{``}$$

The following checks by moments should be observed. Moments about b give:

$$(HB) = \frac{R_1 \times 72 - 19.5 \times 33}{16} = - 8.40 \text{ tons};$$

$$(H'B) = \frac{R_3 \times 72}{16} \qquad\qquad = - 8.415 \text{ tons}.$$

Moments about $\dot{f}$ give :

$$(EF) = \frac{R_1 \times 39}{16} = +\ 17.23 \text{ tons.}$$

The following tables are found by following the same operation for all the weights.

	(dc)	(ef)	(fg)	(gh)	(hb)
w at D	− 12.3	− 8.76	− 5.22	− 1.68	+ 1.86
" " E	− 17.78	− 17.78	− 10.82	− 3.87	+ 3.09
" " F	− 11.49	− 11.49	− 17.23	− 7.13	+ 2.97
" " G	− 6.08	− 6.08	− 9.12	− 12.16	+ 0.65
" " H	− 1.77	− 1.77	− 2.66	− 3.55	− 4.42

	(AD)	(DE)	(EF)	(FG)	(GH)	(HB)
w at D	+ 12.3	+ 8.76	+ 5.22	+ 1.68	− 1.86	− 3.77
" " E	+ 8.89	+ 8.89	+ 10.82	+ 3.87	− 3.09	− 6.94
" " F	+ 5.74	+ 5.74	+ 17.23	+ 7.13	− 2.97	− 8.41
" " G	+ 3.04	+ 3.04	+ 9.12	+ 12.16	− 0.65	− 7.54
" " H	+ 0.89	+ 0.89	+ 2.66	+ 3.55	+ 4.42	− 3.63

	$(d'e')$	$(e'f')$	$(f'g')$	$(g'h')$	$(h'b)$
w at D	+ 0.69	+ 1.38	+ 2.07	+ 2.76	+ 3.45
" " E	+ 1.24	+ 2.49	+ 3.73	+ 4.97	+ 6.22
" " F	+ 1.52	+ 3.04	+ 4.56	+ 6.08	+ 7.60
" " G	+ 1.37	+ 2.70	+ 4.10	+ 5.46	+ 6.83
" " H	+ 0.66	+ 1.32	+ 1.97	+ 2.63	+ 3.29

14

	(CD')	(D'E')	(E'F')	(F'G')	(G'H')	(H'B)
w at D	− 0.69	− 1.38	− 2.07	− 2.76	− 3.45	− 3.82
" " E	− 1.24	− 2.49	− 3.73	− 4.97	− 6.22	− 6.88
" " F	− 1.52	− 3.04	− 4.56	− 6.08	− 7.60	− 8.42
" " G	− 1.37	− 2.73	− 4.10	− 5.46	− 6.83	− 7.56
" " H	− 0.66	− 1.32	− 1.97	− 2.63	− 3.29	− 3.64

Using the main web stresses and the fixed weight chord stresses found in Article 38, the following greatest stresses at once result:

$(dE) = +$ 15.62 tons.
$(Ad) = +$ 12.58 " $\qquad (Ad) = -$ 48.99 tons.
$(dD) = +$ 24.50 " $\qquad (dD) = -$ 12.48 "
$(Dc) = +$ 28.17 " $\qquad (cE) = -$ 24.57 "
$(Ef) = +$ 49.19 " $\qquad (fF) = -$ 40.86 "
$(Fg) = +$ 75.19 " $\qquad (gG) = -$ 61.02 "
$(Gh) = +$ 105.50 " $\qquad (hH) = -$ 84.51 "
$(Hb) = +$ 117.64 " $\qquad (bB) = -$ 217.33 "

$(AD) = -$ 7.92 tons ; $\quad +$ 28.42 tons.
$(DE) = -$ 22.12 " ; $\quad +$ 16.16 "
$(EF) = -$ 42.59 " ; $\quad +$ 18.89 "
$(FG) = -$ 69.34 "
$(GH) = -$ 110.96 "
$(BH) = -$ 153.83 "
$(dc) = +$ 7.92 " ; $\quad -$ 46.98 tons.
$(cf) = +$ 22.12 " ; $\quad -$ 34.72 "
$(fg) = +$ 42.59 " ; $\quad -$ 18.89 "
$(gh) = +$ 69.34 "
$(hb) = +$ 110.96 "

The web members dD, $d'D'$, Ad, Cd', and the portions AF, CF', dg, $d'g'$ of the chords need counterbracing.

The chord stress $(GH) = -110.96$ tons requires the moving load to cover BC and AG.

With the moving load on AB only, there is some ambiguity in the stresses (dc), *compression*, and (DE) *tension*.

In such a case the reaction R_1 is 37.98 tons, and the web member De may be neglected. Under such an assumption, by taking moments about d and E successively, there will result:

$$(AD) = (DE) = \frac{(R_1 - 3) \times 13}{16} = +28.42 \text{ tons};$$

$$(dc) = (cf) = -\frac{(R_1 - 3) \times 26 - 27.23 \times 13}{16} = -34.72 \text{ tons.}$$

This ambiguity cannot be avoided if both the web members dE and Dc exist. It might also have been noticed in the case last treated.

It has already been noticed that the downward reaction of 6.74 tons must be resisted by the latching apparatus.

If there are two or more systems of triangulation, the preceding principles hold true for each. Also, if there is locomotive excess, precisely the same methods are to be employed.

The observations which were made at the end of Article 38 on a pivot or centre-bearing turn-table, over which there are two points of support for the truss, apply, exactly as they stand, to this case. The value of R'_2 must, however, be found by Eq. (3) of this Article.

CHAPTER VI.

Art. 41.—General Considerations.

IN the preceding chapter there was noticed, in detail, the method of prevention of "hammering," by latching down the ends of a swing bridge of two spans. It was also there noticed that the necessity of such an arrangement could only exist in the case of continuity with two spans. For precisely the same reasons given in connection with that case, *the necessity of lifted ends can exist in the event of continuity with two spans only.*

It is plain that if the ends of a swing bridge are pressed upward by forces exceeding the greatest negative reactions determined for latched ends by the formulæ of the last chapter, there can be no hammering, for the ends can never leave their seats or supports.

By a proper device, then, the ends should be pressed upward by forces at least equal to the negative reactions determined for latched ends.

In order to provide for any contingency, however, which may arise, the upward force should somewhat exceed such a value.

Art. 42.—Ends Lifted—One Point of Support Between Extremities—Example.

The figure represents the truss to be taken as an example. The span, depth of truss, and panel lengths, excepting hh',

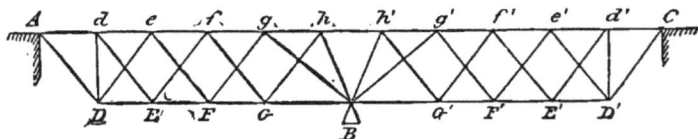

212

are the same as those taken in the two preceding cases; the loading is also the same.

The following are the data to be used :

$$AC = 2\left(Ah + \frac{hh'}{2} \right) = 144 \text{ feet.}$$

Uniform depth of truss = 16 "

Panel length = 13 "

$hh' =$ 14 "

Uniform fixed upper chord panel load = $W =$ 5.00 tons.

" " lower " " " = $W' =$ 2.73 "

" " moving " " = $w =$ 19.50 "

Moving load for unit of length = 1.50 "

The truss is a deck one, as the moving load passes along AC; and as the figure shows, there are two systems of triangulation. It will be assumed, though not strictly true, that the same panel loads are found at h and h' as at the other panel points.

Let the inclination of Gh to a vertical line be denoted by α.

" " Bh " " " β.

" " Bg " " " δ.

Then $\tan \alpha = 0.8125$; $\sec \alpha = 1.29$;

" " $\beta = 0.4375$; " $\beta = 1.09$;

" " $\delta = 1.25$; " $\delta = 1.6$.

Each system of triangulation is to be treated as an independent truss. The fixed weights at D and A will be taken as belonging to the system $ADeF$, etc., while that at d will be assumed to belong to the other system. Similar observations apply to the other arm. As in the preceding case, a fixed load of three (3) tons will be taken at A or C.

The stresses in a swing bridge with ends lifted may be considered as composed of the stresses in two other trusses, one with ends latched down and subjected to the same loads (both fixed and moving), and the other subjected to the action of the upward pressures only, at the ends; the different trusses being supposed of the same form and dimensions in all their parts.

From this, it at once follows that the positions of the moving load for the greatest stresses (*when the ends are lifted*) are exactly the same as those determined in the preceding chapter.

For the stresses in the counters, then, or for those in the members which slope downward from the upper chord and toward the ends, the moving load must extend from the centre to the upper extremities of such members.

These stresses will be compressive, and the member in which such stress is first found is to be determined in the manner already shown.

In order to find the greatest compressive stress in any web member, in one arm, sloping downward from the upper chord, and toward the centre, the moving load must extend from the end of that arm to its upper extremity, and at the same time cover the whole of the other arm.

These conditions of loading are to be taken *while the ends are lifted*, but it will also be necessary to find the web stresses for the *open draw* in the vicinity of the end, as some of these will be the greatest stresses in the web members there located.

It is to be borne in mind that any two web members which intersect in that chord which does not carry the moving load, take their greatest stresses together.

Although positions of moving load for the greatest chord stresses may be assigned, it will probably be the shortest and most labor-saving method to find the chord stresses due to the fixed load and upward pressure together, then find those due to each moving panel load alone, and combine the results. This method will be used.

The example will now be treated.

The following quantities are determined on the supposition that the ends are latched down, by Eqs. (1), (2), and (4) of Article 40.

System *ADdEf*, etc.

$z = 13$ feet . . $M = -\ 61.28$ . . $R_1 = +\ 15.14$ . . $R_3 = -\ 0.851$ tons.
$z = 39$ " . . $M = -134.29$ . . $R_1 = +\ 7.07$ . . $R_3 = -\ 1.87$ "
$z = 65$ " . . $M = -\ 58.59$ . . $R_1 = +\ 1.09$ . . $R_3 = -\ 0.81$ "

$$-\ 3.531 \quad \text{"}$$

System *ADcFg*, etc.

$z = 26$ feet . . $M = -110.18$ . . $R_1 = +\ 10.94$ . . $R_3 = -\ 1.53$ tons.
$z = 52$ " . . $M = -121.22$ . . $R_1 = +\ 3.74$ . . $R_3 = -\ 1.68$ "

$$-\ 3.21 \quad \text{"}$$

Each of these results, it is to be observed, is for a single panel moving load placed at the panel point denoted by the value of z. They have been used in the two preceding cases.

The chord stresses due to each panel moving load *alone* will first be found. As these are all found by exactly the same method, the detailed expressions for two only (one in each system) will be given.

System *ADdEf*, etc.

Panel moving load at f :

$z = 39$ feet ; $\qquad R_1 = +\ 7.07$ tons ; $\qquad R_3 = -\ 1.87$ tons.

(Ad)	$= -\ R_1 \tan \alpha$	$= -\ 5.74$ "
(df)	$= -\ 2\ R_1 \tan \alpha$	$= -\ 11.49$ "
(fh)	$= -\ 11.49 + (12.43 - 7.07) \tan \alpha =$	$-\ 7.20$ "
(hh')	$= -\ 7.20 + 12.43\,(\tan \alpha + \tan \beta) = +$	8.30 "
(DE)	$= -\ (Ad)$	$= +\ 5.74$ "
(EG)	$= 5.74 + 2\ R_1 \tan \alpha$	$= +\ 17.23$ "
(GB)	$= 17.23 - 2 \times 12.43 \times \tan \alpha$	$= -\ 2.96$ "
(Cd')	$= -\ R_3 \tan \alpha$	$= +\ 1.52$ "
$(d'f')$	$= -\ 2$ " "	$= +\ 3.04$ "
$(f'h')$	$= -\ 4$ " "	$= +\ 6.08$ "
(hh')	$= 6.08 - R_3\,(\tan \alpha + \tan \beta)$	$= +\ 8.42$ "
$(D'E')$	$= -\ (Cd')$	$= -\ 1.52$ "
$(E'G')$	$= +\ 3\ R_3 \tan \alpha$	$= -\ 4.56$ "
$(G'B)$	$= +\ 5$ " "	$= -\ 7.60$ "

As numerical checks, the moment method gives the following results:

$$(GB) = \frac{R_1 \times 65 - 19.5 \times 26}{16} = -2.96 \text{ tons.}$$

$$(hh') = -\frac{R_3 \times 72}{16} = +8.41 \text{ tons.}$$

$$(G'B) = \frac{R_3 \times 65}{16} = -7.60 \text{ tons.}$$

System A De Fg, etc.

Panel moving load at e:

$z = 26$ feet; $R_1 = +10.94$ tons; $R_3 = -1.53$ tons.

(Ae)	$= -R_1 \tan \alpha$	$= -8.89$ "
(eg)	$= (Ae) - 2.38 \times \tan \alpha$	$= -10.82$ "
(gg')	$= (eg) + (19.5 - R_1)(\tan \alpha + \tan \delta)$	$= +6.84$ "
(DF)	$= 2R_1 \tan \alpha$	$= +17.78$ "
(FB)	$= (DF) - 2(19.5 - R_1)\tan \alpha$	$= +3.87$ "
(Ce')	$= -R_3 \tan \alpha$	$= +1.24$ "
$(e'g')$	$= -3$ " "	$= +3.73$ "
$(g'g)$	$= (e'g') - R_3(\tan \alpha + \tan \delta)$	$= +6.88$ "
$(D'F')$	$= 2 R_3 \tan \alpha$	$= -2.49$ "
$(F'B)$	$= 4$ " "	$= -4.97$ "

Moments give:

$$(FB) = \frac{R_1 \times 52 - 19.5 \times 26}{16} = +3.86 \text{ tons.}$$

$$(F'B) = \frac{R_3 \times 52}{16} = -4.97 \text{ tons.}$$

All the results for the two systems give the four tables below:

	(Ad)	(df)	(fh)	(hh')	(DE)	(EG)	(GB)
w at d	−12.30	− 8.76	− 1.68	+ 3.77	+12.30	+ 5.22	− 1.86
w at f	− 5.74	−11.49	− 7.20	+ 8.30	+ 5.74	+17.23	− 2.96
w at h	− 0.89	− 1.77	− 3.54	+ 3.63	+ 0.89	+ 2.66	+ 4.43

	(Cd')	(df')	(fh')	(hh')	(D'E')	(E'G')	(G'B)
w at d	+ 0.69	+ 1.38	+ 2.77	+ 3.83	− 0.69	− 2.07	− 3.46
w at f	+ 1.52	+ 3.04	+ 6.08	+ 8.42	− 1.52	− 4.56	− 7.60
w at h	+ 0.66	+ 1.32	+ 2.63	+ 3.64	− 0.66	+ 1.97	− 3.29

	(Ae)	(eg)	(gg')	(DF)	(FB)
w at e	− 8.89	− 10.82	+ 6.84	+ 17.78	+ 3.87
" " E	− 3.04	− 9.12	+ 7.54	+ 6.08	+ 12.16

	(Ce')	(e'g')	(g'g)	(D'F')	(F'B)
w at e	+ 1.24	+ 3.73	+ 6.88	− 2.49	− 4.97
" " g	+ 1.37	+ 4.10	+ 7.56	− 2.73	− 5.46

The open draw stresses due to the fixed weight alone are the following:

$$3 \times \tan \alpha = 2.44 \text{ tons.} \qquad 3 \times \tan \delta = 3.75 \text{ tons.}$$
$$W \times \text{ " } = 4.06 \text{ " } \qquad W \times \text{ " } = 6.25 \text{ " }$$
$$W' \times \text{ " } = 2.22 \text{ " } \qquad W' \times \text{ " } = 3.41 \text{ " }$$

$$IV \times \tan \beta = 2.19 \text{ tons.}$$
$$IV' \times \text{ " } = 1.19 \text{ " }$$

$(Ad) = + 3 \times \tan \alpha$		$= + 2.44$ tons.
$(dc) = (Ad) + IV \tan \alpha$		$= + 6.50$ "
$(cf) = (dc) + \{2 (3 + IV') + W\} \tan \alpha$		$= + 19.88$ "
$(fg) = (cf) + \{2 (IV + IV') + IV\} \tan \alpha$		$= + 36.50$ "
$(gh) = (fg) + (3 + IV + 2W') (\tan \alpha + \tan \delta) + IV \tan \delta$		
		$= + 70.51$ tons.
$(hh') = (gh) + 2 (IV + IV') (\tan \alpha + \tan \beta) + IV \tan \beta$		
		$= + 92.02$ tons.
$(DE) = - (2 \times 3 + IV') \tan \alpha$		$= - 7.10$ "
$(EF) = (DE) - 2IV \tan \alpha - IV' \tan \alpha$		$= - 17.44$ "
$(FG) = (EF) - 2(3 + IV + IV') \tan \alpha - IV' \tan \alpha$	$= - 37.10$ "
$(GB) = (FG) - (4IV + 3IV') \tan \alpha$		$= - 60.00$ "

As a numerical check:

$$(GB) - (3IV + 2IV') \tan \beta - (3 + 2IV + 2IV') \tan \delta =$$
$$- 92.02 \text{ tons} = - (hh').$$

Again, by moments:

$$(hh') = \frac{5 \times 7.73 \times 33 + 3 \times 72 - 7 \times 2.73}{16} = + 92.02 \text{ tons.}$$

The chord stresses resulting from the upward pressure alone still remain to be found.

The total negative reaction, supposing the ends to be latched down, has been shown to be $- (3.53 + 3.21) = - 6.74$ tons. A margin of safety, however, of two tons will be taken; *i. e.*, it will be assumed that the total upward pressure at each end of the bridge has a value of 8.74 tons.

In the example, and in all cases where two or more systems of triangulation have a common point of support at the ends, some ambiguity necessarily arises in regard to the upward pressure. The proportion of the excess carried by either system is indeterminate; and if there is no excess, the proportion of the upward pressure carried by either system, during partial loading of one or both arms, is also indeterminate.

In the absence of anything better, it will be assumed that the excess, in the example, of two tons is equally divided between the two systems. It will farther be assumed that the upward pressure, under all circumstances of loading, is 4.53 tons for the system $ADdEf$, etc., and 4.21 tons for the system $ADeFg$, etc. The chord stresses due to the upward pressures will then be the following:

$$(Ad) = - 8.74 \times tan\,\alpha \qquad\qquad = -\;\;7.10 \text{ tons.}$$
$$(dc) = (Ad) - 4.53 \times tan\,\alpha \qquad = - 10.78 \text{ ``}$$
$$(cf) = (dc) - 2 \times 4.21 \times tan\,\alpha \quad = - 17.62 \text{ ``}$$
$$(fg) = (cf) - 2 \times 4.53 \qquad\text{``} \qquad = - 24.98 \text{ ``}$$
$$(gh) = (fg) - 4.21\,(tan\,\alpha + tan\,\delta) \quad = - 33.66 \text{ ``}$$
$$(hh') = (gh) - 4.53\,(tan\,\alpha + tan\,\beta) \quad = - 39.32 \text{ ``}$$
$$(DE) = 2 \times 4.21 \times tan\,\alpha + 4.53\,tan\,\alpha = + 10.52 \text{ ``}$$
$$(EF) = (DE) + 2 \times 4.53 \times tan\,\alpha \quad = + 17.88 \text{ ``}$$
$$(FG) = (EF) + 2 \times 4.21 \qquad\text{``} \qquad = + 24.72 \text{ ``}$$
$$(GB) = (FG) + 2 \times 4.53 \qquad\text{``} \qquad = + 32.08 \text{ ``}$$

As numerical checks:

$$(hh') = - \frac{8.74 \times 72}{16} = - 39.33 \text{ tons.}$$

$$(GB) + 4.53 \times tan\,\beta + 4.21 \times tan\,\delta = 39.32 \text{ tons} = - (hh').$$

The stresses in the web members, existing with a passing load, will next be found, and those which may be termed counter stresses will first receive attention.

System $ADdE$, etc.

Moving loads at f and h:

$R_1 = 7.07 + 1.09 + 4.53 = + 12.69$ tons. $\therefore \Sigma P = 0$ at f, and (fE) will be the first counter stress.

The shear, s, in fE, is:

$$s = 12.69 - 7.73 = 4.96 \text{ tons.}$$
$$\therefore (fE) = - s \times sec\,\alpha = - 6.40 \text{ tons.}$$
$$\therefore (dE) = (s + 2.73)\,sec\,\alpha = + 9.92 \text{ tons.}$$

Moving loads at d, f, and h :

$$R_1 = 12.69 + 15.14 = + 27.83 \text{ tons.}$$
$$s \text{ (for } AD) = 27.83 \text{ tons.}$$
$$\therefore (AD) = + s \times \sec \alpha = + 35.9 \text{ tons.}$$
$$\therefore (dD) = - s = - 27.83 \text{ tons.}$$

System $ADeF$, etc.

Moving loads at e and g :

$R_1 = 10.94 + 3.74 + 4.21 = 18.89 \text{ tons.}$ $\therefore \Sigma P = 0$ at e, and (De) is the first counter stress.

$$s \text{ (for } De) = 18.89 - (3 + 2.73) = 13.16 \text{ tons.}$$
$$\therefore (De) = - s \times \sec \alpha = - 16.98 \text{ tons.}$$
$$\therefore (AD) = (R_1 - 3) \sec \alpha = + 20.86 \text{ tons.}$$

The main web stresses existing with the moving load are found as follows :

System $ADdE$, etc.

Moving load on Ch' and at d :

$$
\begin{aligned}
R_1 &= 15.14 + 4.53 - 3.53 &&= 16.14 \text{ tons.} \\
(dE) &= - (19.5 + 5 - 16.14) \sec \alpha &&= - 10.78 \text{ ``} \\
(Ef) &= - (dE) + W' \sec \alpha &&= + 14.30 \text{ ``}
\end{aligned}
$$

Moving load on Ch' and Af :

$$
\begin{aligned}
R_1 &= 16.14 + 7.07 &&= 23.21 \text{ tons.} \\
(fG) &= - (2W + 2w + W' - R_1) \sec \alpha &&= - 36.79 \text{ ``} \\
(Gh) &= - (fG) + W' \sec \alpha &&= + 40.31 \text{ ``}
\end{aligned}
$$

Moving load on Ch' and Ah.

$$R_1 = 23.21 + 1.09 = 24.30 \text{ tons.}$$

$$(hB) = - (3W + 3w + 2W' - R_1) \sec \beta = - 59.58 \text{ tons.}$$

System $ADcF$, etc.

Moving load on Ch' and at e:

$$R_1 = 10.94 + 4.21 - 3.21 = 11.94 \text{ tons.}$$

$$(eF) = - (W + w + W' + 3 - R_1) \sec \alpha = - 23.59 \text{ tons.}$$
$$(Fg) = - (eF) + W' \sec \alpha \qquad\qquad = + 27.11 \quad\text{``}$$

Moving load on Ch' and eg:

$$R_1 = 11.94 + 3.74 = 15.68 \text{ tons.}$$

$$(gB) = - \{2(W + W' + w) + 3 - R_1\} \sec \delta = - 66.85 \text{ tons.}$$

A few of the open draw web stresses are the following:

$$
\begin{aligned}
(AD) &= -\ 3 \sec \alpha & &= -\ \ 3.87 \text{ tons.} \\
(dD) &= \quad 0 & &= \quad 0.00 \quad\text{``} \\
(dE) &= -\ W \sec \alpha & &= -\ \ 6.45 \quad\text{``} \\
(De) &= +\ (3 + W') \sec \alpha & &= +\ \ 7.39 \quad\text{``} \\
(eF) &= -\ (3 + W + W') \sec \alpha & &= -\ 13.84 \quad\text{``} \\
(Ef) &= +\ (W + W') \sec \alpha & &= +\ \ 9.97 \quad\text{``}
\end{aligned}
$$

It is unnecessary to give others, as they are not needed; only two of these, it will be seen, are used.

All of the greatest stresses in the truss may now be written by the usual method of combining the results for the different cases of loading.

They are the following:

$$
\begin{aligned}
(AD) &= -\quad 3.87 \text{ tons;} & +\ 56.75 \text{ tons.} \\
(dD) &= -\ 27.83 \quad\text{``} \\
(dE) &= -\ 10.78 \quad\text{``} & +\ \ 9.92 \quad\text{``} \\
(De) &= -\ 16.98 \quad\text{``} & +\ \ 7.39 \quad\text{``} \\
(eF) &= -\ 23.59 \quad\text{``} \\
(Ef) &= -\quad 6.40 \quad\text{``} & +\ 14.30 \quad\text{``} \\
(fG) &= -\ 36.79 \quad\text{``} \\
(Fg) &= & +\ 27.11 \quad\text{``} \\
(gB) &= -\ 66.85 \quad\text{``} \\
(Gh) &= & +\ 40.31 \quad\text{``} \\
(hB) &= -\ 59.58 \quad\text{``}
\end{aligned}
$$

$$(Ad) = + \quad 2.44 \text{ tons}; \quad - 35.52 \text{ tons}$$
$$(de) \;= + \quad 6.50 \quad " \qquad - 38.23 \quad "$$
$$(cf) \;= + \quad 19.88 \quad " \qquad - 39.70 \quad "$$
$$(fg) \;= + \quad 36.50 \quad " \qquad - 20.84 \quad "$$
$$(gh) \;= + \quad 77.15 \quad "$$
$$(hh') = + \; 112.51 \quad "$$
$$(DE) = - \quad 7.10 \quad " \qquad + 46.21 \quad "$$
$$(EF) = - \quad 17.44 \quad " \qquad + 49.41 \quad "$$
$$(FG) = - \quad 37.10 \quad " \qquad + 38.76 \quad "$$
$$(GB) = - \quad 57.52 \quad "$$

The web members AD, dE, De, Ef, and the portions Ag and DG of the chords must be counterbraced. The same treatment must of course be given to corresponding members and portions in the other arm.

The particular form of truss in the figure has been so chosen as to illustrate faults of designs, in general, in consequence of possible ambiguity in the stresses.

If possible, ambiguity should always be avoided. In the present case it would have been far better to have had one system of triangulation, and supported the chords by light verticals, designed to resist compression, extending from the apices.

Precisely the same methods of loading and treatment would be used if there were two *apparent* points of support above B, that point still existing as the *real* point of support of the truss. In fact, the same general observations as those which were made in the last portions of Articles 38 and 40 apply in this case also.

The same methods of loading and treatment would also be used if there were locomotive excess, or if there were one or more than two systems of triangulation.

Art. 43.—Final Observations on the Preceding Methods.

Although particular forms of triangulation have been chosen for the various examples in the different cases of swing bridges, yet the conclusions reached and the principles established are perfectly general. They are applicable to any form of triangulation, and to either the deck or through form of bridge; they also apply whether the two arms are of the same length or of unequal length, the panels being either uniform or irregular. It is only necessary to bear in mind what may be called the " local " circumstances of any given case; these do not, however, affect the general principles. As a single illustration—if the bridge is of the " deck " form, those web members which intersect in the lower chord take their greatest stresses together; if of the " through " form, those which intersect in the upper chord take their greatest stresses together.

CHAPTER VII.

CONTINUOUS TRUSSES OTHER THAN SWING BRIDGES.

Art. 44.—Formulæ for Ordinary Cases — Reactions — Methods of Procedure.

ON account of the doubtful utility of fixed continuous trusses, and the extreme rarity of their occurrence in American practice, general directions and formulæ only will be given. It will be assumed that the moment of inertia (I) and coefficient of elasticity (E) are constant; it will also be assumed that the points of support are all in the same level, as it has been shown in Appendix I to what cases the resulting formulæ apply.

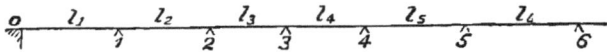

FIG. 1.

Eq. (17) of that Appendix, after introducing these conditions, gives, in connection with the notation of Fig. 1, the following equations:

$$2M_1(l_1 + l_2) + M_2 l_2 + A = 0 \quad . \quad . \quad . \quad . \quad . \quad (1).$$
$$M_1 l_2 + 2M_2(l_2 + l_3) + M_3 l_3 + B = 0 \quad . \quad . \quad . \quad . \quad . \quad (2).$$
$$M_2 l_3 + 2M_3(l_3 + l_4) + M_4 l_4 + C = 0 \quad . \quad . \quad . \quad . \quad . \quad (3).$$
$$M_3 l_4 + 2M_4(l_4 + l_5) + M_5 l_5 + D = 0 \quad . \quad . \quad . \quad . \quad . \quad (4).$$
$$M_4 l_5 + 2M_5(l_5 + l_6) + M_6 l_6 + E = 0 \quad . \quad . \quad . \quad . \quad . \quad (5).$$
$$\text{etc.} + \quad \text{etc.} \quad + \text{etc.} + \quad = 0 \quad . \quad . \quad . \quad . \quad . \quad .$$

The various values of M are the bending moments existing at the supports indicated by the subscripts; the moment at 0 is evidently nothing, since the truss is there simply supported.

224

The quantities A, B, C, etc., have the following values, as Eq. (17), of Appendix I, shows:

$$A = \frac{1}{l_1} \overset{1}{\Sigma} P(l_1^2 - z^2) z + \frac{1}{l_2} \overset{2}{\Sigma} P(l_2^2 - z^2) z \quad . \quad . \quad (6).$$

$$B = \frac{1}{l_2} \overset{2}{\Sigma} P(l_2^2 - z^2) z + \frac{1}{l_3} \overset{3}{\Sigma} P(l_3^2 - z^2) z \quad . \quad . \quad (7).$$

$$C = \frac{1}{l_3} \overset{3}{\Sigma} P(l_3^2 - z^2) z + \frac{1}{l_4} \overset{4}{\Sigma} P(l_4^2 - z^2) z \quad . \quad . \quad (8).$$

$$D = \frac{1}{l_4} \overset{4}{\Sigma} P(l_4^2 - z^2) z + \frac{1}{l_5} \overset{5}{\Sigma} P(l_5^2 - z^2) z \quad . \quad . \quad (9).$$

$$E = \frac{1}{l_5} \overset{5}{\Sigma} P(l_5^2 - z^2) z + \frac{1}{l_6} \overset{6}{\Sigma} P(l_6^2 - z^2) z \quad . \quad . (10).$$

Etc. $=$ etc. $+$ etc. . .

The Eqs. (1) to (5) show, since the end moments are zero, that whatever the number of spans, there will always be as many of those equations as there are unknown bending moments over the points of support. Those moments, therefore, may always be found, and, consequently, the reactions which depend upon them. These reactions are the main objects of search. It will be necessary, then, to determine the bending moments at the points of support.

From Eq. (1):

$$M_2 = -\frac{A}{l_2} - M_1 \frac{2(l_1 + l_2)}{l_2}. \quad . \quad . \quad . \quad (11).$$

By inserting this value of M_2 in Eq. (2), there at once results:

$$M_3 = \frac{2(l_2 + l_3) A - l_2 B}{l_2 l_3} - M_1 \left\{ \frac{l_2^2 - 4(l_1 + l_2)(l_2 + l_3)}{l_2 l_3} \right\} \quad . \quad . \quad (12).$$

15

These values of M_2 and M_3 inserted in Eq. (3), give:

$$M_4 = -\frac{[-l_3^2 + 4(l_2 + l_3)(l_3 + l_4)]A - 2l_2(l_3 + l_4)B + l_2 l_3 C}{l_2 l_3 l_4}$$

$$+ M_1 \left\{ \frac{-8(l_1 + l_2)(l_2 + l_3)(l_3 + l_4) + 2l_2^2(l_3 + l_4) + 2l_3^2(l_1 + l_2)}{l_2 l_3 l_4} \right\} \quad (13).$$

Again, Eq. (4) gives, after inserting in it these values of M_3 and M_4:

$$M_5 = \frac{[-2l_3^2(l_1 + l_4) - 2l_4^2(l_2 + l_3) + 8(l_2 + l_3)(l_3 + l_4)(l_4 + l_5)]A - [-l_3 l_4^2 + 4l_2(l_3 + l_4)(l_4 + l_5)]B + 2l_2 l_3(l_4 + l_5)C + l_2 l_3 l_4 D}{l_2 l_3 l_4 l_5}$$

$$- M_1 \left\{ \frac{-16(l_1 + l_2)(l_2 + l_3)(l_3 + l_4)(l_4 + l_5) + 4l_4^2(l_2 + l_3)(l_4 + l_5) + 4l_3^2(l_1 + l_2)(l_4 + l_5) + 4l_4^2(l_1 + l_2)(l_2 + l_3) - l_3^2 l_4^2}{l_2 l_3 l_4 l_5} \right\} \cdot (14).$$

Any bending moment may thus be found.

It is seen that all moments are given in terms of M_1, which is still unknown. However, the bending moment at the other free end of the truss, from o, Fig. 1, will be zero; consequently its general expression, put equal to zero, will give M_1 in terms of A, B, C, etc., and the lengths of the different spans, *i. e.*, in terms of known quantities. When M_1 is known, all the other bending moments are at once given by Eqs. (11), (12), (13), (14), etc.

As an illustration, if there are five spans, $M_5 = 0$ and Eq. (14) will at once give M_1. Eqs. (11), (12) and (13) then give the other moments desired.

Another method may be followed by which a less number of equations will suffice for a greater number of spans. For example, the Eqs. (11), (12), (13) and (14), with a similar value for M_6 are sufficient for the solution of a case of ten spans.

Let A' be the quantity corresponding to A, which would appear in the equation involving M_9 and M_8, in a continuous truss of ten spans, and corresponding to Eq. (1). Let B', C', D' represent similar quantities in equations corresponding to Eqs. (2), (3) and (4). The following five equations may then be written by the aid of Eqs. (1) to (5):

$$2M_9(l_{10} + l_9) + M_8 l_9 + A' = 0 \quad . \quad . \quad (15).$$
$$M_9 l_9 + 2M_8(l_9 + l_8) + M_7 l_8 + B' = 0 \quad . \quad . \quad (16).$$
$$M_8 l_8 + 2M_7(l_8 + l_7) + M_6 l_7 + C' = 0 \quad . \quad . \quad (17).$$
$$M_7 l_7 + 2M_6(l_7 + l_6) + M_5 l_6 + D' = 0 \quad . \quad . \quad (18).$$
$$M_6 l_6 + 2M_5(l_6 + l_5) + M_4 l_5 + E = 0 \quad . \quad . \quad (19).$$

A value for M_5 may be written by changing, in Eq. (14), M_1 to M_9, l_1 to l_{10}, l_2 to l_9, l_3 to l_8, l_4 to l_7, l_5 to l_6, A to A', B to B', C to C', and D to D'. A value of M_6 in terms of M_1 would be equal to the value of M_4, given by Eq. (13), with exactly the same changes made, in so far as the same quantities appear. These pairs of values of the two quantities M_5 and M_6, equated, would give two equations from which M_1 and M_9 could be immediately deduced. All the other moments would then follow.

The Eqs. (11), (12), (13) and (14), are sufficient in themselves for the solution of a case of nine spans, in the manner just indicated.

The preceding operations represent the most direct method of finding the bending moments over the points of support. All things considered, it is probably as short as anything that can be derived.

Prof. Merriman has, however, given a more elegant method by the use of so-called " Clapyronian numbers." Any method involves sufficient tedium.

The preceding formulæ will be very much simplified if a single weight, only, rests upon some one span, since all the quantities, A, B, C, etc., except two, will then disappear.

The various reactions may be immediately determined by Eqs. (21)–(27) of Appendix I., after the bending moments are found. And when the reactions are known, the stresses in the individual members, for a given condition of loading, are found precisely as for a simple truss supported at each end.

If the ends of the truss are not simply supported, the end moments must be known, else the problem will be indeterminate. In such a case the preceding methods are in no wise changed, but the end moments, instead of being zero, will appear as known quantities.

CHAPTER VIII.

ARCHED RIBS.

Art. 45.—Equilibrium Polygons.

PRELIMINARY to the specific treatment of arched ribs it will be necessary, first to consider some general principles regarding equilibrium polygons for any given system of vertical forces, and then those involved in the theory of flexure.

In the figure below, let AB . . . . . . K be any straight, simple beam, subjected to the action of the vertical forces B, C, D, etc. Let x be measured from any section positive and horizontal toward A, and let P signify any external force such as the reaction at A or any of the forces applied to the beam ;

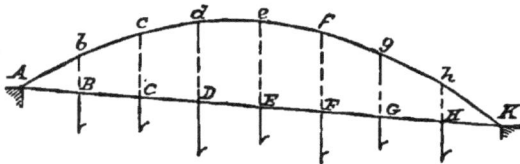

FIG. 1.

then will ΣPx represent the bending moment to which the beam is subjected at the section denoted by x. Now let there be imagined any force T acting parallel to AEK, and let the moments ΣPx be taken at each of the points B, C, D, E, F, G, H. Then if the quotients of those moments divided by the horizontal component of T be supposed represented by the vertical lines bB, cC, dD, etc., respectively, will the polygon $AbcdefghK$ be *one* equilibrium polygon for the given system of loads ; so that if the beam AK were displaced by a tie in which exists the stress T, and the given loads hung from the joints b, c, d, etc., the whole system would be in equilibrium.

229

In order to establish this, it is only necessary to show that no piece of the polygon is subjected to bending; for if that is the case, the line of action of the resultant stress must coincide with its centre line.

Consider any portion of the system, as that lying on the left of the vertical line dD. Those forces which have moments about the point d are the external forces to the left of dD and the stress T in the tie AK; the latter has a lever-arm n, equal to the normal distance from d to AK, and its moment is opposite in sign to ΣPx. Consequently the resultant moment about d will be $M = \Sigma Px - Tn$. *But by construction $\Sigma Px = Tn$,* hence $M = 0$; and the same is, of course, true of every other joint. If T_h is the horizontal component of T, then evidently $Tn = T_h(dD) = \Sigma Px$.

If v be the general representative of the vertical ordinates bB, cC, etc., then, in general,

$$v = \frac{\Sigma Px}{T_h},$$

but T_h is a constant quantity. From these considerations follows this important principle:

The vertical ordinates of the equilibrium polygon of any system of vertical loads are proportional to, and may represent, the bending moments found at the various sections of a beam subjected to the action of the same system of loads, and having the same span.

Since the stress T was taken arbitrarily, it is evident that there may be an indefinite number of equilibrium polygons for any given system of loads; the principle stated above, however, is perfectly general, and is true for all.

Since $vT_h = \Sigma Px =$ constant for any given section, it follows that any variation of T, and therefore T_h, produces an opposite kind of variation in v. Hence *the height of an equilibrium polygon is proportioned to the reciprocal of T or T_h.*

The method of constructing the equilibrium polygon given above is not the most convenient, nor the one commonly used. The method ordinarily used is the usual one for con-

structing the equivalent polygonal frame, and is the following :

Let *AK*, Fig. 2, represent any span, inclined in this case
but ordinarily horizontal, and 1, 2, 3, 4, etc., the vertical loads
acting along their respective lines of action.　In Fig. 3 let the
portions 1, 2, 3, 4, 5, 6, 7, 8, and 9 of the vertical line 1–9 represent those loads taken by any assumed scale.　Since *BC*

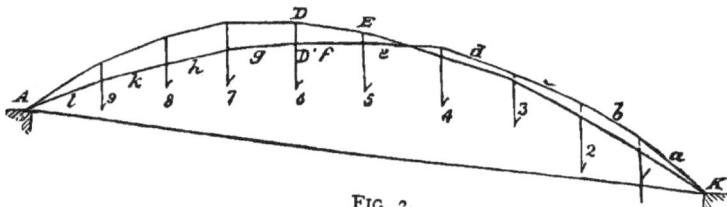

FIG. 2.

represents the sum of all the applied loads, it is also equal to
the sum of the two reactions or shearing stresses at *A* and *K*.
In the case of the simple beam taken, those quantities will of
course be determined by the law of the lever only.

Suppose *A'C* and *A'B* to represent the shearing stresses or
reactions at *A* and *K* respectively.　Then draw *A'P* parallel
to *AK*, and on it take
any point *P*.　From *P*
draw the radial lines
a, b, c, d, ... l, as
shown, a n d starting
from *A* or *K* in Fig. 2,
draw the lines *a, b, c,
d, ... l*, parallel
to the lines denoted
by the same letters in
Fig. 3.　Then will Fig.
2 represent the equi-
librium polygon for the
given span and system
of loading.

The line *AK* or *PA'*
is called the *closing line* of the polygon.　The reaction at *A* is

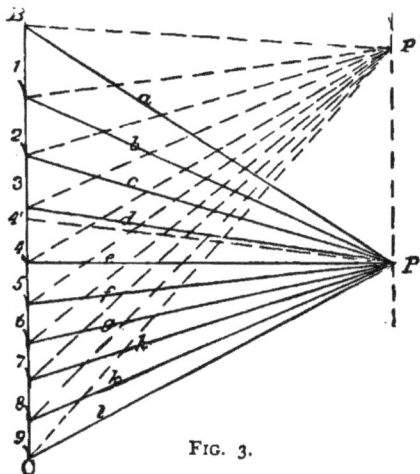

FIG. 3.

evidently composed of the numerical sum of the vertical components in l and AK, while that at K is equal to the numerical difference of the vertical components in a and AK.

The point P, from which the radial lines are drawn, is called
the *pole*, and the normal distance from the pole to the load
line BC, the *pole distance.* The pole distance evidently represents the horizontal component of stress common to all the
members of the polygon.

In order that the equilibrium polygon, constructed according to the principles given above, shall exactly fit the span, it
is only necessary that a proper observance be paid to the
scales used.

From the equation $v = \dfrac{\Sigma Px}{T_h}$ it is seen that the scale for the
forces does not affect the height of any joint of the polygon;
it depends only on the scale according to which x or the horizontal span is drawn.

Let the line PP' be drawn parallel to BC, and let P' be
the pole of a new equilibrium polygon; the pole distance
will, of course, remain the same as before. But the pole distance represents the horizontal component of the stress in the
closing line, and it has already been shown that if T_h remains

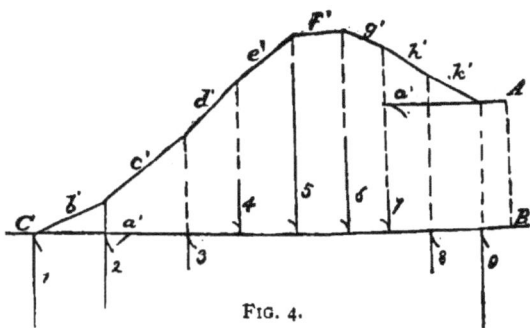

Fig. 4.

the same, v cannot vary. Hence, *any movement of the pole
parallel to the load line does not change the vertical dimensions*

of the equilibrium polygon. *But if the pole distance is changed, the vertical dimensions are changed in the inverse ratio.*

The determination of the deflection polygon of an arched rib with ends fixed, involves the use of an equilibrium polygon, similar to that required for a system of forces whose resultant is a couple. Its method of construction is not at all different from that just given.

In Fig. 4, let the forces 1, 2, 3, 4, 5, 6, 7, 8, and 9, act vertically, *BC* being horizontal, and let the sum of 1, 2, 3, 8, and 9 be numerically equal to the sum of 4, 5, 6, and 7. The double line, *DE*, in Fig. 5, represents the forces shown in Fig. 4.

In Fig. 5, draw the line *a* in a horizontal direction through the upper extremity of force 1, and

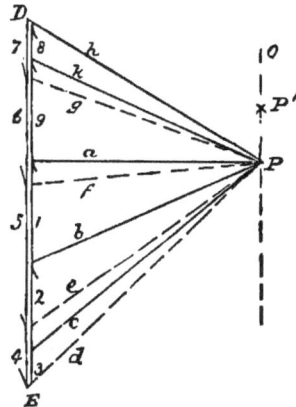

FIG. 5.

take any point on it for the pole *P*. From *P* draw the radial lines in the usual manner as shown.

From *C*, in Fig. 4, draw *b'* parallel to *b* in Fig. 5, until it intersects the line of action of force 2. Then draw the other lines, *c'*, *d'*, etc., parallel to *c*, *d*, etc., until the lines of action of the other forces are intersected. *b'*, *c'*, ... *h'*, *k'*, will then be the equilibrium polygon for the system of forces assumed.

It is seen that the polygon does not close. This simply shows that the resultant of the system is a couple, whose moment is the force *a*, in Fig. 5, multiplied by *AB* (vertical) in Fig. 4.

The following general principle then results: *The equilibrium polygon for any system of parallel forces whose resultant is a couple, is not a closed one.*

This principle is indeed true for any system of forces.

P might have been taken at any other point, as *P'* in *OP*. In that case, however, the equal forces *a*, acting at *A* and *C*, Fig. 4, would be parallel to a line drawn from the upper extremity of force 1 to P'. The vertical dimensions of the polygon, measured from either of the forces a' (in general inclined), will always be the same if the pole remains in the line PP'.

An arched rib is any truss curved in a vertical plane, both of whose chords are convex or concave in the same direction, neither being horizontal ; the ends may be fixed or free.

In Fig. 2 of Art. 45, let $ADEK$ represent an arched rib sustaining the loads 1, 2, 3, 4, ... 9. Now it has already been seen that, so far as equilibrium is concerned, any given system of loading may be sustained by any one of a set of equilibrium polygons consisting of an indefinite number. On the other hand, it is evident that no polygon or arched rib can be drawn, which is not an equilibrium polygon for *some* system of vertical loading; but if that arched rib sustains some other system of loads than that which, it may be said, properly belongs to it, and if its joints be prevented from turning, it will be subjected to bending, which will vary from one section to another.

The arched rib $ADEK$ sustains a system of loading for which AfK is the equilibrium polygon, hence the former will be subjected to varying degrees of bending at various sections. When the rib ADK is subjected to the action of its load, stresses are developed in its different parts, whose horizontal components are all the same because the load is wholly vertical. Now if an equilibrium polygon can be found in which the horizontal component of stress T_h is the same as that developed in $ADEK$, then all the circumstances of stress and bending in the latter can be determined, as will be seen hereafter.

Suppose AfK to be that polygon, then let v' denote the portion of a vertical line intercepted between it and the arched rib, as DD'. The moment about any point D will then be

$$M = \Sigma Px - T_h (v + v').$$

But since AfK is the equilibrium polygon, $\Sigma Px - T_h v = 0$.

$\therefore M = -T_h v'$. When the polygon lies above the rib, v' is negative, and, hence, M positive.

Let the polygon which has the same value of T_h as the arched rib be called the true equilibrium polygon ; then, since T_h is a constant quantity for the same rib, there is established the following important principle:

The bending moments to which the different parts of an arched rib are subjected are proportional to, and may be represented by, the vertical intercepts included between the rib and the true equilibrium polygon.

This principle has been demonstrated for a beam with free ends only, but it is true also for a beam with fixed ends, as will now be shown.

In order to fix the end of the rib it is only necessary to impress upon the rib at A, the point of fixedness, the proper couple whose moment is m; in the fixed rib, as in the free, let T_h represent the horizontal thrust. The true equilibrium polygon for the fixed rib will be that found by increasing the vertical dimensions of a polygon for a free-end rib, formed by using T_h and reactions for fixed ends, by a constant amount equal to $\frac{m}{T_h}$. Again, taking moments about any point of the centre line of the rib, there will result

$$M = \Sigma Px + m - T_h \left(v + \frac{m}{T_h} + v''\right) = -T_h v',$$

as before. This shows that the principle stated above is true for both fixed and free-end ribs. In truth this equation might have been written first, and the special case of the free-end rib deduced by making $m = 0$.

An arched rib, then, when subjected to the action of a load, suffers bending in the same manner as a straight beam, but to a different degree.

In so far as it plays the part of a beam, it must be governed by the general laws of bending or flexure. The formulæ to be given in this connection are those approximate ones based on the common theory of flexure, and found in the ordinary works on that subject.

Art. 47.—General Formulæ.

Let S denote the total shearing stress at any section, P any applied load or external force, M the bending moment at any section, and D the deflection found above; then the six general equations of flexure demonstrated in the Appendix on the Theorem of Three Moments, some of which are made use of in the graphical treatment of arched ribs, are the following:

$$S \quad = \Sigma P,$$

$$M \quad = \Sigma Px,$$

$$P' \quad = \frac{M}{EI},$$

$$\Sigma nP' = \Sigma \frac{nM}{EI},$$

$$D \quad = \Sigma nP'x = \Sigma \frac{nMx}{EI}.$$

$$D_h \quad = \Sigma nP'y.$$

If the beam is originally straight and parallel to the axis of x, n becomes dx and $y = 0$.

As usual, E and I represent the coefficient of elasticity and moment of inertia of the cross-section, respectively.

The limits of the summations are the section considered and any section of reference. The quantities S, M, P' and D then refer simply to that portion of the beam over which the summation extends.

One very important deduction is to be drawn from the above equations, or rather from the second and fifth of them. It is seen from these two equations that $\frac{nM}{EI}$ stands in the same relation to D that P does to M. Consequently if $\frac{nM}{EI}$ be put in the place of P in any graphical construction, D will be represented in the place of M. *If, therefore, an "equilibrium polygon" be constructed for any span by taking $\dfrac{nM}{EI}$ as loads instead of P, the vertical ordinates of the polygon will represent the deflections at the sections denoted by the corresponding values x.*

This polygon may be called the "deflection polygon," and its construction plays a very important part in the determination of the true equilibrium polygon for an arched rib, in the majority of cases.

Art. 48.—Arched Rib with Ends Fixed.

The ends of an arched rib or any girder are considered fixed when the angles formed by their centre lines with any assumed line, at the fixed sections, do not vary under any applied load.

Let BAD, Pl. V., be the centre line of any arched rib ; it will, of course, be considered fixed at the points B and D. This line may be any curve, though for convenience the arc of a circle has been drawn. In the demonstration no attention whatever has been given to the character of the curve, so that it is equally applicable to any other curve.

In the present case the *centre line* of the rib will be divided into equal parts for the application of the load, and each of those parts will be n ; consequently that quantity will have a finite value, and the results will not be strictly accurate, though near enough for all technical purposes.

The piers or points of fixedness are supposed to be immovable, whatever may be the character of the load ; but if that is the case, the summation of the strains at any given distance from the neutral axis of the rib, considered as a truss, taken throughout the whole length of the rib BAD, must be equal to zero. Take that distance as unity, then there results, for one condition, since n is constant, the equation :

$$\Sigma_{\scriptscriptstyle D}^{\scriptscriptstyle B} nP' = \Sigma_{\scriptscriptstyle D}^{\scriptscriptstyle B} \frac{nM}{EI} = \frac{n}{EI} \Sigma_{\scriptscriptstyle D}^{\scriptscriptstyle B} M = 0.$$

$$\therefore \Sigma_{\scriptscriptstyle D}^{\scriptscriptstyle B} M = 0.$$

The quantity $\dfrac{n}{EI}$ is brought outside of the sign Σ, because *the moment of inertia of the cross-section of the rib is supposed to be the same throughout its entire length.* More will be said on this point hereafter.

It has already been shown that the area included between the equilibrium polygon and the curve *BAD* is made up of vertical strips, whose lengths (the vertical intercepts) represent the actual bending moments at the different sections of the rib. Hence ΣM represents *the sum of those vertical lengths or intercepts drawn at the points to which the moments M belong*, and the equation $\Sigma_{\scriptscriptstyle D}^{\scriptscriptstyle n} M = 0$ shows that *the sum on one side of BAD must be equal to that on the other.*

But, as will be seen, there may be an indefinite number of equilibrium polygons which will fulfill this condition; consequently at least one other condition must be obtained.

Since the points *B* and *D* are fixed, the sum of all the deflections, both horizontal and vertical, taken between those two points, must be equal to zero. It has been shown that the vertical deflection at any point, when n and I are considered constant, is $D = \dfrac{n}{EI} \Sigma Mx$; also, from the reasoning applied to the curved girder, that the horizontal deflection is $D_h = \dfrac{n}{EI} \Sigma My$. Now, when these summations extend from *B* to *D*, since those points are fixed, both D and D_h must equal zero. The three equations of condition, then, which must be fulfilled for the rib, are:

$$\Sigma_{\scriptscriptstyle D}^{\scriptscriptstyle n} M = 0; \qquad\qquad \Sigma_{\scriptscriptstyle D}^{\scriptscriptstyle B} Mx = 0;$$
$$\Sigma_{\scriptscriptstyle D}^{\scriptscriptstyle n} My = 0.$$

It has already been stated, and it is evident without much

thought, that any polygon whatever is an equilibrium polygon for *some* load.

Hence consider BAD, Pl. V., an equilibrium polygon for its proper load, and consider it subjected to that load ; denote its moments by M_b.

Again, suppose the polygon a, a', a'', etc., to be the true equilibrium polygon for the given load, and denote its moments, represented by the vertical ordinates drawn from its closing line, by M_a. *

Then, from the principle which precedes the equation $M = -T_h v'$ in the general discussion of equilibrium polygons, there follow the equations :

$$M = M_a - M_b,$$
$$Mx = M_a x - M_b x.$$
$$\therefore \ \Sigma_p^n Mx = \Sigma_p^n M_a x - \Sigma_p^n M_b x = 0.$$
Or, $$\Sigma_p^n M_a x = \Sigma_p^n M_b x.$$

In the same manner, $\Sigma_p^n M_a y = \Sigma_p^n M_b y$. This last equation will be used in fixing the pole distance of the true equilibrium polygon.

It must be remembered that M represents the actual moment to which the rib is subjected at any point.

The application of these two conditions will be shown in the course of the construction of the true equilibrium polygon, as they are needed.

In the figure of Pl. V., let the scale for linear measurements be 10 feet to the inch, and the force scale 15 tons per inch. The curved centre line of the rib is divided into ten equal parts of 10.95 feet each, and that is the constant value of n. The load is not therefore uniformly distributed. The panel length, or horizontal distance between the points of application of the loads, is thus a variable quantity. If the versed sine of the centre line of the rib is small, n may be taken equal to the span divided by the number of panels. But if the versed sine is not larger, even, than in the present case, n

cannot be so taken without sensible error, as will be seen. The other data are as follows :

Span	$= 100$ feet.
Radius	$= 75$ feet.
Angular length of curve	$= 83° 37'$.
Panel fixed load	$= 4$ tons.
Panel moving load	$= 10$ tons.
Centre rise of rib	$= 19.1$ feet.

In the figure BD is the span, and C the centre ; b', b'', b''', etc., are the panel points equidistant in the curve, and through which the loads are supposed to be applied.

Now the actual moment area for the arched rib is supposed, really, to be the difference between the moment area of the true equilibrium polygon for the applied loads and that of the rib itself considered as an equilibrium polygon for the proper load ; both systems of loading being supposed applied to a straight beam fixed at each end.

The first portion of the problem which presents itself, then, is to determine the true equilibrium polygon for the given load. The construction will first be made, and it will then be shown that the two conditions given above are satisfied.

Let the moving load cover the left half, BC, of the span, and suppose the half panel loads at B and D to rest directly on the abutments. According to the scale taken, lay off $B6$ equal to 7 tons, half the total load on the panel Ab^{iv}, and $B5$ equal to half the fixed panel load on Ab^{vi} ; then lay off $6 - 10$, divided into four equal parts, equal to the four equal panel loads on b', b'', b''', and b^{iv}. In the same manner lay off $5 - 1$, equal to the four fixed panel loads on the right half of the span. Assume C as a convenient pole, and draw the radial lines from it to $1, 2, 3, 4, \ldots \ldots 10$. Starting from C, draw $C - 6$ until it intersects a vertical through b^{iv} at a_4 ; from the latter point, $a_4 a_3$ parallel to $C7$ until it intersects the vertical through b''' ; proceed in the same manner until the polygon ECF is drawn. If the ends were free EF would be the closing line, but one must now be found that will satisfy the condition $\Sigma M_a = 0$, or, in other words, the sum of the vertical intercepts drawn from the closing line downwards must be just equal to the sum of those drawn from the same line upwards.

The proper closing line is easily located by trial. If $Cc^{v} = 0.5$ inch and $vv' = 0.96$ inch, there results:

$$\Sigma_D^B M_a = \tfrac{1}{2}vv' + a_1c' - a_2c'' - a_3c''' - a_4c^{iv} - Cc^{v} - a_6c^{vi} - a_7c^{vii}$$
$$+ a_8c^{viii} + a_9c^{ix} + \tfrac{1}{2}v''v''' = 1.88 - 1.89 = -0.01 \text{ inch.}$$

This sum is sufficiently near zero.

The lines vv' and $v''v'''$ are drawn vertically through points b and b^x, distant $\dfrac{n}{4}$ from B and D on the curve BAD, and their halves are taken because in the summation $\Sigma_D^B n M_a$ there would appear terms $\dfrac{n}{2} \times vv'$ and $\dfrac{n}{2} \times v''v'''$, or $n \times \dfrac{vv'}{2}$ and $n \times \dfrac{v''v'''}{2}$.

Similar terms will hereafter appear in similar summations.

The closing line HK then satisfies the condition $\Sigma_D^B M_a = 0$ for the equilibrium polygon ECF.

There still remains the condition $\Sigma_D^B M_a x = \Sigma_D^B M_b x$. This equation will be satisfied by making each of its members equal to zero. The closing line HK must, then, also make $\Sigma_D^B M_a x = 0$. This simply means that the vertical ordinates of the polygon a, measured from HK, multiplied by their horizontal distance, from D, will form a sum equal to zero when their products are added. If the ordinates below HK are taken positive, as $v''v'''$, a_9c^{ix}, a_1c'', etc., and those above, as a_6c^{vi}, negative, and if the ordinates and distances be taken by scale from the drawing, there will result, nearly,

$$v''v''' \times De + a_9c^{ix} \times De' + a_1c' \times De^{v} + \text{etc.} = + \; 99.7$$
$$a_7c^{vii} \times De''' + a_6c^{vi} \times De'' + \text{etc.} \qquad\qquad = - \; 102.0$$

16

The numerical values are nearly enough equal, and the line *HK* will be taken as the proper closing line.

The next step is to find the closing line for the curve *BAD* of the rib, considered as an equilibrium polygon, which will satisfy the same general conditions.

Using precisely the same method of procedure as for the polygon *ECF*, the line $H''K''$ is found to be the one desired, for that line makes the sum of the intercepts above it just equal to the sum of those below it.

Ac_5 is about 0.62 of an inch.

For this curve the summation may be written :

$$\Sigma_D^B M_b = bH' + 2b'c_1 + 2b''c_2 - 2c_3b''' - 2c_4b^{iv} - Ac_5 = 0.$$

Since the curve is symmetrical in reference to *A*, the static moments of the ordinates on one side of $H''K''$, about DK'', will evidently be equal to the same moment of those on the other side.

The second condition may now be applied. That condition is $\Sigma_D^B M_a y = \Sigma_D^B M_b y$. It has already been shown that if M_a and M_b be considered loads applied at distances y from the assumed origin, the ordinates of the equilibrium polygon so constructed will represent the quantities $\Sigma M_a y = D_a$, or $\Sigma M_b y = D_b$.

Through *A*, therefore, draw the horizontal line *RAS*. Assume any line, as *AC*, as the closing line of the deflection polygon, and lay off *At* equal to a half of vv'. Also make tt equal $c'a_1$; $t't'$ equal $c''a_2$; $t't'''$ equal a_3c''', etc. $t''t'''$ is measured in a direction opposite to that of the preceding, because it represents a moment of an opposite sign. In the same way *At* measured to the right of *A* is equal to a half of $v''v'''$; tt_1 equal to $c^{ix}a_9$; t_1t_{11} equal to a_3c^{viii}, etc.* Draw bb^x and note its intersection C'' with *AC*. From C'' draw $C''d_1'$ parallel to *Ct* until it intersects a horizontal drawn through b' in d_1'; draw $d_1'd_1''$ parallel to Ct' until it intersects a horizontal through b'' in d_1''; draw $d_1''d_1'''$ parallel to Ct''

---

* If the drawing had been made accurately, $t^{iv}t_{iv}$ would have been exactly equal to Cc^v.

until it intersects a horizontal through b''', etc. The polygon $C''d_,'d_,''$, etc., will intersect the horizontal RAS at a point distant from A, on the left of it, from which $l_,$ is drawn to a point on the right of A at the intersection of the deflection polygon formed by using $t, t_,, t_{,,}, t_{,,,}$, as before, and which is shown in the figure. The sides of the deflection polygon on the right of CA are parallel to radial lines drawn from C to the points $t, t_,, t_{,,}, t_{,,,}$. The distance $(l_, - l_,)$ represents, in an exaggerated manner, the horizontal deflection of the end of a vertical beam fixed at C, whose length is CA, and which is subjected to the bending moments vv', $c'a_,$, $c''a_2$, etc., at vertical distances from C equal to the heights of b, b', b'', etc., above BD. In the same manner, $l_{,,}$ represents the same quantity for the same beam when subjected to the corresponding moments on the right-hand side of AC.

The line $l_,$ can be determined with less work and more simply when the meaning of the construction is once clearly seen, by laying off, on the left of A, as before, loads represented by the *algebraic* sums $(At + At)$, $(tt' + tt_,)$, $(t't'' + t_,t_{,,})$, etc., and then drawing the equilibrium polygon as usual. The distance from A to the intersection of the polygon with AR will then be equal to $l_,$.

The deflection polygon $d'd''d'''d^{iv}$ is constructed in precisely the same manner as the preceding. Make As equal to a half of bH'; ss' equal to $c_,b'$; $s's''$ equal to c_2b''; $s''s'''$ equal to c_3b''', etc.;* then draw radial lines from those points of division to C. The point d' is at the intersection of $C''d'$, drawn parallel to Cs, with a horizontal line drawn through b'; $d'd''$ is drawn parallel to Cs' until it intersects a horizontal line drawn through b'', and the other sides of the polygon are constructed in the same way.

The polygon cuts the horizontal line RAS in a point distant $\frac{1}{2}l$ from A. There will, of course, be another deflection polygon, precisely the same as the last, on the right-hand side of AC, found by taking $\frac{1}{2}K''D$, c_9b^{ix}, etc., and laying them off from A towards S.

---

* With a sufficient accuracy of construction, As^{iv} would equal a half of Ac_5.

If the intercepts used in the deflection polygons for M_a and M_b, represent those moments by the proper scale, then by the same scale CA will not, in general, represent the true pole distance. But this fact *has the same proportional effect* on both l and l_i. Consequently any result depending on the equality of l and l_i will not be affected.

Instead of using Ac_5 and Cc^v in the manner shown, greater accuracy might have been attained by taking half intercepts at the distance $\dfrac{u}{4}$ on both sides of A and C. Such an operation, however, is unnecessary in all ordinary cases, since moments in the vicinity of A and C have very little effect on the horizontal dimensions of the deflection polygon. Moments, on the contrary, in the vicinity of HH'', have great effect.

Now l_i represents $\cdot\Sigma_D^B M_a y$; and $l_i \Sigma_D^B M_b y$; and in order that the second condition may be satisfied they should be equal. Since l_i is less than l_i it shows that the quantities M_a are too small, or, in other words, the pole distance BC is too large. This last statement is evidently true, if it be remembered that the pole distance is inversely proportional to the vertical ordinates which represent the moments.

Lay off, therefore, on AC produced, the distance CM equal to l_1, and draw through M the horizontal line MN. With a radius CN equal to l, draw the arc of a circle cutting MN in N, then produce the line CP. All moments represented in the lower equilibrium polygon a will have to be increased in the ratio of CN to CM. To make this reduction, draw a horizontal line, for instance, through c^v until it cuts CN in p, then make $c_5 a^v$ equal to Cp; a^v will be one point in the true equilibrium polygon. All other points might be found in the same way, but having found one point, as a^v, a much shorter method may be used.

It has already been shown that the vertical dimensions of two equilibrium polygons for the same loading and span are inversely proportional to the pole distances, the vertical dimensions being measured from the closing lines. In the figure the vertical dimensions of the polygon *ECF* must be increased in the ratio of l_1 to l, or in the ratio of Cc^v to Cp. Hence, on *CN* produced make *CP* equal to *BC*, and draw the horizontal line *OP* cutting *CM* produced in *O*, then will *CO* be the pole distance for the true equilibrium polygon.

In order to find the true pole, make *CL* parallel to *HK*, then draw *LC'* parallel to *BC** and equal to *CO*; the point *C'* will be the true pole.

According to previous principles, the reactions or vertical shearing stresses at *B* and *D* will be $L - 10$ and $L - 1$ respectively, and, since the closing line must be parallel to *BD* in the true equilibrium polygon, *LC'* must be parallel to *BC*. From *C'* draw the radial lines shown to 1, 2, 3, ... and 10; these lines will be parallel to the sides of the true polygon; *i. e.*, draw $a^v a^{iv}$ parallel to *C'* 6 until it cuts a vertical through b^{iv}; $a^{iv} a'''$ parallel to *C'* 7 until it cuts a vertical line drawn through b'''; $a^v a^{vi}$ parallel to *C'* 5, etc. The polygon $a a' a''$ . . . . . . $a^{ix} a^x$ so formed will be the true equilibrium polygon.

As a check some of the points as a' or a'' should also be found by the previous method. Thus make *Cg* equal to $c' a_{,}$, and draw *gp'* parallel to *CD*; *Cp'* should then be equal to $c_, a'$. Other points may be treated in the same manner.

It may now be seen that the polygon $a a' a''$ . . . . . . a^x satisfies the three conditions $\Sigma_D^B M = 0$, $\Sigma_D^B Mx = 0$, and $\Sigma_D^B My = 0$. The first conditions are evidently satisfied by the method of locating the closing lines in the lower polygon $a_1 a_2$ . . . . . . a_9, and the curve *BAD*, for any intercept between the curve and the upper polygon a is the difference between two intercepts each of which belongs to a sum equal to zero, hence the sum of those intercepts is zero. The second condition is satisfied by the location of the point *C'*.

Another check on the degree of accuracy attained in the construction is found in what has just been said, *i. e.*, the

---

* The reason for the true closing line being horizontal is given, though for another purpose, on page 273.

moment area lying above the curve BAD must be just equal
to that lying below it.

By actual measurement LC' is equal to 3.65 inches, hence
the constant horizontal component of stress in any portion of
the rib is 54.8 tons. The resultant stress at any point of the
arched rib is equal to 54.8 tons multiplied by the secant of
the inclinations at that point. The bending moment at any
point to which the rib is subjected is found by multiplying
the vertical intercept between the equilibrium polygon and
the curve by (54.8 tons $= T_h$). For example, the actual moment to which the rib is subjected at b' is ($b'a' \times$ 54.8).

The line of action of T_h (54.8 tons in this case) is, of course,
along the true closing line $H''K''$. This is an important
matter, as will hereafter be seen.

A line drawn through C parallel to EF would cut off on
the load line the reactions which would exist were the ends
free.

The reaction at B, in the present case, is thus seen to be
much greater than would be found in the case of free ends.

A point in the vertical line passing through the centre of
gravity of the load is found at the intersection of the sides a
a' and $a^{1x}\, a^x$ in G. G', at the intersection of Ea_1, and Fa_9,
prolonged, is in the same vertical line.

The pole of the deflection polygons might have been at A,
and the moments laid off from C, in which case the half of
Ac_5, and the same of Cc^v, would have been the moment distances adjacent to C. Precisely the same value for LC' would
probably not be found, because the sum ΣMy is not continuous, and consequently not exact. For this reason it would
be better in an actual case to divide the real panel lengths
into two or more equal parts in order to find the true pole
distance, LC', and consequently the true equilibrium polygon.
T_h can then be used to find the stresses in the members of
the actual rib in a manner that will hereafter be shown.

The diagram should of course be drawn to as large a scale
as possible, and it may often be advisable to exaggerate the
vertical scale so as to make the intersections of the true polygon and curve well defined.

The effect of such an exaggeration may easily be shown since the various steps of the construction remain precisely the same. Suppose AC to be m times as large as it would be made by the scale according to which the span BD is laid off; m denotes the degree of exaggeration. The distance l will be m times as great as it ought to be, and consequently the height of the equilibrium polygon will be increased beyond its true value in the same ratio. But if the true height is only $\dfrac{1}{m}$ of that found, the true pole distance will be $m \times CO$, and LC' must be made equal to that.

The span has been supposed half covered by the moving load, but any other portion might have been taken as well. It is to be noticed that the method is perfectly general, and entirely independent of the character of the curve BAD, or of the loading.

Art. 49.—Arched Rib with Free Ends.

The treatment of the arched rib with free ends is not different in any respect, except one, from that given in the previous case. The exception is this, that the condition $\Sigma M = 0$ must be omitted, since the bending moments at the free ends must disappear.

In this case, again, the centre line is divided into equal parts. Observations made under this head in the preceding Article apply here also.

The expression $\Sigma n P' = \Sigma \dfrac{n.M}{EI}$, sometimes called the bending, is of course based on the common theory of flexure, and denotes simply the difference of inclination of the neutral surface of a straight beam at the two sections indicated by the limits of the summation; with a constant moment of inertia it is usually written $-\dfrac{1}{EI}\int M dx$. The condition of fixedness of the two ends of the ribs requires the position of the neutral surface to remain unchanged at those two sections, consequently $\Sigma n P'$ must be equal to zero between those limits. In the case of free ends, the position of the neutral surface may be any whatever consistent with the elastic properties of the material at those sections. The middle points of the free-end sections must, however, retain their primitive positions, or the summation of the deflections, either horizontal or vertical, between those points must be equal to zero. The only remaining conditions, therefore, are $\Sigma_D^B nMy = 0 = \Sigma_D^B n.Mx$.

But the ends of the rib may have any relative vertical movements, whatever, without changing the circumstances of bending. Consequently the only condition to be fulfilled is, $\Sigma_D^B n.My = 0$.

The same amount and proportion of loading will be taken as in the previous case; the same radius, span, notation, and scale will also be taken. The figure of Pl. VI. represents the construction. The moving load is assumed to cover the half span BC.

As before, take C as the pole for the trial polygon, then make $B\,5$ equal to a half panel fixed load, supposed applied at A, and $B\,6$ a half panel (fixed + moving) load, supposed applied at the same point; also make $5 - 4$ equal to load at b^{vi}, and $6 - 7$ the load at b^{iv}, etc.

Draw radial lines from C to the points of division, 1, 2, 3, 4, 5 . . . . 10, and construct the polygon E, a_1, a_2, a_3, ... F, precisely as before; in truth it is exactly the same polygon that was used in the preceding case. Since there can be no bending moments at B and D, the equilibrium polygon must pass through those points, hence EF is the closing line of the trial polygon, ECF.

As has been seen, the only condition to which the equilibrium polygon is subject is $\Sigma_D^B n.My = 0$; or, as before, as n is constant.

$$\Sigma_D^B My = \Sigma_D^B M_a y - \Sigma_D^B M_b y = 0.$$

Or, $\qquad \Sigma_D^B M_a y = \Sigma_D^B M_b y.$

The method of constructing the deflection polygon is precisely the same as that followed in the previous case, only in the present one a half of the ordinates representing M_a and M_b will be laid off from A in order to keep all the points s and t within the limits of the diagram.

As the half intercepts at the distance $\dfrac{n}{4}$ from B and E are very small, and as their omission will lead to simplicity in the diagram, and not cause much of an error, they will be neglected. In an actual case, however, the omission should be made with caution.

Since the moments at B and D are zero, make As equal to $\frac{1}{2}c_1b'$, ss' equal to $\frac{1}{2}c_2b''$, $s's''$ equal to $\frac{1}{2}c_3b'''$, $s''s'''$ equal to $\frac{1}{2}c_4b^{iv}$; then draw horizontal lines through $b'b''$, $b''b'''$, etc., cutting AC as before. As the end moments are zero, d' will be on AC, then $d'd''$ will be parallel to Cs, $d''d'''$ will be parallel to Cs', and so on until the point d^v on the line RS is reached.

The two portions of the deflection polygons for the moments M_a will not be similar, yet it is only necessary to construct a single deflection polygon, as was shown in the preceding Article. The sums $\sum_{D}^{B} M_a y$ and $\sum_{D}^{B} M_b y$ may each be divided into pairs of terms, each member of the pair having the same value of y; this may be done for any case in which the moments may be taken in pairs. The moments in each pair of terms will of course be located equidistant from AC. Make, therefore, At equal to $\frac{1}{2}(a_1c' + a_9c^{ix})$, $tt_{,} =$ to $\frac{1}{2}(a_2c'' + a_8c^{viii})$, $t_{,}t_{,,}$ equal to $\frac{1}{2}a_3c''' + a_7c^{vii})$, and $t_{,,}t_{,,,}$ equal to $\frac{1}{2}(a_4c^{iv} + a_6c^{vi})$. Draw radial lines as shown, and make $d_{,}'d_{,}''$ parallel to Ct, $d_{,}''d_{,}'''$ parallel to $Ct_{,}$, etc., until the point $d_{,}^v$ is reached.

Lay off on AC produced, CM equal to $Ad_{,}^v$, draw MN parallel to BD, and with a radius CN equal to $2(Ad^v)$ find the point N, and produce CN through that point. In order to find the true equilibrium polygon it is only necessary to increase the ordinates of the trial polygon in the ratio of CM to CN.

Hence draw c^vp parallel to CD, then make Ca^v equal to Cp; a^v will be one point in the true equilibrium polygon. In the same manner make Cg equal a_3c''', and determine p' as before, then make c_8a''' equal to Cp'; a''' will be another point in the true polygon. A shorter way to proceed, however, is the one indicated in the previous case. Draw CL parallel to EF, then a line parallel to BC through L. Make CP equal to BC, and draw OP parallel to BD; OC is the true pole distance. Make, therefore, LC' equal to OC, and C' will be the true pole from which radial lines are to be drawn to 1, 2, 3, 4, 5, ... 10.

Starting from any given point, as B, make Ba' parallel to C' 10, $a'a''$ parallel to C' 9, etc., etc.; the side parallel to C' 1 should pass through the point D.

The two methods should be used to check each other, as was indicated in the previous case.

Since LC' is $3\frac{1}{2}$ inches, $T_h = 3\frac{1}{2} \times 15 = 52.5$ tons, and its line of action is evidently BD.

All those directions of a general character which accompany and follow the construction in the preceding case apply with equal force to the present one and those which follow. It is particularly important in the graphical treatment of all arched ribs to make the polygons approach as nearly their ultimate limits, *i. e.*, curves, as possible; for that reason it will be advisable in most cases to divide the actual panels into two or more equal parts in the search for the true equilibrium polygon.

Art. 50.—Thermal Stresses in the Arched Rib with Ends Fixed.*

Thermal stresses are those stresses which are co-existent with any variation of temperature, in the structure consid-

---

* In reality the deflection which produces stress, in the case of variation of temperature, is not the whole deflection. If the points B and D, in Pl. VII., were free to move, there would be no thermal stresses, but there would be deflection. This deflection, in the present case, would be (if 100 units become 100.12204 for an increase of 180 F.)

$$\left(100 + \frac{165}{180}\,0.12204\right) + 100 \times 19.1 - 19.1 = .021 \text{ ft.}$$

Strictly speaking, the deflection to be used, then, for a change of 165° F. is $0.25 - 0.021 = 0.23$ ft. The difference is so small, however, that it may be neglected, especially since the error is a small one on the side of safety. These observations are general and apply to all cases.

ered, and whose values depend upon that variation. Any variation of temperature in the material of which an arched rib is composed will cause a variation in its length, and consequently a deflection at any given point. Although the temperature is supposed to change, yet the ends of the rib are supposed to remain in their normal positions, so that the general conditions $\Sigma n.M = 0$ and $\Sigma n.My = 0 = \Sigma n.Mx$, hold for thermal stresses as well as for any other, remembering that n is constant.

Any change of form, such as that arising from the application of loading, will cause extra stresses, which are to be determined in precisely the same manner as that used for thermal stresses; in fact, they belong to the same class of stresses.

Every kind of material has its own coefficient of linear expansion; wrought iron, for instance, expanding .12204 units in every 100 for a change of temperature from 32° to 212° F., while tempered steel gives the empirical quantity .12396 for the same conditions. (D. K. Clark, "Rules, Tables, and Data.")

The arched rib shown in Pl. V., and already treated for ordinary stresses, will be supposed to be of such a material and to be subjected to such a change of temperature that the point A will suffer a vertical deflection of 3 inches. It is a matter of indifference in which direction the deflection takes place; it will be supposed upward in the present case.

The effect of the thermal variation is to cause bending moments at the various sections of the rib from which the deflection results. Also, to keep the ends in their original positions requires the existence of a horizontal force, such as the stress which may be supposed to exist in a horizontal tie. The stresses and bending, then, caused by thermal variations are the same as those which would be caused by a horizontal force having the proper line of action; the problem then resolves itself into finding the proper value and line of action of this horizontal force.

The figure to be used, and which will be referred to, is that of Pl. VII., and represents the same rib precisely as Pl. V. For the sake of greater accuracy, n will be taken half as great as in Pls. V. and VI.; its value will then be 5.47 feet.

It will first be assumed that the points b, b_1, b_2, etc., are at a uniform horizontal distance apart of 5 feet, and at the same time equidistant on the curve; that which might be allowable in a very flat curve.

The conditions $\Sigma_D^B nM = 0$ and $\Sigma_D^B nMx = 0$ show $H''K''$ to be the line of action of the horizontal force, which will be called T_h. The line $H''K''$ is the same line as $H''K''$ in Pl. V., since it is located by exactly the same condition, the vertical intercepts between it and the curve BAD representing the moments M. Since the only force acting on the rib is T_h, the bending moment M will be equal to T_h multiplied by the proper vertical intercept. Thus the moment at b_1 will be $T_h \times a_1 b_1$; and that at b_8, $T_h \times a_8 b_8$.

The value of T_h is determined by either of the conditions $\Sigma_A^B nMy = D_h EI$, or $\Sigma_A^B nMx = DEI$; by way of variety the latter will be taken. Assume any point as C for the pole, and make BC the pole distance. Make $B1$ equal to $\frac{1}{4} A a_9$; $1-2$ equal to $a_8 b_8$; $2-3$ equal to $a_7 b_7$; .... and $9-10$ equal to ab, and construct the polygon $Ca'a''$.... a^{xl} in the usual way. $B-10$ is one half $H'b'$, and $a^x II'$ is a vertical line through b', while Bb' is one quarter of Bb. Draw the horizontal line $a^x E$, then the vertical intercepts included between $a^x E$ and the deflection polygon $a^x a^v C$, of the kind (ah), represents in an exaggerated manner the deflections (vertical) of the points in the rib vertically above them.

Now since the actual moment in an equilibrium polygon is equal to the vertical ordinate multiplied by the pole distance, the actual vertical deflection at A is CE multiplied by its pole distance; however, since the real deflections are proportional to the vertical ordinates of the kind (ah), those deflections may be at once found by multiplying those ordinates by a proper ratio. That ratio is a known quantity, because the real deflection at A is known to be three inches. The ordinate CE measures 1.61 inches on the drawing, and represents 16.1 feet on the actual rib; the ratio desired is therefore $\dfrac{3}{12 \times 16.1} = \dfrac{3}{193.2} = \dfrac{1}{64.4}$. To illustrate, the deflection at the section t is $13.9 \div 64.4 = 0.216$ feet. Similarly, the deflection at b_3 is $6.2 \div 64.4 = 0.096$ feet. These quantities will be used farther on.

It will now be necessary to give a little consideration to the general equation $\Sigma nMx = DEI$.

If the ordinates of the kind (ab), measured from $H''K''$, are denoted by y', M, from what has already been said, can be written as $T_h y'$. Hence the general equation may be written,

$$\Sigma nMx = T_h \Sigma ny'x = DEI.$$

Or, since n is constant,

$$T_h = \frac{DEI}{n \Sigma y'x}.$$

This last is the equation from which T_h is to be found.

Now since y' is positive or negative according as it is measured on one side or other of the line $H''K''$, and since n is assumed to be uniform and horizontal, the quantity $n\Sigma y'x$ is the difference between the statical moments of the moment areas on the different sides of that line in reference to the section considered. Written as an integral expression, it would be $\int y'x\,dx$.

The area of the surface Ab_6ma_9 is 1.12 sq. in., or 112 sq. ft. full size. The distance of the centre of gravity of the same area from AC is very simply found by construction. Take any point a_7 for a pole, and a_7m for the pole distance. Make $mm_1 = mm_5 = \frac{1}{2}a_7b_7$, m_1m_2 equal to a_6b_6, and so on, making, however, m_6m_7 equal to a half of Aa_9. Then construct the equilibrium polygon $c, c_1, c_2, \ldots . c_7$. The sides cc_1 and c_6c_7 produced will cut each other in the vertical line GH passing through the centre of gravity of the area Ama_9.

The area $BH''m$ is to be treated in precisely the same manner, taking a_1 as pole, and a_1H'' as pole distance, and making Bn equal to a half of BH''. The vertical line KL passing through the centre of gravity of the area is found as before at the intersection of the sides ff_1 and f_5f_6 prolonged.

The area $BH''m$ is, of course, equal to the area Ama_9, consequently the value of $n\Sigma y'x$ for the section A will be the product of the common area by the horizontal distance between their centres of gravity, *i. e.*, 3.26 inches in the figure, but 32.6 ft. full size.

The cross-section of the rib will be assumed to be of such form that EI has the value of 2,000,000 foot-tons. Hence for the point A, $D = 3$ inches or 0.25 ft., $n\Sigma y'x = 112 \times 32.6 = 3651.2$;

$$\therefore T_h = \frac{2000000 \times 0.25}{3651.2} = 137 \text{ tons.}$$

For any other section, as t, the deflection is $13.9 \div 64.4 = 0.216$ ft. The vertical line passing through the centre of gravity of the area included between m and a vertical line through t passes through the intersection of the sides c_3a_7 and cc_1, prolonged, and the distance between KL and it is, as shown, 3.04 inches in the drawing. For this section

$$n\Sigma y'x = 112 \times 30.4 - 49.82 \times 5.8 = 3116;$$

$$\therefore T_h = \frac{2000000 \times 0.216}{3116} = 139 \text{ tons.}$$

In precisely the same manner, for the point b_3,

$$T_h = \frac{2000000 \times 0.096}{1421} = 135 \text{ tons.}$$

The values should have been the same, except for the errors incident to a small scale and the fact that polygons were used where curves really belonged; yet the difference between the extreme values is only $2\frac{1}{2}$ per cent. of the larger, which is not very much of an error.

If, however, the true value of n (5.47 feet, nearly) be taken along the centre line of the rib, a decidedly different result will be found, since $n\Sigma y'x$ will have a different value.

The quantity x, as before, will be measured from a_9 to the intersections of the dotted lines drawn through the points (b), while y' will represent any vertical ordinate (belonging to any point (b)) from the line $H''K''$, taken positive downward. The point m will be assumed midway between a_3 and a_4.

The following values are measured from the original drawing:

$x = 1.36$	$y' = -6.2$	$\frac{1}{2}(xy') = -\ 4.22$
$x = 5.45$	$y' = -6.1$	$xy' = -\ 33.25$
$x = 10.9$	$y' = -5.5$	$xy' = -\ 60.00$
$x = 16.2$	$y' = -4.6$	$xy' = -\ 74.52$
$x = 21.7$	$y' = -3.2$	$xy' = -\ 69.44$
$x = 26.8$	$y' = -1.35$	$xy' = -\ 36.18$
		-277.61
$x = 32.0$	$y' = 0.7$	$xy' = 22.4$
$x = 36.8$	$y' = 3.2$	$xy' = 117.76$
$x = 41.6$	$y' = 6.0$	$xy' = 249.6$
$x = 46.0$	$y' = 9.2$	$xy' = 423.2$
$x = 49.0$	$y' = 11.98$	$\frac{1}{2}(xy') = 293.5$
		1084.06

The first values of x and y' belong to that portion of the moment area adjacent to Aa_9.

The last values of x and y' belong to that portion of the moment area adjacent to BH''; and half of each product is taken, so that it may be multiplied by the full value of $n = 5.47$ feet.

The formula then gives:

$$T_h = \frac{DEI}{n \Sigma y'x} = \frac{0.25 \times 2000000}{4411.30} = 113.3 \text{ tons (nearly).}^*$$

---

* The method by deflection polygon given in the next Article, produces a result essentially the same as the one above.

BC is the pole distance laid down to a scale of 400,000 foot-tons to the inch,

$$\therefore\ T_h = \frac{1}{4 \times 16.1 \times 5.47} \times \frac{0.62 \times 400000}{6.2} = 113.6 \text{ tons.}$$

The agreement is much closer than can ordinarily be expected with the scales used.

The difference between the results of the two methods is $137.00 - 113.3 = 23.7$ tons.

This difference is by no means small, and shows how carefully the approximate method ought to be used.

With a wrought-iron rib, the deflection taken, 3 inches at the middle of the span, belongs to a change of temperature of about 165° F., a very extreme case, which accounts for the large values of T_h.

This shows, however, in a very marked manner, the importance of putting together an arched rib at about the mean temperature, for then the variation of temperature to be taken in the calculation of thermal stresses will only be about half the variation between the extreme limits.

The effect of T_h is the same as if that force were applied at m and acting toward H'' for the portion Bm fixed at B, but applied at m and acting toward m' for the free-end portion (so considered) mA.

The change of temperature, 165°, changes the radius from 75 feet to 74.536 feet, and increases the length of the curve 0.122 feet.

Art. 51.—Thermal Stresses in the Arched Rib with Ends Free.

The method to be used in the present case is somewhat shorter and simpler than the one used in the preceding, but will probably not give as nearly correct results when the scale used is small.

The figure to be used is that shown in Pl. VIII., and the curve BAD is precisely the same as that shown in Pls. V., VI., and VII.

For the sake of greater accuracy, the curve BA will be divided into ten equal parts. There will then result, $n = bb_1 = b_1b_2 =$ etc., $= 5.47$ (nearly) feet.

Since BD is the true position of the closing line for free ends, the effect of the variation in the temperature will be the same as that of a horizontal stress, T_h, whose line of action is BD, and which will produce a deflection equal to the thermal deflection. What may be called the "thermal moment," therefore, at any point will be equal to T_h multiplied by the vertical ordinate of the curve at that point, or $M = T_h y'$. The moment at b_5, for instance, is $M = T_h \times b_5 n_5$.

Supposing the rib to be of wrought iron, a change of temperature of 137° F. will cause the length to change from 109.454 feet to 109.556 feet, and the radius from 75 feet to 74.67 feet ; the corresponding upward vertical deflection at crown A will be $1\frac{1}{2}$ inches, or $\frac{1}{8}$ of a foot.

Let the two general equations be compared :

$$\Sigma Px = T_h y.$$
$$\Sigma n Mx = \Sigma M'x = EI \,.\, D.$$

From these two equations it is seen that if M' be taken as vertical loading, and EI as pole distance, then the ordinates of the resulting equilibrium polygon will represent the deflections *according to the same scale by which x is measured.* It is important also to notice that M' and EI are of the same denomination (foot-pounds or foot-tons, as the case may be), consequently *M' is to be measured in the same scale as that according to which EI is laid down.*

Since $nM = nT_h y'$ (n being constant), the vertical ordinates of the kind (bn) are proportional to the moments M or M', and they may be taken to represent those moments ; since, however, that would carry the lower extremity of the load line $B - 10$ off the diagram, one third only of those ordinates will be taken in the plate.

As before, EI will be assumed to be 2,000,000 foot-tons, and the scale according to which it is to be laid down for the pole distance at 200,000 foot-tons to the inch. Hence make EHF parallel and equal to BCD. Since BD is ten inches in the diagram, EF is the pole distance, and F will be taken as the pole.

As in the case of the same rib with external loading, the half intercept at the distance $\frac{n}{4}$ from B will be neglected.

Make E 1 equal to $\frac{1}{3}(\frac{1}{4}AC)$, 1-2 equal to $\frac{1}{3}b_8 n_8$, etc., and construct the polygon $a\, a_1\, a_2\, \ldots \ldots\, H$ in the usual man-

17

ner. BA is divided into ten equal parts for the sake of greater accuracy.

The polygon thus constructed will represent the actual deflections to a scale of 10 feet to the inch.

Now Ea is equal to 1.12 (3.32 inches on the original drawing) inches, or 11.2 feet full size, whereas it ought to be but 0.125 foot; and since the pole distance is to remain the same, the moments must be reduced in the ratio of $\frac{1}{8}$ to 13.2; or the moments as actually taken must be multiplied by $\dfrac{1}{8 \times 11.2} = \dfrac{1}{89.6}$. The reduction for the bending moment at A, therefore, is $\frac{1}{4} AC \times 200000 \div 89.6 = \dfrac{0.636666 \times 200000}{89.6}$, since AC is equal to 1.91 inches. Hence,

$$T_h' = \frac{M'}{y'} = \frac{0.636666 \times 200000}{89.6 \times 19.1} = 74.46 \text{ tons.}$$

But since $M' = nM$:

$$T_h = \frac{M}{y} = \frac{M'}{ny'} = \frac{T_h'}{n} = 13.6 \text{ tons.}$$

This operation may be considerably shortened by remembering that $0.636666 \div 19.1 = 1 \div 30$, and that this ratio is constant for all points. If, therefore, the moment at any other point, as b_5, be taken, precisely the same result will be obtained.

The method used in the preceding Article gives, at least approximately, the same result. Taking Bn, Bn_1, etc., for the different values of x, and bn, b_1n_1, etc., for y, there will result:

$x =$ 4.0 feet,	$y' =$	3.7 feet,	$\therefore$	$xy' =$	14.8
$x =$ 8.4 "	$y' =$	6.8 "	$\therefore$	$xy' =$	49.6
$x =$ 13.2 "	$y' =$	9.5 "	$\therefore$	$xy' =$	101.6
$x =$ 18.0 "	$y' =$	12.0 "	$\therefore$	$xy' =$	183.6

$x = 23.1$ feet,	$y' = 14.2$ feet,	$\therefore$	$xy' = 307.2$
$x = 28.2$ "	$y' = 16.0$ "	$\therefore$	$xy' = 425.8$
$x = 33.6$ "	$y' = 17.4$ "	$\therefore$	$xy' = 554.4$
$x = 39.0$ "	$y' = 18.3$ "	$\therefore$	$xy' = 643.5$
$x = 44.4$ "	$y' = 18.9$ "	$\therefore$	$xy' = 799.2$

With these values the partial summation is :

$$\Sigma ny'x = n\Sigma y'x = 16846.00.$$

A product for the moments adjacent to B is nearly :

$$0.92 \times 1 \times 2.73 = 2.51.$$

Another for those adjacent to AC is nearly :

$$19.1 \times 48.64 \times 2.73 = 2536.5.$$

Taking the sum of these results for the complete summation :

$$T_h = \frac{DEI}{\Sigma ny'x} = \frac{0.125 \times 2000000}{19385.00} = 13.0 \text{ tons (nearly).}$$

The polygon $a, a_1, a_2, a_3, \ldots H$ is an exaggerated representation of the movements of the points $b, b_1, b_2,$ etc., when the temperature is changed $137°$ F.

In all cases, as large a diagram as possible must be used, in order to reduce the scale for the pole distance, so that that distance may be the largest possible.

The use to be made of T_h will be shown farther on.

Art. 52.—Arched Rib with Fixed Ends—I and n Variable.

The rib to be taken in this case, and the loading, are precisely the same as taken in the preceding cases. As before, the centre line is divided into ten equal parts of 10.95 feet each. The data are therefore the following :

Span	=	100.00 feet.
Radius	=	75.00 "
Panel fixed load	=	4.00 tons.
Panel moving load	=	10.00 "
Centre rise of rib	=	19.10 feet.

The moving load is supposed to cover the half span BC.

The figure to be referred to is that shown in Pl. IX.

The point h is midway between b' and b'', while k is midway between b'' and b'''.

The moment of inertia I of the cross-section of the rib will be taken as 2,000,000 foot-tons throughout Bh, 1,777,778 foot-tons throughout hk, and 1,600,000 foot-tons throughout kA. The same values hold for similar portions of AD.

1, 2, 3, etc., ... 10, is the load line, and BC the pole distance of the equilibrium polygon $Ea_2 Ca_7 F$. That polygon is drawn in precisely the same manner, in fact, is precisely the same one as that shown in Pl. V. ; the radiating lines drawn from C are, therefore, omitted. As the ends are fixed, EF cannot be the closing line ; but the condition $\Sigma_D^B \dfrac{nM}{EI} = 0$ must first be imposed.

It has already been shown that for any point in the arched rib the moment $M = M_a - M^b$.

Hence if I_1 be the moment of inertia for the portion $2Ak$ of the rib, there may be written :

$$\Sigma_D^B \frac{nM}{EI} = \frac{1}{EI_1}\left(\Sigma_D^B niM_a - \Sigma_D^B niM_b \right) = 0.$$

For Bh, i has the value $\dfrac{I_1}{I} = 0.8$; for kh, the value $\dfrac{I_1}{I} = 0.9$; and for Ak, the value $\dfrac{I_1}{I_1} = 1.0$.

The closing line HK must be so located that:

$$\Sigma_D^B niM_a = 0.$$

But for the equilibrium polygon $Ea_1 a_2$, etc., the summation $\Sigma_D^B niM_a$ has the value :

$\Sigma_D^B niM_a = 0.8n' \times vv' + 0.8n \times a_1c' + 0.9n \times a_2c'' - n \times c''a_3 - n \times c^{iv}a_4 - n \times c^v C - n \times c^{vi}a_6 - n \times c^{vii}a_7 - 0.9n \times c^{viii}a_8 + 0.8n \times a_9c^{ix} + 0.8n' \times v'''v'' = 0.$

On the curve BAD the distance Bb is $\dfrac{n}{4} = \frac{1}{4} \times Bb'$, and $H'v'$ is taken vertically through b; also $Db^{x} = Bb$ and $K'v'''$ is drawn vertically through b^{x}. From the location of b and b^{x} it follows that $n' = \dfrac{n}{2}$, consequently n may be canceled from the series by writing $\dfrac{vv'}{2}$ and $\dfrac{v'''v''}{2}$ instead of the whole quantities themselves.

If vv' be taken at 1.04 inches and $v'''v''$ at 0.66 inches, the above summation (after dropping n in the manner shown) gives a result of $-$ 0.02 of an inch only. This agreement is sufficiently close.

The vertical intercepts and their products by i, are the following:

$$+ 0.8 \times \frac{vv'}{2} = + 0.8 \times \frac{1.04}{2} = + 0.42$$

$$+ 0.8 \times a_1c' = + 0.8 \times 0.56 = + 0.45$$

$$+ 0.9 \times a_2c'' = + 0.9 \times 0.06 = + 0.05$$

$$- 1 \times a_3c''' = - 1 \times 0.30 = - 0.30$$

$$- 1 \times a_4c^{iv} = - 1 \times 0.50 = - 0.50$$

$$- 1 \times Cc^{v} = - 1 \times 0.46 = - 0.46$$

$$- 1 \times a_6c^{vi} = - 1 \times 0.31 = - 0.31$$

$$- 1 \times a_7c^{vii} = - 1 \times 0.10 = - 0.10$$

$$+ 0.9 \times a_8c^{viii} = + 0.9 \times 0.15 = + 0.13$$

$$+ 0.8 \times a_9c^{ix} = + 0.8 \times 0.42 = + 0.34$$

$$+ 0.8 \times \frac{v''v'''}{2} = + 0.8 \times \frac{0.66}{2} = + 0.26$$

The other condition for the closing line HK is $\Sigma_D^B n M_a ix = 0$.

Taking the products of the ordinates (nMi) of different signs, by their horizontal distances from DK'', as was done in Art. 48, there will result the sums $+91.8$ and -90.5. The algebraic sum is only $+1.3$, which is near enough to zero. HK will therefore be taken as the proper closing line.

The condition which locates the closing line $H''K''$ is similar to that which placed HK; it is the following:

$$\Sigma_D^B niM_b = 0.8 \times 2n' \times H'b + 0.8 \times 2n \times c_1b' + 0.9 \times 2n \times c_2b''$$
$$- 2n \times b'''c_3 - 2n \times b^{IV}c_4 - n \times Ac_5 = 0.$$

Since the curve BAD is symmetrical in reference to AC, the line $H''K''$ will evidently be horizontal. For the same reason, $2n'$ and $2n$ are written in the summation in all its terms except the last. As before, n' is one half n, and by so writing it, n may be dropped from the series.

By making $Ac_5 = 0.58$ inch, the summation (after dropping n) gives $2.18 - 2.16 = 0.02$ only; $H''K''$ will therefore be assumed to be the proper closing line for BAD.

The vertical intercepts and their products by i are the following:

$$2 \times 0.8 \times \frac{bH'}{2} = 2 \times 0.8 \times \frac{1.13}{2} = 0.90$$

$$2 \times 0.8 \times b'c_1 = 2 \times 0.8 \times 0.66 = 1.06$$

$$2 \times 0.9 \times b''c_2 = 2 \times 0.9 \times 0.12 = 0.22$$

$$2 \times b'''c_3 = \qquad 2 \times 0.27 = 0.54$$

$$2 \times b^{IV}c_4 = \qquad 2 \times 0.52 = 1.04$$

$$Ac_5 = \qquad 0.58 = 0.58$$

As before, since the curve BAD is symmetrical in reference to A, the two conditions $\Sigma_D^B niM_b = 0$, and $\Sigma_D^B niM_b x = 0$, are equivalent.

The equation expressing the condition that the horizontal deflection of D in reference to B is nothing, is the general one already given:

$$\Sigma_D^B \frac{n.My}{EI} = \frac{n_1}{EI_1} \Sigma_D^B ri.My = \frac{n_1}{EI_1} \left(\Sigma_D^B ri.M_a y - \Sigma_D^B ri.M_b y \right) = 0.$$

Or, as in preceding cases :

$$\Sigma_D^B ri.M_a y = \Sigma_D^B ri.M_b y.$$

The quantity n_1 is any standard value of n, just as I_1 is a standard value of I, and r is such a variable ratio that for any section $n = rn_1$ or $r = \dfrac{n}{n_1}$. In the present case n_1 will have the value 10.95 feet; consequently r will be unity for all sections except b, and for that one it will be $\frac{1}{2}$.

The ratio r might have been used in the previous summations of the present Article in exactly the same manner and with exactly the same values as in the present one.

The ratio i has the same value as before.

The principles on which the remaining constructions are based are precisely the same as those shown in the preceding Articles ; the difference in the construction itself is simply this, that $ri.M_a$ and $ri.M_b$ are taken instead of M_a and M_b. In other words, r and i, in general, in this case, have values different from unity, while in the preceding cases $r = i = 1$.

The following are the values of $(ri.M_a)$:

$$
\begin{aligned}
\tfrac{1}{2} \times 0.8 .(\quad vv' \; + v''v''') &= \quad At; \\
1 \times 0.8 .(\quad a_1 c' \; + a_9 c^{ix}) &= \quad tt'; \\
1 \times 0.9 .(\quad a_2 c'' \; - a_8 c^{viii}) &= \quad t't''; \\
1 \times \quad 1 .(- a_3 c''' \; - a_7 c^{vii}) &= - t''t'''; \\
1 \times \quad 1 .(- a_4 c^{iv} \; - a_6 c^{vi}) &= - t'''t^{iv}; \\
1 \times 1 \times C c^{v} &= \quad t^{iv}A.
\end{aligned}
$$

Draw radial lines from C to the points t; then draw the horizontal lines bC'', $b'd_1'd'$, $b''d_1''d''$, etc. $C''d_1'$ is parallel to Ct; $d_1'd_1''$ is parallel to Ct'; $d_1''d_1'''$ is parallel to Ct'', etc. $CC''d_1'd_1''d_1'''d_1^{iv}d_1^{v}$ is then the deflection polygon for the moments M_a.

The following are the values for (riM_b):

$$\tfrac{1}{2} \times 0.8 \times 2H'b \quad = \quad As;$$
$$1 \times 0.8 \times 2b'c_1 \quad = \quad ss';$$
$$1 \times 0.9 \times 2b''c_2 \quad = \quad s's'';$$
$$1 \times \quad 1 \times (-2b'''c_3) = -s''s''';$$
$$1 \times \quad 1 \times (-2b^{iv}c_4) = -s'''s^{iv};$$
$$1 \times \quad 1 \times (-Ac_5) \quad = -s^{iv}A.$$

The point A belongs to the curve BAD.

The sides of the deflection polygon $C''d'd''d'''d^{iv}d^v$ are parallel to radiating lines drawn from C to the points s.

Ad_1^v represents $\sum_{D}^{B}riM_ay$, and Ad^v represents $\sum_{D}^{B}riM_by$. Since the first is less than the second, the moments M_a must be increased in the ratio of Ad_1^v to Ad^v. Hence on AC prolonged, make CM equal to Ad_1^v; draw MN parallel to BD and with a radius CN equal to Ad^v find the point N. Prolong the line CN; this line will enable the true moments M_a to be determined in the manner already shown in the other cases.

Draw c^vp parallel to BD, c_5a^v equal to Cp will give a point a^v in the true equilibrium polygon. Again, Ck'' equals KF, hence $K''a^x$, equal to Ck''', gives the true point a^x.

Also, Cg is equal to HE; and $H''a$, equal to Cp', gives the true point a. All points in the true polygon might be thus determined, but it is advisable to check by the other method already shown.

For this purpose draw CL parallel to HK, and LC' parallel to BD. Then make CP equal to BC and draw OP parallel to BD. Take LC' equal to CO. C' is the pole and LC' the pole distance of the true equilibrium polygon. Finally draw radiating lines from C' to the load points 1, 2, 3, 4, 5, 6, etc. Starting from any point already determined, as a, draw aa' parallel to $C'10$; $a'a''$ parallel to $C'9$; $a''a'''$ parallel to $C'8$, etc. The different points found by the two methods ought to coincide.

The polygon $aa'a''a'''a^{iv}a^va^{vi}a^{vii}a^{viii}a^{ix}a^x$ is the true equilibrium polygon, which was to be found.

$C'L$ is 3.86 inches; hence T_h, whose line of action is $H''K''$ is equal to $3.86 \times 15 = 57.90$ tons.

In the determination of the thermal stresses the same figures will be used, and there will be supposed such a change of temperature that the point A will suffer a vertical deflection of 1.5 inches or 0.125 of a foot; the same, in fact, as was supposed in a previous case.

As the ends of the rib are fixed, the general conditions, $\Sigma_D^B riM = 0$; and $\Sigma_D^B riMy = \Sigma_D^B riMx = 0$ hold as well for thermal stresses as others. Consequently $H''K''$ will be the line of action of the horizontal stress T_h, induced by the variation of temperature.

As before, let y' denote the vertical ordinate of any point in the curve BAD from the line $H''K''$, then there may be written:

$$\frac{n_1}{EI_1}\Sigma_D^A riMx = D = \frac{n_1 T_h}{EI_1}\Sigma_D^A riy'x$$

$$\therefore \quad T_h = \frac{DEI_1}{n_1 \Sigma_D^A riy'x}.$$

In order to save confusion in the figure, n_1 will be taken at its previous value 10.95 feet.

Also,

$$EI_1 = 1600000 \text{ foot-tons};$$
$$D = 0.125 \text{ foot}.$$

A half of the moment at A will be supposed applied at a point c distant $\frac{n_1}{4}$ from A on the curve BAD, and none at all at A.

The co-ordinate x will be measured from c_5 towards K''. The following values then result:

$x = 2.72$ ft.	$riy' = \frac{1}{2} \times 1 \times 5.7 = 2.85$	$riy'x =$	7.752
$x = 10.90$ "	$riy' = 1 \times 1 \times 5.2 = 5.2$	$riy'x =$	56.68
$x = 21.7$ "	$riy' = 1 \times 1 \times 2.7 = 2.7$	$riy'x =$	58.59

123.022

$x = 3.2$ ft. $riy' = -1 \times 0.9 \times 1.2 = -1.08$ $riy'x = - 34.56$
$x = 41.6$ " $riy' = -1 \times 0.8 \times 6.6 = -5.28$ $riy'x = -219.648$
$x = 47.9$ " $riy' = -\frac{1}{2} \times 0.8 \times 11.3 = -4.52$ $riy'x = -206.508$

$$-460.716$$

Hence,
$$\Sigma_D^A riy'x = 123.022 - 460.716 = -337.694.$$

The negative sign will be dropped hereafter, as it refers simply to the direction in which y was measured.

Making the substitutions:

$$T_h = \frac{0.125 \times 1600000}{10.95 \times 337.694} = 54.1 \text{ tons.}$$

The method by the deflection polygon gives nearly the same result, as will now be shown.

Comparing the two equations:

$$\Sigma Px = T_h.y,$$
$$\Sigma n_1 ri.Mx = EI_1 . D,$$

it is seen that if EI_1 be taken as the pole distance in the deflection polygon, $n_1 ri.M$ must be the general expression for the load at any point.

Since EI_1 is 1,600,000 foot-tons, CD will represent it at 320,000 foot-tons per inch; and that will be taken as the pole distance. $D - 3$, measured downwards, will be the load line. If the loads were taken at $10.95 \times ri.M$, or $10.95\,riy'$, the lower limit, 3, of the load line would not be on the diagram. The loads will therefore be taken as $4\,riy'$ and a proper reduction will be made afterwards. Hence, make

$$D - 1 = \quad 4\,riy' = \quad 4 \times 0.285 \text{ inches.}$$
$$1 - 2 = \quad 4\,riy' = \quad 4 \times 0.52 \quad \text{"}$$
$$2 - 3 = \quad 4\,riy' = \quad 4 \times 0.27 \quad \text{"}$$
$$3 - 4 = -4\,riy' = -4 \times 0.108 \quad \text{"}$$
$$4 - 5 = -4\,riy' = -4 \times 0.528 \quad \text{"}$$
$$5 - D = -4\,riy' = -4 \times 0.452 \quad \text{"}$$

The values of riy' are taken from the table immediately above.

By drawing radial lines from C to the points 1, 2, 3, 4, 5, the polygon $Cd_1 d_2 d_3 d_4 d_5 d_6 d_7$ is formed in the usual manner.

The deflection Dd_7 measures 2.67 inches or 26.7 feet, full size. Since $\dfrac{n_1}{4} = 2.74$, the deflection with the true moment-loads would be $26.7 \times 2.74 = 73.16$ feet, whereas it should be but one-eighth of a foot. Hence, measured by the same scale, the quantities $n_1 riM$ must be $\dfrac{1}{8 \times 73.16}$ of those taken in the figure, the pole distance CD remaining the same.

Since $M = T_h y'$, there may be written for the point A:

$$10.95 \times ri T_h y' = 10.95 \times 0.285 \times 320000 \div 8 \times 73.16.$$

As $r = \tfrac{1}{2}$, $i = 1$ and $y' = 5.7$ feet there results:

$$T_h = 54.7 \text{ tons.}$$

The deflection at other points might be used in the manner already shown in a preceding Article.

It is thus seen that the constructions are equally simple in principle whether r and i are constant or variable.

If the ends had not been fixed, it would only have been necessary to use the condition

$$\frac{n_1}{EI_1} \sum_{D}^{B} ri M y = 0.$$

Art. 53.—Determination of Stresses in the Members of an Arched Rib— Example—Fixed Ends—Consideration of Details.

It has been shown, in the preceding Articles, how to determine the horizontal tension T_h in the various cases which may arise; the method of using this horizontal tension in the determination of the stresses in the individual members of an arched rib remains to be shown.

For this purpose there will be taken the rib shown in Pl.

X., Fig. 3, having ends free, *i. c.*, free to turn about the points M and L.

The curve which has hitherto been used, and called the "centre line" of the rib, is the centre line of the neutral surface of the arched rib considered as a beam ; consequently the centres of gravity of the various cross sections of the rib must be found in this "centre line," *in all cases.* In other words, the "centre line" which has been used in the preceding articles is the locus of the centres of gravity of the normal cross sections of the actual rib.

In the rib taken as the example, Pl. X., Fig. 3, the apices in the upper and lower chords lie in the concentric circumferences of circles having radii of 78 and 72 feet, respectively ; and the centres of gravity of the normal cross sections will be supposed to lie on the circumference of a circle having the same centre, whose radius is 75 feet. Thus the centre line is precisely the same as has been used in the preceding Articles. The same span and loading will also be taken.

The extremities of the span, or points M and L, at which the horizontal tension or force is applied, *must lie in the centre line.*

All the loading will be assumed to be applied at the apices of the upper chord, although the operations would be exactly the same if the fixed load were divided in any proportion between the two chords.

The apices of the triangles of the web system were located as follows: As was done in finding T_h, the centre line was divided into ten equal parts. The upper chord panel points are vertically over those points of division. The upper chord panels were then bisected, and radii were drawn through these points of bisection. The lower chord panel points were taken at the intersections of those radii with the circumference of the circle whose radius was 72 feet.

There is only one point to be observed in forming the chord panels, *i. c.*, the panel points must be so located that the load will act exactly as was supposed in determining T_h.

The following loads will then be assumed to act through the upper chord panel points :

7 tons at the intersection of 1 and 3 ;

14 " " " intersections of 4 and 5, 6 and 7, 8 and 9, 10 and 11 ;

9 " " " intersection of 12 and 13 ;

4 " " " intersections of 14 and 15, 16 and 17, 18 and 19, 20 and 21 ;

2 " " " intersection of 22 and 24.

As the ends are free, $L - 10 = 50.7$ tons in Pl. VI., gives the reaction R at the left end of the span, or at L in the example. Also $L - 1$ in Pl. VI., gives the reaction as 30.3 tons at M in the example. In Art. 49 it is found that $T_h = 52.5$ tons for this case, and the two methods in Art. 51 give the thermal stresses in the horizontal tie as 13.6 and 16.7 tons. The thermal tension will be taken at 15 tons. The total tension in the tie will then be $52.5 + 15.00 = 67.5$ tons, as shown.

It is a matter of no consequence whether the tie exist or not. If it does not exist, the abutments at M and L must then supply the horizontal force of 67.5 tons.

Fig. 4 of Pl. X. is the complete diagram for the stresses with the load taken ; it is drawn to a scale of 20 tons to the inch, nearly. The lines indicated by letters or figures in the diagram are parallel to the members of the rib indicated by the same letters or figures, though the parallelism is not exactly shown in the plate for all the lines.

With the following explanations relating to the diagram, little more is needed :

$$a'b' = b'c' = c'd' = d'f' = 14 \text{ tons.}$$
$$o'n' = n'm' = m'k' = k'h' = 4 \quad \text{"}$$
$$h'f' = 9 \quad \text{"}$$

$$q'a' = (1) + 7.$$
$$p'o' = (24) + 2.$$
$$PS = (E).$$

$a'e'$ is the reaction R, and $c'o'$ is the reaction at M, while

$c'P$ is the horizontal tension 67.5 tons. Although the diagram appears very complicated, yet, it is really composed of very simple *five*-sided figures, as may easily be seen. Let the rib be divided through c, 7, B, and T; then the portion of the rib between that surface of division and L is held in equilibrium by the action of the stresses in the members divided (considered as forces external to that portion), the applied loads, (T), and the reaction R. The resultant of the loads and R is a vertical shear represented by $e'c'$ in the diagram. The forces acting upon the portion of the rib in question are then represented by the lines $c'c'$, (T), (B), (7), and (c) in Fig. 4, and these constitute a simple five-sided figure. The arrow heads show the direction of action of these forces, and enable the *kind* of stress to be recognized at a glance.

The whole diagram is thus composed of just such pentagons.

As a check on the accuracy of the construction of the diagram, if it is worked continuously from L to M, the four-sided figure involving (T), (23) and (24) should exactly close. It is far more conducive to accuracy, however, to work up the diagram from both ends, and if the work has been accurately done, the diagrams will give the same stress in that member which becomes common to both where they meet.

The stresses, as determined by the original diagram from which Fig. 4 was constructed, are written in Fig. 3.

If an arched rib is subjected to a load, advancing panel by panel, the stresses due to the fixed load alone may first be determined and then tabulated. The stresses in all the members of the rib due to each panel moving load may then be found and tabulated also. The greatest stress of either kind in any member may then be determined by a combination of these results in the usual manner.

Some of the stresses found by diagram should be checked by moments in the following manner. The horizontal distances of the panel points, in the left half of the rib holding 14 tons each, from a vertical line bisecting the span and passing through the intersection of 12 and 13, are 10.9 feet, 21.7 feet, 32 feet, and 41.5 feet. The normal distance from the intersection of (12) and (13) to E, is 6.15 feet (by scale). Hence, taking moments about that intersection,

$$(E) = \frac{50.7 \times 50 - 14(10.9 + 21.7 + 32 + 41.5) - 67.5 \times 22.1}{6.15} =$$
$$- 72 \text{ tons.}$$

The diagram gave 72.5 tons, and the agreement is sufficiently close.

On account of the ill-defined intersections of the prolonged chord sections in any panel, the method of moments for the web stresses is not satisfactory unless one chord stress in the panel is known. The web stress can then be found by moments in a manner to be presently illustrated.

Again, let (G) be determined by taking moments about the intersection of 16 and 17. Draw a vertical line through that point. The horizontal distances of the three upper chord panel points on the right of that line, from the same, are 10.3 feet, 19.8 feet, and 28.5 feet (by scale). In the same manner the vertical distance of the point above T is 19.25 feet. Hence,

$$(G) = \frac{30.3 \times 28.5 - 4(10.3 + 19.8) - 67.5 \times 19.25}{6.15} =$$
$$- 90.5 \text{ tons.}$$

The diagram gave 95 tons, and the agreement is not close. This illustrates in a marked manner the great fault of the graphical method. In constructing the original diagram, shown by Fig. 4 of Pl. X., the rib was drawn to a scale of 5 feet per inch, and the diagram itself to a scale of 10 tons per inch, and although the greatest care was taken, yet the stresses found for the right half of the rib may, in some members, be wrong to the extent of even twenty per cent. The method requires the largest and most accurate figures possible, and the very nicest instruments, for extended diagrams.

By far the most accurate, and, all things considered, the most satisfactory method, is the combination of moments and diagram, so freely used in the treatment of bowstring trusses. In this method a stress in either chord is found by moments, the other two stresses (one a chord and the other a web) in the same panel are then immediately found by a simple five-sided figure or diagram.

The chord stresses (E) and (G) have just been found. In Fig. 1 take bd equal to 67.5 tons and parallel to T, and make dc (parallel to E as well as T) equal to $(E) = - 72.00$ tons. A section is supposed to be taken through the panel in which f, 13 and E are found; consequently the vertical shear $S = 50.7 - 65.00 = - 14.3$ tons. The remainder of the diagram needs no explanation. It gives:

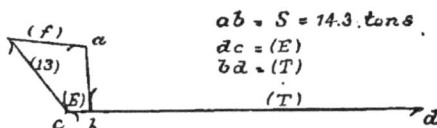

FIG. 1.

$(13) = - 20.3$ tons; $(f) = + 17.4$ tons.

Fig. 2 is drawn in precisely the same manner by using $(G) = - 90.5$ tons, which has already been determined by

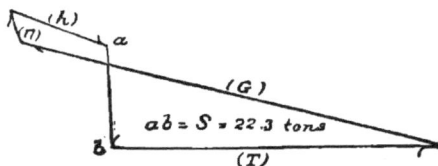

FIG. 2.

the moment method. The section is taken through G, 17 and h.

Fig. 2 gives:

$(17) = - 5.5$ tons; $(h) = + 23.0$ tons.

All the other stresses may be found in the same manner. The difference in the case of (h) between the results of the two methods is four tons, or about 17 per cent. of the smallest result.

By the method of Figs. 1 and 2 accurate results may be obtained by taking a scale of even twenty tons to the inch, but a larger diagram is preferable.

If the ends of the rib are fixed, the shortest method of finding the stresses is in no way different from that given in connection with Figs. 1 and 2, excepting this: the chord stress which is found by moments will have a different value. If the ends are fixed, *the reaction R, at the left end of the span, will be L — 10 of Pl. V.; and the horizontal tension (T) will be taken as* 54.8 + 113.3 = 168.1 *tons, from Arts.* 48 *and* 50.

The line of action of T_h for both external load and thermal stresses is $H''K''$ of either Pl. V. or Pl. VII.; let Pl. V. be considered.

The action of T_h through H'' (taken as acting toward K''), so far as the rib *BAD* is concerned, *is equivalent to T_h acting through B toward D, combined with a right hand couple whose force is T_h and whose lever arm is BH''.* Let the moment of this couple be called M. This moment will cause compression throughout the upper chord of the rib and tension throughout the lower.

Let M_1 represent the moment (of the external forces and T_h) about any panel point of the rib, as F, Fig. 3 (the reaction and T_h being taken for the particular case, as just indicated). Then any chord stress, as (BE), will be

$$(BE) = \frac{M + M_1}{n_1};$$

n_1 being the normal depth of the rib, as shown. Particular care is to be taken in regard to the signs of M and M_1, *i. e.*, it is to be noticed whether they tend to produce the same or different kinds of stress in BE.

After (BE) is found, the diagrams are to be drawn precisely like Figs. 1 and 2, and the resulting stresses, scaled from the diagrams, are the ones desired.

The following, but longer methods, may also be used:

18

The first portion of the operation is simply the application of the method by diagram, or the combination of moments and diagram, already given in connection with the case of ends free.

All the individual stresses in the rib are to be found in this manner; *those for the web members are the true web stresses desired* if the rib is of uniform normal depth. The chord stresses thus found are, however, in all cases, to be modified.

Let the normal depth of the rib at any section be n_1; the distance BH'', Pl. V., h''; and the general expression for any chord stress due to the moment $M = T_h h''$, c.

Then will result:

$$c = \frac{M}{n_1} = \frac{T_h h''}{n_1}$$

Then let (c) be the general expression for any chord stress already found without considering the moment M; the numerical value may be either positive or negative. Finally, the resultant chord stress desired will be, for the upper chord,

$$(c) - c = (c) - \frac{T_h h''}{n_1},$$

and for the lower,

$$(c) + c = (c) + \frac{T_h h''}{n_1}.$$

If the rib is of uniform normal depth, *i.e.*, if n_1 is constant, the web stresses will not be affected by the moment M, for it (the moment M) will cause uniform chord stresses throughout the rib.

If the normal depth, however, is not constant, the moment

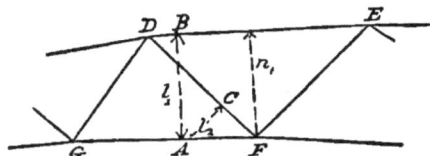

FIG. 3.

M will cause web stresses which may be determined very accurately in the following manner:

Let it be desired to determine the stress in the web member *DF* of a portion of an arched rib, shown in Fig 3, and let *w* denote that stress. The stress in *DE* is *c*, found by the method just given.

In Fig. 4, take $(DE) = c$ and parallel to *DE* in Fig. 3. The lines (GF) and (DF) in Fig. 4, are then parallel to *GF* and

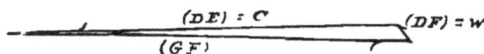

FIG. 4.

DF in Fig. 3, and they are the stresses in those members. All the web stresses and the chord stresses in one chord may be thus found. This operation is simply the method of Figs. 1 and 2 applied to this case.

The web stresses may be found by using moments, only, in the following manner :

Take *A* as any convenient point in *GF*. Let l_1 represent *AB*, and l_2, *AC* ; these lines are normal to *DE* and *DF* respectively.

Let *M* be still considered right-handed and positive, and let it first be assumed that *c* is compression. Moments about *A* give :

$$w = - \frac{-cl_1 + T_h h'}{l_2}.$$

If *c* is tension, or belongs to a panel in the lower chord, then (w) will be the stress in a member like *DG*. There will then result :

$$w = \frac{-cl_1 + T_h h''}{l_2}.$$

In either of these formulæ, (w) will represent tension or compression according as the result is positive or negative.

Let (w) represent any web stress already found by neglecting *M*, then will any resultant web stress desired be :

$$(w) + w;$$

the signs of both these quantities being implicit.

As the lever arms l_1, l_2, and n_1 are scaled from the drawing, the rib should be laid down as accurately and to as large
scale as possible.

In important cases these different methods should be used
as checks.

A very common system of bracing for arched ribs, although
a very unsatisfactory one, is that shown in Fig. 5. The web

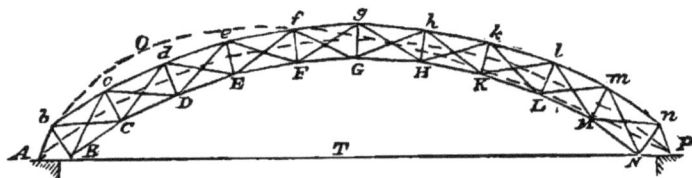

FIG. 5.

members, bB, cC, dD, etc., are normal to the centre line of
the rib, and are designed for *tension* only. The other web
members are for compression only

Let the ends be supposed free, and take AOP for the true
equilibrium polygon for a given load.

For any given loading the stresses in the different members
are indeterminate, unless about half of the compression web
members are neglected.

With the assumed position of the equilibrium polygon AOP,
for instance, it is seen that compression will increase in the
upper chord from b to d (nearly); from that point to g
(nearly) it will decrease. The compression in the lower chord
will increase from G to L (nearly) and then decrease from
that point to N. The points of greatest chord stresses of
either kind are those at which the polygon and centre line of
rib are parallel.

From these considerations it results that the web members
bC, cD, De, Ef, Fg, Gh, Hk, Kl, LM, and mN may be omitted;
they must be omitted, in fact, if the stresses are to be determinate. Having made these omissions, the stresses are to
be found by the methods already given, as those methods
are *perfectly general.*

Precisely the same observations apply to the case of fixed ends, or to that in which the normal members are in compression and the others in tension.

It is by no means certain that the stresses thus found will really exist in the rib, but the assumption is the best that can be made. This system of bracing is, at best, very unsatisfactory.

The free ends of an arched rib are sometimes arranged, in regard to support, as shown in Fig. 5, Pl. X. There are two ties or sets of ties, T' and T'', instead of one, T_h. A and B are the points at which these ties take hold of the rib. E is the intersection of AB and the centre line, EF, of the rib. *The span to be used in finding T_h for either external loads or thermal variation is the horizontal distance between E and the corresponding point at the other end of the rib.* T', T_h, and T'' are parallel to each other, and ac is normal to the three.

Now if T' and T'' are determinate, there may be written:

$$T' = \frac{bc}{ac} T_h;$$

$$T'' = \frac{ab}{ac} T_h.$$

In such a case the systems of triangulation in the rib may be separated, T' will belong to one and T'' to the other. The stresses may then be found in each system separately, and the results combined for the resultant stresses of the rib. If the web members may be counterbraced, the resultant stresses are thus determinate. It is not certain, however, that the tensions T' and T'' will have the values given above.

For the determination of the stresses in the rib, however, it is not necessary to resolve T_h into T' and T'', except for the panel $ABDC$.

In the case of a design, if the dimensions required by the calculations of this Article give a value to the moment of inertia I very different from that assumed in the determination of T_h, for either thermal variations or external load, it will be necessary to make an entirely new set of calculations with another value of I. This must be done until the agreement between the assumed and required values of I is sufficiently close.

Art. 54.—Arched Rib Free at Ends and Jointed at the Crown.

Suppose the rib to be represented in the figure.

Since there is a joint at A, the bending moments must be zero at that point; consequently the equilibrium polygon for any load must pass through that point. This fact furnishes a

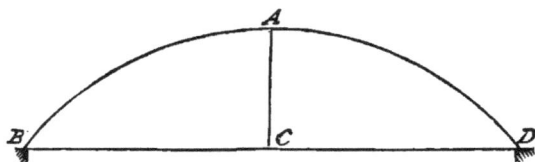

FIG. 6.

very simple method of determining T_h. Denote by ΣPx the moment of all the external forces about the joint A; then, since M must be equal to zero,

$$\Sigma Px - T_h(AC) = 0. \quad \therefore \ T_h = \frac{\Sigma Px}{AC}.$$

In this rib variations of temperature produce no variations of stress in BD, except that due to the slight change of AC, as shown by the formula above. This, however, is a very small quantity, and would ordinarily be neglected. If necessary, it would be allowed for by taking the value of AC at the lowest temperature to which the rib would be subjected.

Other arched ribs are seldom constructed, but they are to be treated by the same general methods, precisely, as those used in the preceding cases.

CHAPTER IX.

SUSPENSION BRIDGES.

Art. 55.—Curve of Cable for Uniform Load per Unit of Span—Suspension Rods Vertical—Heights of Towers, Equal or Unequal—Generalization.

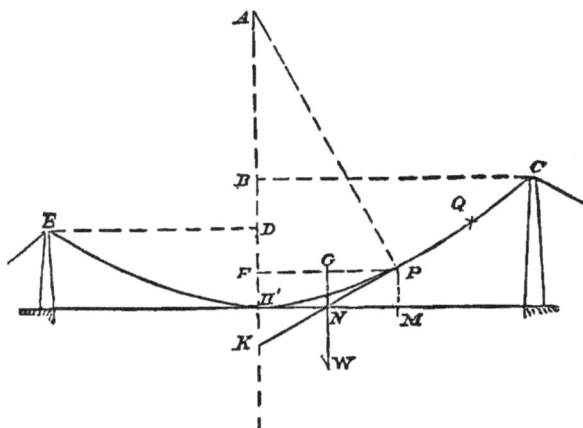

FIG. I.

IN the figure, let $EH'C$ represent the cable of a suspension bridge carrying a load extending over the whole span. In the ordinary experience of an engineer, the load carried by a suspension bridge cable is nearly uniform in intensity in reference to a horizontal line; so nearly uniform per foot of span, in fact, that it is assumed to be exactly so, and such an assumption will be made in the present instance.

The use of the stiffening truss, to be presently noticed, makes this assumption essentially true.

Let $(ED + BC) = l = $ span; $BH' = h_1$; $DH' = h_2$; $w = $ load for horizontal foot, and let x be measured hori-

279

zontally from H', the lowest point of the cable. The height of the highest tower is, of course, h_1, and that of the other h_2.

The ordinate of any point P is x, the load on $H'M$ is, consequently, $W = wx$. Draw PK tangent to the curve at P, then by the first principles of statics, it is known that the direction of the cable tensions at P and H' and the direction of W must intersect in one point N. Since, however, w is uniform along x, the resultant direction of W passes through N, half way between H' and M. Hence $FH' = H'K$; or, since FK is the subtangent, *the abscissa, FH', of the curve is equal to half the subtangent,* consequently the curve is the ordinary parabola.

Again, it is known that the horizontal component of the tension of a cable will be a constant quantity if the loading (as in the present case) be wholly vertical; let that component be denoted by H.

Let GNP be taken for the triangle (right angled) of forces at P, in which NP represents the cable tension at P, GN the load, $W = wx$ and GP the constant horizontal component H.

Then let AP be drawn normal to the curve at P; the triangles AFP and GNP will be similar. There can now be at once written the relation :

$$\frac{AF}{GP} = \frac{FP}{GN} = \frac{x}{wx} = \frac{1}{w};$$

but

$$GP = H \quad \therefore \quad AF = \frac{H}{w} = \text{constant} \quad . \quad . \quad (1).$$

Now AF is the subnormal of the curve of the cable, and since it is constant, the curve is the ordinary parabola.

The preceding results may be generalized in a very simple and easy manner.

If any two points, as P and Q, be considered fixed, and if the portion PQ of the cable carry the same intensity of load w as before, there will at once result the general case of a flexible cable carrying a load whose intensity, along a straight line, and direction are uniform. There may then be stated the general principle : *If a perfectly flexible cable carry a load uniform in direction and intensity in reference to a straight line, the cable will assume the form of an ordinary parabola whose axis will be parallel to the direction of the loading.*

This principle finds its application in the case of a suspension bridge with inclined, but parallel, suspension rods.

Art. 56.—Parameter of Curve—Distance of Lowest Point of Cable from either Extremity of Span—Inclination of Cable at any Point.

Attending to the figure and notation of the previous Article, the equation of the curve, the origin of co-ordinates being taken at H', is :

$$x^2 = 2py ;$$

in which $2p$ is the parameter.

Let $BC = x_1$ and $ED = x_2$, then there may be written $x_1^2 = 2ph_1$, $x_2^2 = 2ph_2$ and $2x_1 x_2 = 4p\sqrt{h_1 h_2}$.

Hence

$$(x_1 + x_2)^2 = l^2 = 2p(\sqrt{h_1} + \sqrt{h_2})^2 = 2p(h_1 + 2\sqrt{h_1 h_2} + h_2) \ . \ . \ . \ (1).$$

$$\therefore p = \frac{l^2}{2(\sqrt{h_1} + \sqrt{h_2})^2} = \frac{l^2}{2(h_1 + 2\sqrt{h_1 h_2} + h_2)} \ . \ . \ (2).$$

If the towers are of the same height, then $h_1 = h_2 = h$ and

$$p = \frac{l^2}{8h} \quad . \quad . \quad . \quad . \quad . \quad . \quad (3).$$

Now x_1 is the horizontal distance from the lowest point of the cable to that end of the span at which h_1 is found, *i. e.*, BC in the figure, while x_2 is the other segment of the span, and by the equations immediately preceding :

$$x_1 = \frac{l\sqrt{h_1}}{\sqrt{h_1} + \sqrt{h_2}} \quad . \quad . \quad . \quad . \quad . \quad (4).$$

$$x_2 = \frac{l\sqrt{h_2}}{\sqrt{h_1} + \sqrt{h_2}} \quad \cdots \cdots \quad (5).$$

$$\text{If } h_1 = h_2; \ x_1 = x_2 = \frac{l}{2} \quad \cdots \cdots \quad (6).$$

Referring to the figure, since $KH' = H'F = y$, if i is the inclination to a horizontal line, of the curve at any point P, then $x \ tan \ i = 2y$;

$$\text{hence, } tan \ i = \frac{2y}{x} \ \therefore \ sec \ i = \sqrt{1 + \frac{4y^2}{x^2}} \quad \cdots \quad (7).$$

At the summits of the towers:

$$tan \ i_1 = \frac{2h_1}{x_1} \text{ and } tan \ i_2 = \frac{2h_2}{x_2} \quad \cdots \cdots \quad (8).$$

$$\text{If } h_1 = h_2, \ tan \ i_1 = tan \ i_2 = \frac{4h}{l} \quad \cdots \quad (9).$$

Art. 57.—Resultant Tension at any Point of the Cable.

In the first Article of this Chapter there was recognized the general principle that if the loading on a cable is uniform in direction, the component of cable tension normal to that direction will be constant at all points of the cable. In the present case the resultant tension at the lowest point of the cable will be this constant component H.

From Eq. 1, Art. 55, $H = wAF$. But AF is the subnormal of the curve, and, from Analytical Geometry, it is known to be equal to one half of the parameter, or equal to p (using the same notation as before).

Hence, after taking the value of p from the previous Article:

$$H = wp = \frac{wl^2}{2\left(\sqrt{h_1} + \sqrt{h_2}\right)^2} = \frac{wl^2}{2\left(h_1 + 2\sqrt{h_1 h_2} + h_2\right)} \quad \cdot \ (1).$$

Let R denote the resultant tension at any point, then by the triangle of forces GNP, in the figure:

$$R = H \sec i = H \sqrt{1 + \frac{4y^2}{x^2}} \quad . \quad . \quad . \quad . \quad (2).$$

Eq. (2) gives the tension at any point. At the summits of the towers there are found:

$$R_1 = H \sqrt{1 + \frac{4h_1^2}{x_1^2}} \quad . \quad . \quad . \quad . \quad (3).$$

$$R_2 = H \sqrt{1 + \frac{4h_2^2}{x_2^2}} \quad . \quad . \quad . \quad . \quad (4).$$

If $h_1 = h_2$, consequently $x_1 = x_2 = \frac{l}{2}$, then:

$$H = \frac{wl^2}{8h}, \quad R_1 = R_2 = H \sqrt{1 + \frac{16h^2}{l^2}} \quad . \quad . \quad . \quad (5).$$

Art. 58.—Length of Curve between Vertex and any Point whose Co-ordinates are x and y, or at which the Inclination to a Horizontal Line is i.

The usual expression for the length of a part of one branch of a parabola, beginning at the vertex, as determined by the integral calculus, may easily be put in the following form, denoting by c the length in question:

$$c = \frac{x^2}{4y} \left\{ \frac{2y}{x} \sqrt{1 + \frac{4y^2}{x^2}} + \text{hyp. log.} \left(\frac{2y}{x} + \sqrt{1 + \frac{4y^2}{x^2}} \right) \right\} (1).$$

Or, using the values for *tan i*, *sec i*, and p, determined in the preceding Articles:

$$c = \frac{p}{2} \{ tan\ i\ sec\ i + \text{hyp. log. } (tan\ i + sec\ i) \} \quad . \quad . \quad . \quad (2).$$

The total length of the cable will, of course, be found by putting x_1 and h_1 for x and y in Eq. (1), or i_1 for i in Eq. (2); then x_2 and h_2 for x and y, or i_2 for i, and adding the results.

Denoting those results by c_1 and c_2 the total length will then be :

$$c_1 + c_2.$$

An approximate formula sometimes used is determined as follows. In the figure of the first Article of this chapter consider $H'P$ to be the arc of a circle, and let x and y be taken as heretofore ; also let R be the radius of the circle. The ordinary expression for the length of a circular arc, in the integral calculus, is :

$$\int \frac{dx}{\left(1 - \frac{x^2}{R^2}\right)^{\frac{1}{2}}} = \int \frac{dx}{1 - \frac{1}{2}\frac{1}{R^2}x^2} \text{ (nearly),}$$

if x is small compared with R. Again, performing the division indicated and omitting all terms in the quotient after the second, there will result :

$$\int_0^x dx\left(1 + \frac{1}{2R^2}x^2\right) = x\left(1 + \frac{x^2}{6R^2}\right) \quad . \quad . \quad (3).$$

If y^2 be omitted in the expression $x^2 = 2Ry - y^2$, and the resulting value of R be inserted in Eq. (3), there will at once be found :

$$x\left(1 + \frac{2y^2}{3x^2}\right) \quad . \quad . \quad . \quad . \quad . \quad (4).$$

As before, to find the total length of the curve, x_1 and h_1, and x_2 and h_2 must be inserted in succession in Eq. (4), and the results added.

If the heights of the towers are equal to each other and to h, the total length will be

$$l\left(1 + \frac{8h^2}{3l^2}\right) \quad . \quad . \quad . \quad . \quad . \quad (5).$$

The expressions (4) and (5) are evidently not close approximations except for very flat curves, in which case the nature of the curve is a matter of indifference.

Art. 59.—Deflection of Cable for Change in Length, the Span Remaining the Same.

The approximate formula (4) of the preceding Article is usually used in determining the deflection.

The total length of the cable is:

$$c_1 + c_2 = x_1 + x_2 + \frac{2}{3}\left(\frac{h_1^2}{x_1} + \frac{h_2^2}{x_2}\right),$$

Differentiating:

$$d(c_1 + c_2) = \frac{4}{3}\left(\frac{h_1}{x_1} + \frac{h_2}{x_2}\right) dh * \quad \cdots \quad (1).$$

$$\therefore \quad dh = \frac{3d(c_1 + c_2)}{4\left(\dfrac{h_1}{x_1} + \dfrac{h_2}{x_2}\right)} \quad \cdots \cdots \quad (2).$$

The variation in the length of the cable, whether arising from variation in temperature or any other cause, is to be put for $d(c_1 + c_2)$ in Eqs. (1) and (2), then dh will be the corresponding deflection of the lowest point of the cable.

If the towers are of the same height, and, consequently, $c_1 = c_2$, $h_1 = h_2$, $x_1 = x_2 = \dfrac{l}{2}$:

$$2dc_1 = \frac{16}{3}\,\frac{h_1}{l}\cdot dh \quad \cdots \cdots \quad (3).$$

$$dh = \frac{3l}{16h_1}\cdot 2dc_1 \quad \cdots \cdots \quad (4).$$

---

* Since $h_1 - h_2 = $ constant, $dh_1 = dh_2 = dh$.

It is assumed in Eqs. (1) and (2) that the lowest point of the cable remains at the same horizontal distance from the ends of the span, though such is not really the case.

The true deflection can only be found by trial by the use of Eq. (1) of the previous Article.

Let $(c_1 + c_2)$ be the known length of the cable before variation in its length takes place ; then let h_1, h_2, x_1 and x_2 be the original heights of towers and segments of span, also known. Let y_1 and y_2 be the heights of towers above the lowest point of the cable after the variation in its length has taken place ; and let it be assumed, as before, that x_1 and x_2 remain the same whatever the deflection.

Let v be the variation in length of the cable.

Then, since $v = -(c_1 + c_2) + (c_1 + c_2 + v)$:

$$v = \frac{x_1^2}{4y_1}\left\{ \frac{2y_1}{x_1}\sqrt{1 + \frac{4y_1^2}{x_1^2}} + \text{hyp. log.}\left(\frac{2y_1}{x_1} + \sqrt{1 + \frac{4y_1^2}{x_1^2}} \right) \right\}$$

$$+ \frac{x_2^2}{4y_2}\left\{ \frac{2y_2}{x_2}\sqrt{1 + \frac{4y_2^2}{x_2^2}} + \text{hyp. log.}\left(\frac{2y_2}{x_2} \right.\right.$$

$$\left.\left. + \sqrt{1 + \frac{4y_2^2}{x_2^2}} \right) \right\} - (c_1 + c_2) \quad . \quad . \quad . \quad (5).$$

But there is also the equation of condition :

$$y_1 - y_2 = h_1 - h_2 = \text{constant} \quad . \quad . \quad . \quad (6).$$

The value of y_1 or y_2 may be taken from Eq. (6) and put in Eq. (5), there will then be but one unknown quantity in the right member of that equation, and its value must be found by trial. The first value of y_1 or y_2 taken may be h_1 or h_2 increased or decreased, as the case may be, by dh given by Eq. (2).

The deflection sought is, of course :

$$y_1 - h_1 = y_2 - h_2.$$

If the new heights, y_1 and y_2 are given, the variation of length, v, will be at once given by Eq. (5).

If heights of towers are the same, Eq. (6) will not be needed; for making $x_1 = x_2 = \dfrac{l}{2}$, $c_1 = c_2$, and $y_1 = y_2 = h$, there results:

$$v = \frac{2l^2}{16h} \left\{ \frac{4h}{l} \sqrt{1 + \frac{16h^2}{l^2}} + \text{hyp. log.} \left(\frac{4h}{l} + \sqrt{1 + \frac{16h^2}{l^2}} \right) \right\}$$
$$- 2c_1 \quad \cdots \quad \cdots \quad (7).$$

In Eq. (7) h is then to be found by trial, as before, if v is given; or if h is given, v at once results.

The deflection of the middle point of the truss will be:

$$h - h_1.$$

It is to be noticed that in Eqs. (5) and (7) all the quantities, y_1, y_2, and h, increase in the same direction with v. This materially simplifies the approximation by trial.

The determination of v in Eq. (5) might be made without assuming x_1 and x_2 to remain constant, for there are two other equations of condition:

$$x_1' + x_2' = l,$$
and
$$\frac{x_1'^2}{y_1} = \frac{x_2'^2}{y_2}.$$

These, with Eqs. (5) and (6) would be sufficient in order to find the four unknown quantities, y_1, y_2, x_1' and x_2'.

Such a degree of extreme accuracy, however, is unnecessary.

Art. 60.—Suspension Canti-Levers.

In the figure, ABD represents a suspension canti-lever. The cable BC goes over either to another span or to an anchorage, while A is the end of the canti-lever. The cable AB is in precisely the same condition as the half of a cable belonging to a span equal to $2AD$; consequently its tension R at any point and its inclination at the same point are to be found by the formulæ already given. In fact, all the circumstances are precisely the same except this, the platform is subjected to a thrust, uniform throughout its whole length, and equal to the constant horizontal component, H, of the tension R.

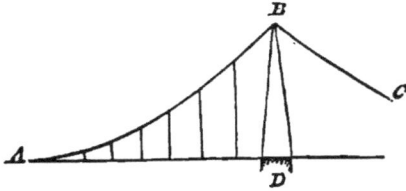

FIG. I.

Art. 61.—**Suspension Bridge with Inclined Suspension Rods—Inclination of Cable to a Horizontal Line—Cable Tension—Direct Stress on Platform—Length of Cable.**

In this case the suspension rods, or suspenders, are all supposed to be equally inclined to a vertical or horizontal line, and, consequently, are parallel to each other; they are also supposed to take hold of the platform or stiffening truss at points equidistant from each other. These conditions cause the cable to be subjected, in each of its parts, to the action of parallel loading of uniform intensity in reference to the span. As was shown in the first Article of this Chapter, the curve of the cable will be composed of common parabolas having axes parallel to the suspension rods.

In the figure, A is the lowest point of the cable, while BD and FG represent suspension rods on either side of A. The

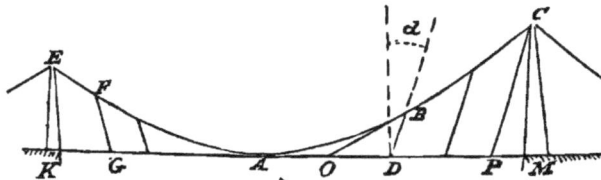

FIG. I.

angle α is the common inclination of all suspenders to a vertical line, it also represents the inclination of the axis of either of the parabolas *AC* or *AE* to the same line.

The vertex of the parabola *AC* is on the left of *A*, and the vertex of *AE* is on the right of the same point.

Let *x* be horizontal in direction and measured from *A*, and let the length of any suspender, as *BD* be denoted by *y*, but let *CP* be designated by y_1. Either parabola, as *AC*, will then be referred to oblique co-ordinates in the usual manner.

If *OB* be drawn tangent to the curve at *B*, *AO* will be equal to *OD*, or $\frac{x}{2}$. If *i* represents the inclination of the curve at any point, as *B*, to a horizontal line, the triangle *OBD* will give:

$$\frac{BD}{OD} = \frac{2y}{x} = \frac{\sin i}{\cos (\alpha + i)} = \frac{\sin i}{\cos \alpha \cos i - \sin \alpha \sin i},$$

$$\therefore \quad \tan i = \frac{\dfrac{2y}{x} \cos \alpha}{1 + \dfrac{2y}{x} \sin \alpha}. \quad . \quad . \quad (1).$$

In the usual manner, $\sec i = \sqrt{1 + \tan^2 i}$, or,

$$\sec i = \sqrt{\frac{1 + \dfrac{4y}{x} \sin \alpha + \dfrac{4y^2}{x^2}}{1 + \dfrac{2y}{x} \sin \alpha}} \quad . \quad . \quad (2).$$

At the point *C*, if $CM = h_1$, $AP = x_1$, and $AM = a$, in Eqs. (1) and (2), $h_1 \sec \alpha$ is to be put for *y*, and $(a - h_1 \tan \alpha)$ for *x*.

Exactly similar equations apply to the other portion of the span.

For the point *A*, Eqs. (1) and (2) apparently become indeterminate, but only apparently, for the relation,

$$\frac{y}{x^2} = \frac{h_1 \sec \alpha}{(a - h_1 \tan \alpha)^2}, \quad . \quad . \quad . \quad (3).$$

gives,

$$\frac{2y}{x} = \frac{2h_1 \sec \alpha}{(a - h_1 \tan \alpha)^2} x,$$

and when $x = 0$, consequently, $\frac{2y}{x}$ becomes zero, making $\tan i$ equal to zero also.

If OBD be taken as a triangle of forces, OB will be the cable tension R at B; while OD will be the horizontal component H, and BD will represent $wx \sec \alpha$. w is the total load per unit of span on AD.

From the triangle in question,

$$\frac{H}{wx \sec \alpha} = \frac{\cos (\alpha + i)}{\sin i} = \frac{x}{2y}, \quad \therefore H = \frac{wx^2 \sec \alpha}{2y},$$

$$\therefore H = \frac{w (a - h_1 \tan \alpha)^2}{2h_1}. \quad . \quad . \quad . \quad (4).$$

As was to be expected, Eq. (4) shows H to be a constant quantity, but it is not a rectangular component in this case.

The same triangle gives for the resultant tension at any point:

$$R = \sqrt{(wx \sec \alpha)^2 + H^2 + 2wx \, H \tan \alpha}. \quad . \quad (5).$$

For the point C, x becomes $(a - h_1 \tan \alpha)$.

If l is the span, these equations apply to the other portion of it, by taking h_2 for h_1, and $(l - a)$ for a.

If the towers are of equal heights, h_1 becomes equal to h_2, and $a = l - a = \frac{l}{2}$.

Let p be the horizontal distance between any two suspenders, then the tension, t, in that suspender will be:

$$t = wp \sec \alpha \quad . \quad . \quad . \quad . \quad . \quad (6).$$

The direct stress in the platform is caused by the horizontal component of the tension in the suspension rods. This stress may exist as tension in the platform, in which case it will exert no action on the towers. Remembering that all the suspension rods must, at any instant, be subjected to a uniform stress, it is evident that the direct tension in the platform will have its greatest value at the centre, and will be equal to

$$nt \sin \alpha = nwp \tan \alpha;$$

in which n is the number of suspension rods in each half of the span, supposing towers to be of equal heights. If n' be the number of suspenders between the end of the span and any point, the tension in the platform at that point will be

$$n't \sin \alpha = n'wp \tan \alpha.$$

If the towers are of unequal heights, there will be a greater number of suspenders on one side of the lowest point of the cable than on the other. Let n_1 be the number in that portion of the span adjacent to the highest tower, and n_2 the number in the other portion; n_1 will be greater than n_2. In this case, then, the platform at the foot of the highest tower will sustain a thrust given by the expression

$$(n_1 - n_2) t \sin \alpha = (n_1 - n_2) wp \tan \alpha.$$

If the platform is to sustain a direct thrust only, at the feet of the two towers it will have to sustain thrusts given by the expressions

$$n_1 t \sin \alpha = n_1 wp \tan \alpha$$
$$n_2 t \sin \alpha = n_2 wp \tan \alpha.$$

If n' represents the number of suspension rods between the centre and any point, the thrust at that point will be

$$n't \sin \alpha = n'wp \tan \alpha.$$

In the case of a suspension cantilever, in addition to the thrust given above there will be one denoted by H, uniform throughout its length. Other calculations for a suspension cantilever are precisely the same as those already given.

The length of the cable from the lowest point to any other point at which the inclination to a horizontal is i, is readily found by means of the formula used for the cable with vertical rods. In the present case the inclination of the cable at any point to a line perpendicular to the axis of the parabola is $(i+\alpha)$; consequently there is simply to be found the length of the parabolic arc between the points at which the inclinations to the axis are $(90-(i+\alpha))$ and $(90-\alpha)$.

The formula mentioned then gives

$$c = \frac{p}{2}\Big(tan\ (i+\alpha)\ sec\ (i+\alpha) - tan\ \alpha\ sec\ \alpha +$$

$$\text{hyp. log.} \frac{tan\ (i+\alpha)+sec\ (i+\alpha)}{tan\ \alpha+sec\ \alpha}\Big) \quad . \quad . \quad (7).$$

It is known from analytical geometry that p takes the following form in terms of the oblique co-ordinates used in this case :

$$p = \frac{x^2\ cos^2\ \alpha}{2y} = \frac{(a-h_1\ tan\ \alpha)^2\ cos^3\ \alpha}{2h_1}.$$

Eq. (7) is, of course, to be applied to both branches of the curve to obtain the total length.

From what was said in the demonstration of the approximate formula, it may be seen that it can be applied to the present case by changing x to $(x+y\ sin\ \alpha)$ and, y to $y\ cos\ \alpha$. The formula then becomes :

$$c = x + y\ sin\ \alpha + \frac{2y^2\ cos^2\ \alpha}{3\ (x+y\ sin\ \alpha)} \quad . \quad . \quad (8).$$

Art. 62.—Suspension Rods; Lengths, and Stresses.

In the following calculations it is virtually assumed that the cable lies in a vertical plane, and that the suspension rods are vertical. This, however, does not affect the generality of the results obtained, for in all cases the suspension rods are supposed parallel to each other, and the lengths found by the formulæ of this Article are to be taken as the vertical projections of the true or actual lengths. The true lengths are therefore to be found by multiplying the values of h_0, h_1, h_2, etc., by the secant of the common inclination to a vertical line, of the suspension rods.

Since a flat parabola nearly coincides with a circle, the camber may be supposed to be formed by a parabolic arc. Let the co-ordinate x be measured from A toward B, in Fig. 14, Pl. XII., and y perpendicular to it; also let $AB = x_1 =$ half span. Then since the curve of the cable is supposed to be a parabola in a vertical plane:

$$y' = y_1 \frac{x^2}{x_1^2}.$$

In the same manner for the camber:

$$y'' = \delta \frac{x^2}{x_1^2}.$$

Then the total length of any suspender **is:**

$$h = y' + y'' + c.$$

When the suspenders are separated by a constant distance, d, simpler formulæ may be found.

Each suspender is composed of the sum of two variable lengths (y' and y'') and a constant length, c. Now if ($y_1 + \delta$) be written for y_1:

$$y = (y_1 + \delta) \frac{x^2}{x_1^2},$$

will evidently be the sum of the two variable lengths referred to. Hence, if h_0, h_1, h_2, h_3 * * *. h_n represent the lengths of the suspenders as shown in the figure:

$$h_0 = c,$$

$$h_1 = c + \frac{d^2}{x_1^2}(y_1 + \delta),$$

$$h_2 = c + \frac{4d^2}{x_1^2} (y_1 + \delta),$$

$$h_3 = c + \frac{9d^2}{x_1^2} (y_1 + \delta),$$

$$h_{n-1} = c + \frac{(n-1)^2 d^2}{x_1^2} (y_1 + \delta)$$

$$h_n = c + \left(\frac{n^2 d^2}{x_1^2} = 1\right)(y_1 + \delta) = c + y_1 + \delta.$$

Having computed the lengths of the suspenders for one half the span, the results may be used for the other half if the piers are of the same height; otherwise the lengths must be computed separately. .

The stress in any suspension rod is the vertical load which it carries, multiplied by the secant of its inclination to a vertical line.

Art. 63.—Pressure on the Tower—Stability of the Latter—Anchorage.

Let P_v = vertical component of pressure on tower head.
" P_h = horizontal " " " " " "
" R = resultant " " " " " "
" T_p and T_p', Fig. 13, Pl. XII., = tensions of the cable on different sides of the pier head.

α, α', and θ represent inclinations to the vertical as shown. When friction on the saddle is considered:

$$P_v = T_p \cos \alpha + T_p' \cos \alpha' ; \quad P_h = T_p \sin \alpha - T_p' \sin \alpha' ;$$

$$R = \sqrt{P_v^2 + P_h^2} ; \quad \cos \theta = \frac{P_v}{R}.$$

When friction on the saddle is not considered, $T_p = T_p'$;

$$\therefore P_v = T_p (\cos \alpha + \cos \alpha'); \quad P_h = T_p (\sin \alpha - \sin \alpha') \quad . \quad (1).$$

$$R = \sqrt{P_v^2 + P_h^2}; \quad \cos \theta = \frac{P_v}{R}.$$

In the same case if $\alpha = \alpha'$;

$P_v = 2W$, $P_h = 0$, $R = 2W$, $\theta = 0$. ($W = \frac{1}{2}$ weight of load and structure.)

There are two cases in which the resultant pressure on the tower, caused by the tension in the cables, may be vertical in direction. Both, however, are founded on the single condition that the horizontal components of the cable tension, on each side of the tower head are equal to each other.

This condition will exist if $\alpha = \alpha'$ in Eq. (1), making $P_h = 0$; or if the saddle be supported on rollers and roller friction be omitted. In the latter case $P_h = 0$ because $T_p \sin \alpha = T_p' \sin \alpha'$, and not because α necessarily equals α'.

In discussing the stability of position of masonry towers, let the distance of the centre of pressure from the centre of figure of the section of the pier be denoted by q'. If this latter does not exceed q (the limit of safety for q'), which may be ascertained by determining the line of resistance for the pier, stability of position will be secured.

It is supposed, of course, that θ has some value greater than zero; otherwise $q' = 0$.

The stability of friction for masonry towers will be secured, at any joint, if the obliquity of the resultant pressure be less than the angle of repose.

Iron and timber towers are to be treated, each as a whole, as long columns, by Gordon's formula.

If the anchorage is a mass of masonry, the stabilities of position and friction are to be considered.

Let $W =$ weight of mass and q' the normal distance from a vertical line through its centre of gravity to the centre of figure of its base. Let T_p equal the tension in anchor chains and p the normal distance from the centre of pressure to its line of direction; and q the distance from the centre of pressure of the base of the foundation to its centre of figure.

Then, in order that stability of position may be secured :

$$T_p p \angle W (q + q')$$

Stability of friction is secured if the greatest obliquity of the resultant pressure on any section (including the base) is less than the angle of repose for the surfaces in contact.

If it were not for friction between the anchor chain and its supports, on the circular part of the chain (see Fig. 15, Pl. XII.), the tension would be the same throughout its whole length; but on account of friction, the tension diminishes on the circular part, from link to link downward, according to the law of friction between cords and cylinders, and is, therefore, the least at the bottom.

The diminution of tension of the anchor chain is computed by this formula:

$$T_p = T_p' E^{f\theta}$$

in which T_p = tension of anchor chain before friction takes effect; T_p' = tension of any point below the first point of support; E = base of *Napierian system of logarithms;* f = coefficient of friction; θ = length of arc considered. The anchorage can be ruptured only by the breaking of chain, or bolt, or plate, or pulling out the whole masonry. The probability of the latter can be determined by comparing the tension of the chain, at the upper surface of the masonry, with the weight of the whole masonry.

Art. 64.*—Theory of the Stiffening Truss—Ends Anchored—Continuous Load—Single Weight.

It has been seen that when a suspension bridge cable carries a load covering the entire span, of uniform intensity

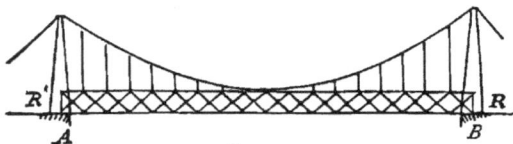

FIG. 1.

per horizontal unit, its centre line forms a parabolic curve. When, however, such a cable carries an isolated weight, or a partially uniform load, it is evident that the centre line of the

---

* This and the following two Arts. form the substance of a paper presented to the Pi Eta Scientific Society, in June, 1879.

cable will assume a form different from that of the preceding parabola, unless such a change is prevented by some special device. Such a special device is the stiffening truss.

The objections to a change of form in the cable of a suspension bridge are of two kinds. Not only would destructive undulations result, but, also, the determination of stresses would become exceedingly complicated and uncertain.

Two cases may arise: the stiffening truss may be securely anchored at its ends; or its ends may simply rest upon supports and be free to rise, in which case there can be no negative or *downward* reaction.

The former case will be considered first.

It is desired to have the cable retain, for all positions of the moving load, the same parabolic form. Now, it has already been seen that such a result can be attained only by assuming a uniform pull on the suspension rods from end to end of the span. Let T be the general expression for this uniform pull for any suspension rod, and let t be its intensity per unit of span, so that if p be the panel length of the stiffening truss, $T = pt$. Let w be the weight per unit of span of the fixed load sustained by the cables. This will, of course, be composed of the weights of the truss, suspension rods and cable or cables. Let w' be the moving load per unit of span; l the span; R the reaction at B; R' the reaction at A, and let the moving load pass on the bridge from B.

Also let x_1 be the distance from B to the head of the moving load; the latter being supposed continuous from B.

Since all the forces acting on the stiffening truss are vertical in direction, there are only two general conditional equations of equilibrium, and those simply indicate that the sum of all the external vertical forces, as well as the sum of the moments of the same, about any point, must be zero.

Those two equations are the following:

$$wl + w'x_1 - tl - R - R' = 0 . \quad . \quad . \quad . \quad (1).$$

$$(w + w')\frac{x_1^2}{2} - t\,\frac{x_1^2}{2} - Rx_1 + (t - w)\frac{(l - x_1)^2}{2} +$$
$$R'(l - x_1) = 0. \quad . \quad . \quad (2).$$

Eq. (2) can be at once written by taking moments about the point x_1 at the head of the moving load.

Eqs. (1) and (2) are the only equations of condition necessary for equilibrium, but they hold three unknown quantities, *i. e.*, t, R, and R'; hence, any one of those three quantities may be assumed at pleasure, and the other two determined from Eqs. (1) and (2). This indetermination simply means that unless another condition be imposed, it cannot be ascertained how much the truss will carry *as a simple truss*, and how much in connection with the cable.

This other condition is virtually the following: *the stiffening truss must act wholly in connection with the cable, and carry no load whatever as an ordinary truss.*

The direct consequence from this condition of the problem is, that the sum of all the uniform upward forces, $T = pt$, must be equal in amount to the sum of all the loads of the kinds w and w'. But the line of action of the resultant of the latter is not, for a partial moving load, the line of action of the resultant of the former: consequently, the truss will be subjected to the action of a couple. In order that equilibrium may be assured, therefore, another couple of equal moment, but opposite sign, must be applied to the truss; the forces of this couple must act at the extremities A and B, and they are nothing more than the reactions R and R'. From this there at once results:

$$R = -R'.$$

This condition, in Eq. (1), gives:

$$t = w + w'\frac{x_1}{l} \quad . \quad . \quad . \quad . \quad . \quad (3).$$

Eq. (2) gives:

$$R = -R' = \frac{w'x_1}{2}\left(1 - \frac{x_1}{l}\right). \quad . \quad . \quad (4).$$

Eq. (4) shows that both reactions, R and R', are zero for $x_1 = l$, or for $w' = 0$.

It is also seen *that R and R' are always numerically equal, but have opposite directions, hence R' is a downward reaction, and its maximum value will indicate the amount of anchorage required at each end of the truss.*

Using Eq. (4):

$$\frac{dR}{dx_1} = \frac{w'}{2} - \frac{w'x_1}{2l} - \frac{w'x_1}{2l} = 0, \qquad \therefore \; x_1 = \frac{l}{2}.$$

Putting $x_1 = \dfrac{l}{2}$ in Eq. (4):

$$R = \frac{w'l}{8} \quad . \quad . \quad . \quad . \quad . \quad . \quad (5).$$

Eq. (5) shows the maximum value of $(-R')$ and gives the amount of anchorage required at either end of the truss; it also shows the greatest shear to be provided for at either end of the truss.

The general value for the shear at any section of the portion of the truss covered by the moving load is:

$$S = R + tx - wx - w'x \quad . \quad . \quad . \quad (6).$$

Or,

$$S = \frac{w'x_1}{2} - \frac{w'x_1^2}{2l} + x\left(\frac{x_1}{l} - 1\right)w' \quad . \quad . \quad (7).$$

This value of S shows it to be positive near the end of the bridge; it then decreases as x increases, passes through the value zero, and then increases as a negative quantity. As a negative quantity it attains its maximum value for $x = x_1$; it then becomes:

$$S_1 = -\frac{w'x_1}{2}\left(1 - \frac{x_1}{l}\right) = R' \quad . \quad . \quad (8).$$

Hence the two reactions at the ends of the truss and the shear at the head of the moving load are always numerically equal.

Eq. (8), consequently, takes its maximum value for $x_1 = \dfrac{l}{2}$, and that value is given in Eq. (5); this last equation, therefore, gives the maximum shear which is to be provided for at the head of the moving load.

Now since this maximum shear is to be provided for at both ends and at the middle of the truss, it would probably be advisable in all ordinary cases to design all the web members of the truss to be of uniform size, and capable of carrying this maximum shear, although there would then be a little waste of material in the vicinity of the quarter points of the span on each side of the centre.

This supposes, of course, that the chords of the stiffening truss are parallel and horizontal. If the chords are not parallel the amount of shear carried by the web members will depend on the inclination of one or both the chords.

For all values of x and x_1, for the portion of the span covered by the moving load, the total shear will be given by Eq. (7).

For the portion of the span not covered by the moving load the general value for the shear is (*measuring x from A*):

$$S = -R' + w x - t x = \frac{w' x_1}{2} - \frac{w' x_1^2}{2l} - \frac{w' x_1 x}{l} \quad . \quad . \quad (9).$$

This expression attains its greatest values for $x = 0$ and $x = l - x_1$. In the first case the shear becomes $-R'$, and in the second R. These results show nothing new.

Since the maximum shear in a simple truss, of the span l and uniform loading of intensity $(w + w')$, is $\frac{1}{8}(w + w')l$, it is seen that the maximum shear in the stiffening truss of same span is only one-fourth of that due to the moving load alone in the case of the simple truss.

The general value of the bending moment to which the truss is subjected, for the portion covered by the moving load, is:

$$M = Rx - (w + w' - t)\frac{x^2}{2} \quad . \quad . \quad . \quad (10).$$

Eq. (10) shows that if $x = x_1$, the bending moment is equal to zero. Hence, *at the head of the moving load, for all its positions, there is a section of contraflexure or no bending, and consequently the loaded and unloaded portions of the stiffening truss are each in the condition of a simple beam supported only at each end, and loaded uniformly throughout its length.*

If the values of R and t from Eqs. (3) and (4) be inserted in Eq. (10), and if $\dfrac{dM}{dx}$ be put equal to zero, it will be found that the bending moment has its maximum value for $x = \dfrac{x_1}{2}$; as might have been anticipated.

Putting $x = \dfrac{x_1}{2}$ in Eq. (10):

$$M = \frac{w' x_1^2}{8}\left(1 - \frac{x_1}{l}\right) \quad \ldots \ldots \quad (11).$$

By differentiating in respect to x_1 it will be found that M has its maximum value for $x_1 = \tfrac{2}{3}l$. Denoting this value of M by M_1, there results:

$$M_1 = \frac{w' l^2}{54} \quad \ldots \ldots \ldots \quad (12).$$

Eq. (12) shows the maximum bending moment to which any loaded portion of the truss can be subjected.

If a simple truss supported at each end be subjected to the action of a uniform load, of the intensity $(w + w')$, throughout its entire length, the greatest bending moment will be:

$$M' = \frac{(w + w')\, l^2}{8} = \frac{w' l^2}{8}\,(\text{if } w = 0);$$

$$\therefore \quad M_0' = \tfrac{4}{27}\, M_0' > \tfrac{1}{4}\, M_0' \quad \ldots \quad (13).*$$

From what has already been shown it is evident that the

---

* The subscript o indicates that $w = 0$ in M'.

greatest bending moment for the portion of the truss not covered by the moving load will occur at the distance $\left(\dfrac{l-x_1}{2}\right)$ from the reaction R'. The general value, therefore, for the greatest moment for that portion will be :

$$M = \tfrac{1}{2} R' (l - x_1) + (t - w)\frac{(l - x_1)^2}{8} \quad . \quad . \quad (14).$$

Putting the differential coefficient of M in respect to x_1 (after inserting the values of R' and t) equal to zero, there results :

$$x_1 = \tfrac{2}{3}l \pm \sqrt{\tfrac{4}{9}l^2 - \tfrac{3}{9}l^2}$$

$$\therefore \quad x_1 = \tfrac{2}{3}l \pm \frac{l}{3} = l \text{ or } \tfrac{1}{3}l \quad . \quad . \quad . \quad . \quad (15).$$

The latter value $(\tfrac{1}{3}l)$ gives a maximum, and inserted in Eq. (14):

$$M_1' = -\frac{w'l^2}{54} = -M_1 = -\tfrac{1}{4}M_0' \text{ (nearly)} \quad . \quad . \quad (16).$$

Eqs. (12) and (16) show that *the greatest bending moments, to which the stiffening truss is subjected, are equal, but of opposite kinds ; and it is seen that they occur when one-third or two-thirds of the span are covered by the moving load.* The chords, therefore, of the stiffening truss must be designed to resist both tension and compression.

Eqs. (10) and (14) are general expressions for all the bending moments to which any portion of the truss can possibly be subjected, but in all ordinary cases it would probably be best to make the chords uniform in section (supposing the depth of the truss to be constant) from end to end, and capable of resisting the moments given by Eqs. (12) and (16).

Eq. (13) shows that the greatest bending moment to which a stiffening truss can be subjected is only $\tfrac{1}{4}$ of that found in a simple truss supported at each end and loaded with a uniform load equal in intensity to that of the moving load on the stiffening truss.

It may be interesting to notice that the resultant load on the portion x_1 of the truss is *downward*, since $(w + w') >$ $\left(t = w + \dfrac{w' x_1}{l} \right)$: but that that on the portion $(l - x_1)$ is *upward*, since $w < t$.

If the bridge is traversed by a single concentrated load or weight, W, the general method of procedure is precisely the same as before. Let the weight W' pass on the bridge from the end B, in the figure, and let x_1 denote its distance from that point; also measure x from the same point.

The general equations of condition are:

$$R + R' + tl - wl - W = 0 \quad . \quad . \quad . \quad (17).$$

$$- R'(l - x_1) + (w - t) \frac{(l - x_1)^2}{2} - (w - t) \frac{x_1^2}{2} + R x_1 = 0 . \ (18).$$

Eq. (18) is written at once by taking moments about the point of application of W.

By precisely the same method as before, there may be found the result, $R = - R'$.

In Eqs. (17) and (18) let there be put $R = - R'$, then there results :

$$t = w + \frac{W}{l} \quad . \quad . \quad . \quad . \quad . \quad . \quad (19).$$

$$R = - R' = W\left(\tfrac{1}{2} - \frac{x_1}{l} \right) \quad . \quad . \quad . \quad (20).$$

When $x_1 = \dfrac{l}{2}$, $R = 0$ and $R' = 0$

Eq. (20) shows that *the reaction nearest the weight W will always be positive, or upward; and that the other will be negative, or downward.*

The maximum value of R or R' is found (by making $x_1 = 0$ in Eq. (20)) to be $\dfrac{W}{2}$, and *that is the amount of anchorage required for the weight W, alone, at each end of the truss.*

The point of application of the weight W divides the span into two segments.

The general value of the shear for the shorter segment is:

$$S = tx - wx + R = \frac{W}{l}x + W\left(\tfrac{1}{2} - \frac{x_1}{l}\right) \cdot \cdot \ (21).$$

This has its greatest value for $x = x_1$; it then becomes:

$$S' = \frac{W}{2}.$$

Hence S' is the uniform maximum upward shear that must be provided for, throughout the whole length of the truss.

The general value of the shear in the longer segment of the span is:

$$S = R' - w(l - x) + t(l - x) = W\left(\tfrac{1}{2} + \frac{x_1}{l} - \frac{x}{l}\right) \cdot \cdot \ (22).$$

This expression attains a positive maximum for $x = x_1$, that being the least positive value of x admissible; the resulting value of S is $\tfrac{1}{2}W$, which shows nothing new.

The negative maximum for $x = l$ is simply the reaction R'. Putting $S = 0$ in Eq. (22) there results:

$$x = x_1 + \frac{l}{2}.$$

Hence for all points of the span between $\dfrac{l}{2} + x_1$ and l there will be negative or downward shear. The maximum negative shear, however, to be provided for, is shown by Eq. (22) to exist in one-half of the truss *when W rests on the opposite end;* or, when in that equation $x_1 = 0$ and $x > \dfrac{l}{2}$. The negative or downward shear to be provided for has, then, for its general expression :

$$S_1 = - W\left(\frac{x}{l} - \tfrac{1}{2}\right) \quad . \quad . \quad . \quad (23).$$

Eq. (23) is to be applied to each half of the truss, and it is also seen that the web members which take up S_1 should increase, in ultimate resistance, uniformly from the centre to the ends of the truss ; supposing the chords to be parallel and horizontal.

In many cases, however, it may be best to design them of uniform dimensions belonging to those at the ends.

The bending moment for any point of the smaller segment of the truss is :

$$M = Rx + (t - w)\frac{x^2}{2} = W\left(\tfrac{1}{2} - \frac{x_1}{l}\right)x + \frac{Wx^2}{2l} \quad . \quad . \quad . \quad (24).$$

Since both terms of this moment are positive, it will attain its greatest value for $x = x_1$; it then becomes :

$$M_1 = \frac{W}{2}\left(x_1 - \frac{x_1^2}{l}\right) \quad . \quad . \quad . \quad . \quad (25).$$

Eq. (25) gives the general value of the greatest positive bending moment to which any point of the truss will be subjected, *for which case x_1 must never be made greater than $\tfrac{1}{2} l$.*

The bending moment M_1 will evidently cause compression in the upper chord, and tension in the lower.

For the longer segment of the truss the general value of the bending moment is :

$$M = R'(l - x) + (t - w)\frac{(l - x)^2}{2}.$$

Or,

$$M = - W\left(\tfrac{1}{2} - \frac{x_1}{l}\right)(l - x) + \frac{W}{l}\frac{(l - x)^2}{2} \quad . \quad (26).$$

20

There is evidently a point of contra-flexure for the longer segment of the truss, for if the second member of Eq. (26) be put equal to zero, there results $x = 2x_1$. Hence *all the portion $l - 2x_1$, of the truss, will be subjected to a negative bending moment causing tension in the upper chord and compression in the lower.*

In Eq. (26), putting $\dfrac{dM}{d(l - x)} = 0$, there results :

$$x = x_1 + \frac{l}{2} \quad . \quad . \quad . \quad . \quad . \quad (27).$$

This value of x, in Eq. (26) gives :

$$M' = -\frac{Wl}{8} + \frac{Wx_1}{2} - \frac{Wx_1^2}{2l} \quad . \quad . \quad . \quad (28).$$

Eq. (28) gives the general value for the maximum negative bending moment at any point in the entire truss, for any position of the weight W. In Eq. (28), it is to be remembered, x_1 must always be less than $\dfrac{l}{2}$; also, that the point at which M' exists will be given by the value of x in Eq. (27).

The formulæ for a continuous load, taken in connection with those for a single weight, will give all the circumstances of bending or shearing which can exist with any condition or position of loading.

When the "shear" has been determined for any section, the stress in the web member which is to carry it (if the chords are parallel and horizontal) will be found at once by multiplying that "shear" by the secant of inclination of the web member to a vertical line. If there are two or more systems of triangulation in the truss, then each system is to be treated as a single truss in the usual manner.

If desirable, after the reactions R and R' and the upward load $T = pt$ are known, the stresses in the individual members of the stiffening truss can be traced as in the case of an ordinary truss supported at each end.

Art. 65.—Theory of the Stiffening Truss—Ends Free—Continuous Load —Single Weight.

In this Article the notation of the previous one will be continued, and the same figure will be referred to.

The case of a continuous load will first be treated, and, as before, it will be supposed to pass on the bridge from B.

Since the ends are not anchored, in this case there can be no negative or downward reaction, *consequently R' will be zero.*

As before, putting the sum of all the vertical forces acting on the truss equal to zero, and taking moments about the head of the moving load, there result the two general Equations of condition :

$$wl + w'x_1 - tl - R = 0 \quad \cdots \cdots \quad (1).$$

$$(w + w' - t)\frac{x_1^2}{2} - (w - t)\frac{(l - x_1)^2}{2} - Rx_1 = 0 \quad \cdots \quad (2).$$

Since there are now but two unknown quantities, R and t, the problem is perfectly determinate. Eqs. (1) and (2) give :

$$R = w'x_1\left(1 - \frac{x_1}{l}\right) \quad \cdots \cdots \quad (3)$$

$$t = w + \frac{w'x_1^2}{l^2} \quad \cdots \cdots \quad (4)$$

The general value for the shear at any section of the truss, for the portion covered by the moving load, is

$$S = R + (t - w - w')x = w'(x_1 - x) - w'\frac{x_1^2}{l}\left(1 - \frac{x}{l}\right) \quad \cdot \quad (5).$$

Evidently S has its maximum positive value for $x = 0$; its greatest negative value for $x = x_1$, and the value zero for $x = \frac{lx_1}{l + x_1}$.

In order to find the head of the moving load for that position which makes R a maximum, let $\dfrac{dR}{dx_1}$ be put equal to zero.

There is then found $x_1 = \dfrac{l}{2}$. Hence *the moving load covering half the span gives the maximum reaction R.*

Placing $x_1 = \dfrac{l}{2}$ in Eq. (3), the greatest value of R becomes $R_1 = w'\ l \div 4$.

In order to determine the greatest web stresses it is necessary to find the greatest shear at any point whose abscissa is x. This maximum shear at once results by placing the first derivative of S in respect to x_1, from Eq. (5), equal to zero. That operation gives:

$$ 1 - \frac{2x_1}{l}\left(1 - \frac{x}{l}\right) = 0 \therefore x_1 = \frac{l^2}{2(l - x)}. $$

By the introduction of this value of x_1 in Eq. (5), the greatest shear for any section located by x becomes:

$$ max.\ S = w'\left\{\frac{l^2}{4(l - x)} - x\right\}. $$

It is clear that this value is a maximum for the reason that $\dfrac{d^2 S}{dx_1^2}$ is a negative quantity in which x_1 does not appear.

It is further evident that *max. S* is a positive, or upward shear, from the fact which was observed at the bottom of the preceding page, that the greatest negative shear occurs at the head of the moving load. By making $x = x_1$ in Eq. (5) that greatest negative shear becomes:

$$ S_1 = -w'\frac{x_1^2}{l}\left(1 - \frac{x_1}{l}\right). $$

It will be necessary to apply *max. S* and S_1 to a half of the span, regarding the shear for a positive direction on one side of the centre as negative for the other. These two values of the shear will enable all the greatest web stresses to be determined.

Since $R' = 0$, the whole truss will be subjected to bending moments of the same sign; such bending moments, in fact, as will put the upper chord in compression and the lower one in tension.

The general value of the bending moment for that portion of the truss covered by the moving load is,

$$M = Rx - (w + w' - t)\frac{x^2}{2},$$

$$\therefore\quad M = w'x_1x - \frac{w'x_1^2x}{l} - \frac{w'x^2}{2} + \frac{w'x_1^2x^2}{2l^2} \quad . \quad . \quad (6).$$

Since the shear S is zero for $x = \dfrac{lx_1}{l + x_1}$, that value of x in Eq. (6) will give the maximum value of M. This latter is :

$$M_1 = w'x_1^2\frac{l - x_1}{2(l + x_1)} \quad . \quad . \quad . \quad . \quad (7).$$

Putting $\dfrac{dM}{dx_1} = 0$, there is found,

$$x_1 = \frac{l}{2}(-1 \pm \sqrt{5}) = +0.618l.$$

The absolute maximum bending moment exists, therefore, when the moving load covers 0.618 of the span. That moment has for its value :

$$0.0451w'l^2 \text{ (nearly).} \quad . \quad . \quad . \quad . \quad (8).$$

If $x_1 = 0.618l$ be put in the expression $x = \dfrac{lx_1}{l + x_1}$, there will result $x = 0.382l$, nearly.

Eq. (6) gives the general value of the bending moment for any position of the load, but it would probably be the most convenient to make the moving load cover $0.618l$, and design the chords for the distance $0.382l$ from each end to resist the bending due to that position of the load, and then design all of the middle $2(0.5 - 0.382)l = 0.236l$ to resist the moment given by the expression (8). By this arrangement there would be a little surplus of material at the middle of the truss.

If a single weight rests upon the bridge, the two general equations of equilibrium, obtained in precisely the same manner as heretofore, are :

$$wl + W - tl - R = 0 \quad \ldots \quad (9).$$

$$(w - t)\frac{x_1^2}{2} - (w - t)\frac{(l - x_1)^2}{2} - Rx_1 = 0 \ \ldots \ (10).$$

These equations then give :

$$t = w + \frac{2Wx_1}{l^2}. \quad \ldots \quad (11).$$

$$R = W\left(1 - \frac{2x_1}{l}\right). \quad \ldots \quad (12).$$

If $x_1 = \frac{l}{2}$, $t = w + \frac{W}{l}$, and $R = 0$.

The general values of the shear S, and moment M, are the following :

$$S = R + (t - w)x,$$

$$\therefore \ S = W\left(1 - \frac{2x_1}{l^2}\right) + \frac{2Wx_1x}{l^2}. \quad \ . \ (13).$$

$$M = Rx + (t - w)\frac{x^2}{2}$$

$$\therefore \ M = W\left(1 - \frac{2x_1}{l}\right)x + \frac{Wx_1x^2}{l^2} \quad \ . \ . \ (14).$$

Eqs. (13) and (14) show, since x_1 must not be taken less than x, that if the maximum shear and bending moment are desired for any section, the weight, W, must be placed at that section, and x be made x_1 in those equations.

That section at which the bending moment will attain its absolute maximum value is found by putting $x = x_1$ in Eq. (14), then taking the first differential coefficient of M in respect to x_1, equating to zero and solving. There results:

$$x_1 = \tfrac{2}{3}l \pm \tfrac{1}{3}l = \tfrac{1}{3}l.$$

This value in Eq. (14), when $x = x_1$, gives:

$$M = \tfrac{4}{27}Wl.$$

The equations for the continuous and single moving loads, used in combination, will give moments and shears for any character and position of loading whatever.

The general observations in regard to finding web and chord stresses, at the close of the last Article, apply equally well to this case.

As was to be anticipated, R and R' in the two preceding Articles, are independent of the fixed load w.

Art. 66.—Approximate Character of the Preceding Investigations—Deflection of the Truss.

In the two preceding Articles it has been virtually assumed that the deflection of the cable, due to its lengthening under stress, is just sufficient to allow the truss to take the deflection due to the loads $T = pt$, w', and W in the different cases.

Such, however, is not really the case.

In all ordinary cases, the cable does not deflect to that extent. The result is, as is evident, that the truss is not subjected to the amount of bending assumed. The error, however, is only a small one, and on the safe side, as should be the case.

It has been found by experiment that if the ends of the truss are anchored, the stiffening truss will be subjected to a maximum moment equal to that existing in an ordinary truss supported at each end, of about one-eighth the span, and carrying load over its entire length of the same amount as that of the moving load on the suspension bridge.

This same case, treated analytically, as was seen, gave about $\frac{1}{4}$ instead of $\frac{1}{8}$.

Approximate values of the deflection of the stiffening truss can be found by the ordinary formulæ used for solid beams in the subject of resistance of materials, in the different cases, where t, R, R', and the moving load are known.

CHAPTER X.

Art. 67.—Classes of Bridges—Forms of Compression Members—Chords Continuous or Non-continuous.

REGARDING the systems of construction, truss-bridge structures are divided into two classes at the present time, *i. e.*, bridges with "pin connections" and bridges with "riveted connections." In the former class the connection of web members with the chords or with each other is made by a single pin only, as in Fig. 3, Pl. III. (the figure shows simply the upper chord and tension web members together with one end post). The pins are shown at 1, 2, 3, 4, and 5, where the tension members join the chord. In the latter class the connections mentioned are made by means of rivets, as shown in Fig. 6, Pl. XI. In that figure A is a tension, and B a compression web member, while C is a portion of the lower chord.

Screw connections for tension web members and simple abutting connections for compression ends have been used, but are not usually employed at present.

A screw connection is formed by passing a tension member through the chord at one of the joints, and placing upon the end of it a nut ; this method, therefore, can only be conveniently used when the tension members are of circular cross-section.

An abutting connection for a compression member is formed by simply abutting either end against the chord, which is properly formed for the purpose at the joint. The end of the post or strut is inserted in the chord, or else a projection of the chord passes into the end of the post ; or, again, some simple device is employed for the purpose of

313

keeping the ends of the post in position, and for nothing else, as the entire compressive stress in the post or strut is transmitted through the abutting surfaces.

Occasionally screw, abutting, and pin connections are combined in a single bridge.

Without regarding systems of construction, truss-bridges are divided into:

" Deck " bridges, *i. e.,* the applied load is on the chord in compression.

" Through " bridges, *i. e.,* the applied load is on the chord in tension.

" Pony " trusses are through truss-bridges when the trusses are not sufficiently high (or deep) to need overhead cross-bracing, and they are seldom put up for spans of over eighty feet, although there are examples of one hundred feet (and even more) in length.

The lateral stability of such long pony trusses, however, is very precarious.

Figs. 1 and 2 of Pl. XI. show the ordinary forms of plate girder, stringers and floor beams, with plate hangers at the ends of the latter.

The various forms of cross-sections of upper chords and posts, and compression members generally, are almost innumerable, and subject only to the fancy of the engineer or builder. The principle which should always be kept in view is this: That the material should be as far as possible from the neutral axis. Figs. 3, 4, 5, and 6 of Pl. III. show methods of building up the upper chord. It consists of riveting plates to a pair of channel bars, or to a pair of channel bars and an ⏉ beam.

The bars or beams are frequently built of plates and angles.

The blackened portions represent sections.

Chords are continuous or non-continuous according as they are built up in a continuous manner from end to end, or built up of panels abutting against each other at the panel points. The former are principally used at present.

Tension members are always of rectangular or circular section. Fig. 5, Pl. XII., is a lower chord "eye-bar." In writing of channel bars, **I** beams, angle-irons, iron bars and rods, they are indicated as follows: **C, I, L, ⊏⊐, O**; that is, by skeletons of their sections.

Art. 68.—Cumulative Stresses.

Stresses are said to be cumulative in any part of a structure when they are transmitted through that part to other parts, whose whole duty is to sustain them, the part in question being subject at the same time to its own stress. The member in which the stresses are cumulative is, therefore, overstrained to some extent in some one or more portions of it. The two channel bars in Fig. 3, Pl. III., are the portions of that upper chord which are subjected to cumulative stresses. If $C'C'$ is supposed to be the centre line of the bridge, then the compressive stress in the chord increases as $C'C'$ is approached from the end, in consequence of the components in the direction of the centre line of the chord of the stresses in the inclined ties, A, B, C, etc. This increase of stress is provided for by riveting plates to the upper flanges of the two **C**s, as shown in the figure. It is evident that the plates do not receive the stress which they are intended to bear, except indirectly through the **C**s and the rivets which connect the latter with the plates. Now since the **C**s are supposed to have their own share of direct compressive stress to sustain, it is plain that the material in the vicinity of the front of the pin (looking from the centre of the pin toward the centre of the bridge) is subjected to a much greater intensity of compressive stress than should exist in the structure. This relates only to the material in front of the pin, and that is the only vicinity in which cumulative stresses would exist if the plates could be so securely riveted to the **C**s that the whole chord could be depended upon to act as one piece. In practice, however, no such riveted work exists. The **C**s must inevitably yield to some extent before they bear sufficiently on the rivets to give to the plates their proper share of the stress. The result is that not only the material in front of the pin but the whole of the [s are overstrained by these cumulative stresses. The only remedy is to so proportion the chord that those parts which are designed to sustain stress shall receive it immediately, and not indirectly through some other part.

The Fig. 5, Pl. III., shows a method of accomplishing this object. The plate *ab* is a light one riveted to the top flanges of the [s, and extends throughout the whole length of the chord. The increase of the areas of cross-sections are obtained by riveting plates to the flat sides of the [s, and by adding an I, if necessary, as shown. The parts of the chord thus receive stresses immediately from the pins, and cumulative stresses are obviated.

It is also evident that if the stresses are applied to the centres of gravity of the cross-sections, or parts of the cross-section, no cumulative stresses will exist.

Cumulative stresses are as liable to occur in riveted connections as in pin connections ; in fact, more so. It may be said to be impracticable to so construct riveted work that cumulative stresses will not exist, and in this respect pin connections have the advantage of riveted connections. Fig. 6, Pl. XI., illustrates the matter for a riveted chord. The plate *C* is common to the whole chord, and all web members are riveted to it, as shown by *A* and *B*, so that before the rivets can take their share of the stress in transferring it to the plate and Ls *ED*, it (the plate *C*) will necessarily yield to such an extent that cumulative stresses will exist throughout its whole length to a greater or less degree.

Art. 69.—Direct Stress Combined with Bending in Chords.

If direct stress is not applied to the centres of gravity of the ends of a piece subjected to compression, it is clear that bending must take place.

In Figs. 3 and 4 of Pl. III., the horizontal components of the oblique forces in the ties *A*, *B*, etc., do not act through the centres of gravity of the sections of the chord, hence there must be a bending in those chords. If the chords were perfectly straight, and if the centres of the pin-holes were all at the same distance from the centres of gravity of the different cross-sections, as well as in the same straight line, then the total direct stress to which the chord is subject at any section would produce bending at that section, and the lever-arm would be the same for all sections. Camber and deflection from loading, however, so complicate the matter that it is quite impossible to make even a satisfactory approximate computation of the chord bending arising from this cause. All chords in compression, therefore, should be so designed that the axes of the pins may traverse the centres of gravity of their sections, even though the ties rest directly on the upper chord. It is clear that when this bending exists, the proper distribution of direct stress is greatly disturbed, though to an indeterminate extent ; and is, except in most rare cases, a very faulty construction.

If it be supposed that the total direct stress in the chord acts as if the latter were perfectly straight, so that it all produces flexure ; and if it then be supposed that the increment only, at each pin or panel point produces flexure in the adjacent panel ; it is evident that the first supposition will make the flexure the greatest possible, while the second will make it the least possible.

It is farther evident that if cumulative stresses occur, flexure must necessarily exist, for the simple reason that the direct stress is not uniformly distributed over the cross-section of the chord.

Although this flexure is indeterminate in amount and shows a faulty design, the attempt to utilize it has sometimes been made, and the analysis on which the practice was based will now be given, it being premised that the ties are supposed to rest directly on the chords, as shown in Figs. 4 of Pls. III. and I.

Suppose, in Pl. III., Fig. 6 to be an enlarged cross-section of the chord in Fig. 4, and let fg pass through the centre of gravity of the cross-section, being parallel to ab and cd'; then, since the increment of the chord stress transmitted through the pin from the ties AA is applied to the cross-section of the chord at a distance h below the centre of gravity, there will be an excess over the uniform intensity of stress in the cross-section at the lower side EE, and a deficiency at the upper side ab.

This excess or deficiency (the same in amount, of course, for certain sections only) is found in a very simple manner, as follows:

Let P be the increment of direct compressive stress given to the chord by the ties AA, and let P_1 be the total direct compressive stress in the section. Put S for the area of the cross-section.

Now the variation of the intensity of stress from the mean is due to the moment Ph, and since this moment is constant for all points between any two pins, as 1—2 or 2—3, Fig. 4, Pl. III., the variation in intensity is also constant between these points.

The moment $Ph = \dfrac{RI}{d_1}$, in which d_1 equals the distance from fg to EE, and R the intensity of stress at EE due to bending, gives

$$\therefore\quad R = \frac{Phd_1}{I}.$$

The intensity of stress at ab due to bending is, of course, equal to

$$\frac{d - d_1}{d_1}\,R.$$

Now the total intensity of stress at EE, Fig. 6, Pl. III., is equal to $\dfrac{P_1}{S} + R$; and that at ab, $\dfrac{P_1}{S} - R\dfrac{d - d_1}{d_1}$; the intensity at $fg = \dfrac{P_1}{S}$, whatever may be the figure of the cross-section. The variation of intensity at any point in the section may be easily found from R by a simple proportion, and the total intensity by adding that to $\dfrac{P_1}{S}$.

As a first case, let the chord be a non-continuous one, so that each panel, so far as the panel moving load is concerned, is a simple beam supported at each end.

If the load rests on the upper chord immediately, as shown in Fig. 4, Pl. III., it will produce tension at the lower side of the chord and compression at the upper by simple flexure, an opposite tendency to that exerted by the moment Ph.

The moments due to the moving load on any panel vary (that is, increase) from the joints to the middle point of the panel, where, of course, the moment is maximum. Denote by R' the greatest intensity of the tensile stress caused by the moment of the moving load, then

$$R' = \frac{d_1 \Sigma w x_1}{I},$$

in which $\Sigma w x_1$ expresses the greatest moment of the applied load.

For a uniform load :

$$\Sigma w x_1 = \frac{w l^2}{8},$$

and it exists at the centre of the panel. Now, ordinarily, the chord would be required to resist the bending moment expressed by $\Sigma w x$, but h may be so chosen that for its maximum value $R = R'$, and then no extra metal will be required on account of the flexure produced by the moving load. This value of h is found as follows: put

$$\frac{P h d_1}{I} = \frac{d_1 \Sigma w x_1}{I}; \quad \therefore \quad h = \frac{\Sigma w x_1}{P}.$$

In all ordinary cases of uniform load

$$\Sigma w x_1 = \frac{w l^2}{8},$$

in which w is the intensity of uniform load.

$$\therefore \quad h = \frac{wl^2}{8P}.$$

The value of h, therefore, is independent of the form of cross-sections. If $R < R'$, additional material will be needed in order to prevent an excess of compressive stress at the upper part of the chord, and a deficiency at the lower side. When $R > R'$, there is an excess of compressive stress at EE, Fig. 6, Pl. III.

When the lower chord sustains the moving load directly, as in Fig. 4, Pl. XI., the only change arises from this: That it is in tension instead of compression, and h is measured above the centre of gravity of the cross-section, instead of below it ; also, the section is rectangular.

In the case of the lower chord, however, if the stress in one panel is given to the adjacent one through the medium of a pin, then the total stress in the panel under consideration must be put for P in the formulæ above. This must also be done in every case where the total chord stress produces bending. The chord stress used may be taken as the maximum (that which exists when the moving load covers the whole truss), for in all other cases there is a surplus of material with which to resist the bending.

This method of neutralizing the flexure produced by the direct application of the moving load to the chords is very unsatisfactory in many ways and should never be used. At all places in the panel except the centre the metal is still over-strained by the flexure due to its own stress, and in the vicinity of the panel points this condition exists to a very serious extent. When this consideration is coupled with the great uncertainty attached to the hypothesis on which the analysis is based, the unsatisfactory character of the method is sufficiently evident to effect its exclusion from the best practice.

If the chord is continuous, the objections to the method already mentioned gather considerably increased force. At and near the ends of the panels the fixed and moving load produce flexure in the same direction as the direct chord stress, and to twice the amount of that at the centre. It is not necessary, therefore, to consider this case farther.

A considerable saving of material can be effected by placing the ties directly on the upper chord of a deck bridge, and with a proper design it in no manner conflicts with the best practice. *In all cases the axis of the pin should traverse the centre of gravity of the chord section*, or as nearly so as practicable, in order, if possible, to eliminate all flexure due to the direct chord stress. The chord section should then be so formed that the combined stresses due to flexure in the exterior fibres, and the direct chord stress shall at no point exceed a proper value per square unit. This value may be taken at 8,000 to 9,000 pounds per square inch for wrought-iron upper chords, or 10,000 to 11,000 for mild steel members of the same kind. These values may be taken comparatively high for the reason that an indefinitely small portion only of the material is subjected to these intensities, and that small portion is well supported against fatigue by the material about it, which is considerably understrained.

If the notation previously used in this Art. be still maintained, the maximum external moment Σwx_1 will develop in the most remote fibres at the distance d_1 from the neutral axis the intensity.

$$R' = \frac{d_1 \Sigma wx_1}{I}. \quad \ldots \ldots \ldots \ldots \quad (1).$$

If, on the other hand, P_1 is the total direct stress in the chord and S the area of cross section, while p is the greatest allowable combined stress, then there will result : ·

$$p = \frac{d_1 \Sigma wx_1}{I} + \frac{P_1}{S}. \quad \ldots \ldots \ldots \quad (2).$$

Eq. (2) shows that in the most efficient design the moment of inertia I of the section must be the greatest possible. At the same time considerations affecting the joint details render

21

it advisable that the centre of gravity should lie not far from the mid-depth of the section. These two conditions are fulfilled by placing large quantities of the material, and as nearly as possible in equal amounts, at the top and bottom of the chords, as shown in Fig. 18 of Pl. XII. The cover plate *bc* and angles *dd* are made as light as the circumstances of proper design will permit, but the angles *aa* are made as heavy as possible. An unequal-legged angle is a very good one for *aa* with the longest leg horizontal, and it is sometimes necessary to rivet a narrow plate to those horizontal legs in order to properly balance the section. The centre of gravity line *fg* will usually lie a little above the centre of figure.

As the centre of the span is approached from the end the chord section must be rapidly increased but *in no case should that increase be made by thickening the cover plates or increasing their number*, as such an operation inevitably means cumulative stresses or flexure by direct stress. In rare cases it may be admissible to slightly thicken a cover plate, if there is but one, but, as a rule, Fig. 6 of Pl. III., shows a design to be carefully avoided. All increase of section should be obtained by thickening the side or web plates, or increasing their number; or, again, by increasing the angles, or, finally, by introducing an interior eye-beam, as shown in Fig. 5, Pl. III.

If the chord is non-continuous, Σwx_1 is simply the bending for a span equal in length to a panel and due to the track load, own weight and superimposed moving load, and is easily determined. If the chord is continuous, on the contrary, the analysis for the moving load bending is not simple. The total bending for this case, however, may properly and safely be taken at three-fourths its value for a non-continuous chord.

The upper chord section required in the case of combined bending and direct stress is readily found by the aid of Eq. (2). The radius of gyration r can be easily and with sufficient accuracy predetermined; so that Sr^2 can be put for I in that equation. After that substitution is made, there at once results:

$$S = \frac{1}{p}\left(\frac{d_1 \Sigma w x_1}{r^2} + P_1\right) \quad . \quad . \quad . \quad . \quad . \quad . \quad . \quad . \quad (3).$$

This is a very convenient formula for practical use.

Art. 70.—Riveted Joints and Pressure on Rivets.

In riveted bridge work the pitch of rivets (*i. e.*, the distance from centre to centre) should not be less than three diameters, although it sometimes is; if possible it should be from four to eight diameters of the rivet, provided that value does not exceed about fourteen or sixteen times the plate thickness. The diameter of the rivet is determined by the amount of stress which the joint is to carry, so that the intensity of pressure against the surface of the rivet in contact with the plate shall not exceed a given value. If the rivet and hole were in ideally perfect contact, this intensity could easily be found, having given the amount of stress which the rivet is to carry. But such is never the case. The only resort left, therefore, is to assume that the rivet does fit perfectly, and fix a low enough value for the intensity of pressure against its surface to make the joint safe.

In Fig. 1, Pl. XII., suppose *EF* to be a part of a plate in which is drilled or punched the rivet-hole *ADBK*, and suppose the stress to be exerted on the plate in the direction of the arrow at *K*, then the surface of contact between the plate and rivet will be projected in *ADB;* contact will not take place throughout the whole semi-circumference when the plate is not subjected to stress, unless the rivet fits the hole with absolute accuracy.

Since all material is elastic to some degree, there will be a surface of contact when the plate is subject to stress, even when the rivet does not accurately fit the hole, and this surface will evidently increase with the stress in the plate.

But suppose that the rivet fits the hole exactly, then the pressure on the surface of contact, *ADB*, will be of uniform intensity, and the case will be similar to that of fluid pressure on a cylinder. Let p denote this intensity whose direction is normal to *ADB* at every point of it (friction is omitted from

consideration), then the total pressure in the direction of the
arrow at K exerted by the plate on the rivet for each unit of
length of the latter is equal to pAB.

Let t be the thickness of the plate, as shown, and put d for
the diameter AB, then the total pressure against the rivet in
the direction of the arrow is

$$P = ptd.$$

The quantity p is the greatest mean value of the intensity
of compressive stress which it is desirable to put upon the
material under the given circumstances. It is usually taken
as high as 12,000 lbs. per square inch, although 10,000 is
a safer value. Of course, the actual maximum value of the
intensity immediately in front of the centre, C, is much
greater than either 10,000 lbs. or 12,000 lbs. If T' is the
amount of stress which the joint is required to carry, then
the number of rivets, so far as the previous consideration is
concerned, is equal to

$$\frac{T'}{ptd} = n.$$

The riveted joint itself, as shown in Fig. 5, Pl. XI., may
now be examined. The distance c should be at least $2\frac{1}{2}$
diameters of the rivet. By the arrangement of the rivets
shown, when the pitch is from four to eight diameters, the
strength of the plate of the width w will only be decreased
by about the amount of metal taken out in one rivet-hole,
although experiments to settle this point definitely are
wanting.

After having determined the pitch and distance of c, as
above, there are five methods of rupture of the joint only
which need serious attention. These five are: (1) tearing of
the plate though the rivet-hole E, (2) tearing of the cover-
plates through the rivet-holes at the middle of the joint, two
in the figure, (3) shearing of the rivets, (4) and (5) rupture by
compression at the surface of the contact between the rivets
and the plates. The safe shearing stress to which rivets are
subjected in bridge structures is usually taken at 7,500 lbs.
This gives a safety factor of from 5 to 6.

Put S for the intensity of the maximum safe shearing stress on rivets (7,500 lbs. for wrought iron), p for the intensity of the maximum compressive stress (10,000 to 12,000 for wrought iron), and T for the maximum working tensile stress; also, n' for the number of rivets on the line through the middle of the joint (two in the figure). Let t and t' represent the thickness of the plate and covers as shown. Then equal liability to rupture in the five ways mentioned is expressed as follows:

$$Tt\,(w-d) = 2\,Tt'\,(w-n'd) = \frac{\pi n d^2}{4}.\ 2\ .\ S = ntdp =$$
$$2nt'dp = T'.$$

It almost always happens that these quantities are not each equal to T', but none of these should be less. If only one cover-plate is used, the $2Tt'$ should be replaced by Tt', $2S$ by S, and $2nt'$ by nt'. The form of this Equation of condition may be somewhat changed by piling of the plates, etc., but it will remain essentially the same, and serves to illustrate the principle which must govern in all cases.

We see, therefore, that in obtaining the amount of pressure which should be put upon a rivet, p ought to be multiplied by its diameter, and not by the semi-circumference ADB, Fig. 1.

Art. 71.—Riveted Connections between Web Members and Chords.

When web members, as A and B, Fig. 6, Pl. XI., are riveted to the chord, the centre line (*i. e.*, the line joining the centres of gravity of the sections of the members) should pass through the centre of gravity of a system of points situated at the centres of the rivet-holes; otherwise the intensity of stress in any section of the member will not be uniform, and it (the web member) will be subjected to flexure. It is supposed, of course, that each rivet carries the same amount of stress, which, however, is probably seldom true, but it is the best assumption that can be made. Fig. 6 represents a proper distribution of rivets in reference to the centre lines Ac and Bc.

It will be observed that the strut, composed of two unequal legged angles, has its connection with the chord through both legs by means of the angle lugs. This should always be done in similar cases, for in no other way can an angle-brace develop its full strength. The practice of riveting single legs, only, of angle-braces to chords is highly objectionable, for the reason that the actual resistance of the brace is far below the nominal.

When three or more pieces are riveted together at the same joint, all the centre lines of stress should intersect at one point, if flexure is to be avoided. It is frequently impracticable to do this in riveted connections, and the impracticability constitutes a serious objection to that character of work.

If T is the total tension in the member A of Fig. 6, Pl. XI., and C the compression in B, then there will be developed at a the bending moment :

$$T \times ac \sin \theta ;$$

and at b the bending moment :

$$C \times bc \sin \theta ;$$

it being supposed that ac, bc, and ab are the centre lines of stress of the two members and chord.

It is very true that the metal is well supported in the vicinity of the joint, but unless provision is made for the flexure, as shown in Art. 69, which is seldom or never the case, some of the metal will be over-strained. Hence, this flexure should always be made a minimum, and reduced to zero if possible.

In pin connections this bending at the joints is, of course, entirely obviated when all centre lines of stress intersect at the centre of the pin.

Art. 72.—Floor-Beams and Stringers.—Plate Girders.

The load applied to a bridge rests immediately on the floor-beams, generally speaking, and is transferred through them to the joints of the truss. In railway bridges the track and ties lie on stringers, which rest on the floor-beams. There are two or more stringers for each track.

In highway bridges, the floor-beams support stringers, say

two feet apart (sometimes less and sometimes a little more), running parallel to the centre line of the bridge, which carry the floor.

In railway bridges, the moving load is applied to the beams at the ends of the stringers, but in highway bridges the greatest moving load, for which the beam is to be designed, may be taken as uniformly distributed over the entire length of the beam, or rather that part of it between the points of support.

Floor-beams should always be supported at the ends by a single hanger, or by some equivalent arrangement, which rests at the *centre* of the pin, or centre of the chord in riveted connections. Double hangers may be made tolerable by some equalizing device, usually of an expansive character, but as a rule they cannot be too strongly condemned, for in such cases the deflection of the beam will throw the greater part or all of the weight of the beam and its load on the inner hangers. The result will be not only a great overstraining of the latter, but a prejudicial redistribution of stresses in both web members and chords. The excessive load on the inner hangers will cause an overstrain in the inner tension braces which will extend to the inner lower chord members, and even to the posts and upper chord.

Floor-beams are frequently built into vertical posts. In such cases a concentrated central bearing on the pin should be provided.

Plate girder stringers for railway bridges are either supported in between the floor-beams, or partially so and partially above, or are supported wholly on the top of the floor-beams. With proper designing there is little difference in cost in the various methods. The first, however, is far preferable, for the reason that it gives the greatest stiffness to the floor system, other things being equal.

The depths of plate girder stringers is usually found between one-ninth and one-twelfth of their spans, *i. c.*, the panel lengths. The floor-beam depth should be as great as economical considerations will permit, in order that the deflections may be the smallest possible.

The webs of stringers and floor-beams slightly aid resistance to flexure, but rivets in stiffness and splice plates, if such exist, decrease this resistance to some extent. Hence, it is the best practice to disregard the resistance of the web to flexure, and to assume that it resists the shear only. This is the more advisable when it is remembered that the rivets, in giving stress to the flanges, produce a flexure in the latter which is always neglected. This flexure arises from the fact that the rivet holes never pass through the centre of gravity of the flange angle section.

The true depth of a plate girder is the vertical depth between the rivet hole centres, but by a curious confusion between rolled and built sections, it is commonly taken as the vertical distance between the centres of gravity of the flange angles.

All stress is given to the flanges by, or through, the rivets, binding them to the web, hence their proper distribution becomes a matter of importance ; it will be shown by two examples.

The exact analytical determination of the web thickness cannot be reached, but the following approximate analysis is frequently used.

It is shown by the theory of elasticity that if two planes at right angles to each other and to a plane normal to the neutral surface of a bent beam, be so taken that their intersection shall be found in that neutral surface while their common inclinations to it is 45°, then there will exist at the neutral axis the same intensity of stress on the two planes, but one stress will be tension and the other compression. It is farther shown that the common intensity of the two stresses is the same as that of either the transverse or longitudinal shear at the same point, which is also known to be $\frac{3}{2}$ the mean for the whole section in the case of a solid rectangular beam.

Now, since the intensity of shear at the neutral surface of such a beam is a maximum and zero at the top and bottom surface, and since it has been assumed that the entire web takes the shear only, it follows that if the shear be assumed to be uniformly distributed throughout any transverse section of the web, that the latter may be supposed to be composed of an indefinitely great number of columns, each of which is an indefinitely thin strip of the web, making an angle of 45° with the axis of the beam. In a direction normal to these columns an equal intensity of tension will, of course, exist.

One of the preceding assumptions is an error on the side of danger, by making the shear at the neutral surface only two-thirds of its actual value; while the other, by making the shear at the top and bottom surfaces equal to the mean, instead of zero, is an error on the side of safety, and its influence largely predominates over the former.

The elementary columns of the web may be assumed to have their ends fixed at, or by, the flange rivets of a built beam, and if d' is the vertical depth, between rivet hole centres of the two flanges, the length of the elementary columns will be:

$$l = d' \sec 45° = 1.414 \, d' \quad . \quad . \quad . \quad . \quad . \quad . \quad (1).$$

If S is the greatest total shear at any transverse section, A the area of that section of the web; then taking the depth as d', and s the mean shear, or:

$$s = \frac{S}{A};$$

these elementary columns will be subjected to an intensity of compression equal to s. Hence if t, the thickness of the web, is sufficiently great, there may be taken by Gordon's formula:

$$s = \frac{f}{1 + \dfrac{l^2}{a\,t^2}} \cdot \quad . \quad . \quad . \quad . \quad . \quad . \quad (2).$$

Or, $$t = l \sqrt{\frac{s}{a\,(f - s)}} \quad . \quad . \quad . \quad . \quad . \quad . \quad (4).$$

If, for wrought iron, $a = 3000$ and $f = 8000$, there will result:

$$t = 0.0183 \, l \sqrt{\frac{s}{8000 - s}} \quad . \quad . \quad . \quad . \quad . \quad (4).$$

The empirical constants for steel are yet to be determined.

In applying Eq. (4) to wrought iron plate girders, it will be found that the resulting values of t are excessive for large beams. The approximations already indicated, and the additional fact that the elementary columns are held in place throughout their whole length by the tension in the web, equal in intensity to the compression, and at right angles to the latter, are sufficient to justify the anticipation of such results. Eq. (3), therefore, has its chief value as the basis of an empirical formula for the web thickness.

Although the web resists "shear," it is evident, from the preceding analysis, that the method of failure of a web will be that of buckling, in which the corrugations will be at right angles to the elementary columns. Hence, if the web is so held that these corrugations are prevented, its resistance will be very materially increased. This is accomplished by riveting angles, usually in pairs, on each side of the web at proper intervals, so that the web plate is securely held between them. The office of these "stiffeners," it is to be remembered, is simply to *stiffen* the web, and prevent its buckling. If they are assumed to act as struts, the transverse shearing strain in the web at the section considered, must in so much exceed the compressive strain in the stiffener-struts, that the rivets can transfer to it its proper load, at the same time presupposing a perfect condition of riveting. In reality, neither of those conditions can possibly exist.

These stiffeners are ordinarily riveted to the web at right angles to the axis of the beam. If no elementary column is to be without the support of, at least, one of these stiffeners in some portion of its length, they must be placed at a distance apart, measured along the axis of the beam not greater than the vertical depth between rivet hole centres ; and that limit is very commonly given in specifications, although it is sometimes placed at once and a half that depth.

About the same amount of stiffening would be secured by placing the stiffeners at 45° with the axis of the beam, and at intervals of twice the depths, but the difficulties of construction would be increased.

A safe rule, and one frequently used, though purely conventional, is to introduce stiffeners when the mean shear, *i. e.*, $S \div A$, exceeds 4000 pounds per square inch for wrought iron.

The function of the rivets holding the flanges to the web is next to be considered. In reality, these rivets may have two offices to perform. If the load of the girder rests on one of its flanges, the flange rivets will sustain it directly; but their chief office is to give the flanges their proper stress. If then the stresses be determined for any two points between the end of a girder, and the point of greatest flange stress, the shearing or bearing resistance of all the rivets between those points must be equal to, or not less than the resultant of the difference between the determined flange stresses and the load resting on the flange between the same points. If no load rests on the flange, but is carried directly by the web, the "resultant" is evidently the simple difference between the determined flange stresses.

These elementary considerations constitute the entire method of finding the pitch and number of rivets in the flanges of built beams, and will be applied to two examples.

The complete design of the truss, treated in Art. 11, is to be given in subsequent pages, and in the present connection the stringers and floor-beams will be discussed. The stringers will be placed 7 feet apart centres, and the rails will be laid on 8 inch by 8 inch ties 9 feet long, spaced 16 inches from centre to centre. The ties, rails, guard rails, splices, spikes, etc., will then weigh about 325 pounds per lin. ft. The depth of stringers will be taken at 27 inches throughout their lengths, and the iron of each stringer will be assumed to weigh 100 pounds per lineal foot. The total fixed load will then amount to 263 pounds per lineal foot of each stringer, while the moving load is one-half of the concentrations given in the engine diagram of Art. 11. The principles established in Art. 7 show that the four 10,000 pound driving wheel loads will produce the greatest bending moment when either wheel *A* or *B* is at the distance 1.062 feet from the centre of the panel, and will be found under the wheel in question.

With that position, the following moving load bending moments will exist :

Maximum . . . . 119,200 ft. lbs.
7½ feet from the end . . . . . . 105,400 " "
5 " " " " . . . . 85,600 " "
2½ " " " " . . . . . . 44,800 " "

It is evident that the last three of these values are not the greatest moments for those points, but as the flanges are to be of uniform section, it is not necessary to seek them, as might easily be done by the aid of the principles of Art. 7.

The vertical depth between the rivet hole centres of these stringers will be taken at 2 feet. The flange stresses at the various points will then be :

At centre . . . $\dfrac{263 \times (20.55)^2}{8 \times 2} + \dfrac{119,200}{2} = 66,540$ lbs. *CD.*

7½ ft. from end $\dfrac{263 \times 7.5 \times 13.05}{2 \times 2} + \dfrac{105,400}{2} = 59,144$ " . *EF.*

5 " " " $\dfrac{263 \times 5 \times 15.55}{2 \times 2} + \dfrac{85,600}{2} = 47,920$ " . *GH.*

2½ " " " $\dfrac{263 \times 2.5 \times 18.05}{2 \times 2} + \dfrac{44,800}{2} = 25,367$ " . *KL.*

The allowed working stresses in the flanges of the stringers will be taken at 7,000 pounds per sq. in. of gross section in compression and 8,000 pounds per sq. in. of net section in tension. The diameter of rivets in the stringers and floor beams of railway bridges is chiefly a matter of judgment ; it usually ranges three-quarters to seven-eighths of an inch. In the present instance, rivets of the latter diameter, before being driven, will be taken. The metal punched out for a rivet should leave a hole not more than one-sixteenth of an inch greater in diameter than that of the cold rivet. But the metal immediately about the edge of the hole is materially injured for tensile purposes, and in the tension-chord angles the disc of metal rendered valueless should be taken one-eighth of an inch greater in diameter than that of the cold rivet, *i.e.*, for the present case, one inch. In the compression flange no metal need be deducted for the rivet holes.

The upper flange section at the centre will be :

$$66,540 \div 7,000 = 9.5 \text{ sq. in.}$$

The net lower flange section will be :

$$66,540 \div 8,000 = 8.3 \text{ sq. in.}$$

The following flanges will satisfy the conditions :

Upper flange . . .	$\begin{cases} 2 - 5'' \times 4'' \text{ 30 lb. angles.} \\ 1 - 10'' \times \frac{3}{8}'' \text{ cover plate.} \end{cases}$
Lower flange	$2 - 5'' \times 4''$ 47 lb. angles.

The 62 lb. angles have a thickness of three-quarters of an inch, so that the excess of gross section exactly covers the metal destroyed by the punch.

It is the best practice to run a plate the whole length of the upper flange in order to make it act as a unit in resisting compression. The lower flange needs no cover plate, and, again, the additional rows of rivet holes would produce more dead metal.

Although there is a little waste of metal near the ends in small plate girders, it is economy to save labor by making the flanges of uniform section throughout their lengths. This economy ceases when the flanges become so heavy that one or more cover plates become necessary in the tension flange and more than one in the compression flange.

Cover plates should be carried at least a foot beyond the section at which the additional metal is required, and rivets should be closely pitched in that portion. If rivets pierce both legs of an angle in tension, metal should be deducted for all the rows in both legs.

The location of the preceding sections is shown in Fig. 1, of Pl. XI. The increments of stresses for the different segments of the flanges will be :

$$EC \text{ or } FD = 66,540 - 59,144 = 7,396 \text{ lbs.}$$
$$GE \text{ or } HF = 59,144 - 47,920 = 11,224 \text{ ``}$$
$$GK \text{ or } HL = 47,920 - 25,367 = 22,553 \text{ ``}$$
$$KA \text{ or } BL = = 25,367 \text{ ``}$$

The extent of flange over which the weight of each driving wheel weighing 10,000 pounds will be distributed is indeterminate, but as the ties are 16 inches apart centres, it will be sufficiently accurate to take that distance as $2\frac{1}{3}$ feet. As each driving wheel passes over the entire stringer, the resultant stresses which the rivets in the various sections into which the beam is divided will be obliged to carry, are as follows :

$$EC \text{ or } FD = \sqrt{(7,396)^2 + (10,000)^2} = 12,440$$
$$GE \text{ or } HF = \sqrt{(11,224)^2 + (10,000)^2} = 15,000.$$
$$GK \text{ or } HL = \sqrt{(22,553)^2 + (10,000)^2} = 24,700.$$
$$KA \text{ or } BL = \sqrt{(25,367)^2 + (10,000)^2} = 27,300.$$

In order to determine the number of rivets in any of these sections, it will be necessary to fix the thickness of web plate. The minimum thickness permissible is, to some extent, a matter of judgment, but it is safe to say that no built beam for railroad purposes should have a less thickness of web than $\frac{5}{16}$ of an inch, and a limit of $\frac{3}{8}$ is still better practice. The latter limit will be used here.

With the position of moving load already determined in Art. 11, the greatest shear at the end of the stringer is $24,830 + 2,700 = 27,530$ pounds $= S$. The sectional area A of a 27 by $\frac{3}{8}$ inch plate is 10.125 sq. ins. $= A$. Hence $s = S \div A = 2,720$ pounds ; also $l = 24 \times 1.414 = 33.94$ inches. Hence Eq. (4) gives :

$$t = 0.45 \text{ inch.}$$

Since this result is greater than the assumed value of t, the hypothetical elementary columns are not capable of sustaining their loads without exceeding by a little the proper working stress; but as the hypothesis involves a considerable safety error, the assumed value of t is probably ample. However, as the additional metal is very small in amount, a pair of 3 by 2½ 16-pound L stiffeners will be introduced at the distance of 27 inches from each end as shown.

The working resistance to shearing offered by rivets in the truss will be taken at 7,500 pounds per sq. in., and the limiting pressure between rivets and walls of holes will be fixed at 12,000 pounds per sq. in. under the same circumstances. The floor of a bridge is subject to shocks due to track imperfections, and the above values should be reduced by 25 per cent., making the working resistance to shearing 5,625 pounds, and to pressure 9,000 pounds per sq. in. As the web is embraced by an angle iron on each side, each rivet will be subjected to double shear, and the thickness of bearing surface of the web will be much less than that of the two flange angles. In the determination of the shearing and bearing resistances of rivets, the cold diameter, before being driven, should, as a margin of safety, be considered. Hence, those resistances for the seven-eighths rivets under treatment, are:

$$2.\tfrac{1}{4} \cdot \pi \, (0.875)^2 \cdot 5,625 = 6,750 \text{ pounds; and,}$$
$$0.875 \cdot 0.375 \cdot 9,000 = 3,000 \quad \text{``}$$

The latter quantity is much the smaller, and will govern the number of rivets in each section, as follows:

EC or FD.....12,440 ÷ 3,000 = 4 rivets required.
GE or HF.....15,000 ÷ 3,000 = 5   "   "
GK or HL.....24,700 ÷ 3,000 = 8   "   "
KA or BL.....27,300 ÷ 3,000 = 9   "   "

The nearest whole number is taken in each case.

These results give at and near the end about nine rivets to each two and a half feet. A uniform pitch of three inches, therefore, will be assumed throughout each flange. If the load is uniform in intensity, it is well known that the variation of flange stress is very little for a considerable distance either side of the centre. This example shows, however, that concentrated loads may require just as close centre riveting, *i.e.*, just as small pitch, as at the ends. The number of sections into which a beam must be divided is a matter of judgment in each particular case.

The greatest shear at the end of the stringer has already been seen to be 27,530 pounds, hence the end stiffeners at A must transfer that amount to the floor-beam web. An intensity of 4,000 pounds per square inch of normal section is a safe and proper value for those members. The end stiffeners will then be assumed to be $2 - 4'' \times 4''$ 35 lb. angles, one being on each side of the web at each end of the beam and extending the full depth between the legs of the flange angles of the latter. Fillers whose thickness is just a little less than that of the heaviest flange angle will be required under these end stiffeners. Light intermediate stiffeners may be bent to fit if the flange angles are not too heavy; otherwise fillers must again be used.

The number of rivets required to transfer the greatest reaction at the stringer ends to the floor-beam (see Fig. 2, Pl. XI.) is found by taking the reaction of the locomotive load (found in Art. 11 to be 34,600 pounds) and adding to it the fixed load, or, $20.55 \times 263 = 5,405$ pounds, then dividing the result by the bearing capacity of seven-eighths rivet in the three-eighths web of the floor-beam, as that is less than the resistance of the same rivet in double shear. The number of rivets needed will then be $(34,600 + 5,405) \div 3,000 = 13$. Seven rivets will be placed in each 4×4 end angle, as shown in Fig. 2, of Pl. XI., making 14 for the end of each stiffener. If the loads were very great, it would be necessary to reinforce the web plate of the floor-beam where it receives the ends of the stiffeners, by riveting a plate on each side. In the present instance, however, it is unnecessary.

The bill of material, with the weights of one stringer, may now be written as follows:

1	27 × $\frac{3}{8}$ Plate		20.55 ft. long			........695 lbs.	
2	5 × 4 30 lb. angles	"	"	"	411	"
2	5 × 4 47 "	"	"	"	"	644	"
1	10 × $\frac{3}{8}$ Plate		"	"	"	257	"
4	4 × $\frac{11}{16}$ Filling plates 19 ins.		"	 60	"
4	3 × $\frac{1}{2}$ "	"	19 "	"	 32	"
4	4 × 4 35 lb. angles 27 "		"	105	"
4	3 × 2$\frac{1}{2}$ 16 "	"	27 "	"	 48	"
	$\frac{3}{4}$ Rivets					120	"

Total weight of stringer........2,372 lbs.

The actual weight per foot is thus about 15 lbs. more than was assumed. Although this makes an insignificant difference in the flange section, and is amply provided for in the present instance, in practice the actual weight should be a little under the assumed, and not over it.

The arrangement of connecting the stringers to the floor-beams shown in Fig. 2 of Pl. XI., has the merit of making a very stiff floor system. It is proper to say, however, that many other methods are used. The small angle brackets seen under the ends of the stringers and riveted to the lower flange angles of the floor-beam are simply for convenience in erection, and are not considered as essential to the resistance of the joint.

According to the preceding bill of material, the actual maximum weight concentrated at the stringer ends will be:

$$34,600 + 20.55 \times 300 = 40,800 \text{ nearly.}$$

The depth of the floor-beam will be taken at 36 inches throughout its entire length, so that the vertical distance between rivet centres in the two flanges will be about 33 inches.

If the weight of the floor-beam be assumed at 150 lbs. per lineal foot, the total flange stresses at the centre, at the stringer points, and at points 2$\frac{1}{2}$ feet distant from the ends, will be:

22

$$\frac{40,800 \times 5}{2.75} + \frac{150 \times (17)^2}{8 \times 2.75} = \qquad 76,170,$$

$$\frac{40,800 \times 5}{2.75} + \frac{150 \times 5}{2 \times 2.75} \quad \times \ 12 \ = 75,840,$$

$$\frac{40,800 \times 2.5}{2.75} + \frac{150 \times 2.5}{2 \times 2.75} \quad \times \ 14.5 = 38,090.$$

Using the same working stresses in the flanges as were fixed for the stringers, there will be found for the upper flange section at the centre:

$$76,170 \div 7,000 = 10.9 \text{ sq. ins.};$$

and for the net lower flange section:

$$76,170 \div 8,000 = 9.5 \text{ sq. ins.}$$

The following flanges will satisfy these requirements:

Upper flange ...... $\begin{cases} 2 - & 5'' \times 4'' \ 36 \text{ lb. angles,} \\ 1 - 12 & \times \tfrac{5}{16} \text{ cover plate.} \end{cases}$

Lower flange ....2 — $5'' \times 4''$ 55 lb. angles.

The thickness of the 55 lb. angle is about 0.7 inch, so that if seven-eighth cold rivets are used the metal destroyed by the punch is just about equal to the excess of the total section over the net.

The total shear at the end of the floor-beam is 40,800 + 8.5 $\times$ 150 = 42,075 lbs. = S. If the web thickness be assumed at three-eighths of an inch, $A = 36 \times \tfrac{3}{8} = 13.5$ sq. ins. Hence, $s = S \div A = 42,075 \div 13.5 = 3,120$ lbs.

$l = 33 \times 1,414 = 46.66$. These quantities inserted in Eq. (4) give:

$$t = 0.68 \text{ inch.}$$

This value shows that if the hypothetical elementary columns are to sustain 3,120 lbs to the sq. in., the web must be 0.68 inch thick. But such a thickness is plainly excessive, and shows how the formula errs in the direction of safety. A web thickness of three-eighths of an inch will be assumed, and 3″ × 2½″ 16 lb. angle stiffeners placed half-way between the stringer supports and the ends, as shown at *EF*, Fig. 2, Pl. XI. These stiffeners will require 3″ × ⅝″ filling plates 28 inches long.

If shearing and bearing resistances of 5,625 and 9,000 lbs. per sq. in., respectively, be taken, as was done in designing the stringers, the bearing value of one rivet in the three-eighths web will be, as before, $0.875 \times 0.375 \times 9,000 = 3,000$ pounds.

Hence the number of seven-eighths rivets required between the three sections of the beam will be :

$$Centre \ and \ C \ldots \ldots (76,170 - 75,840) \div 3,000 = 1,$$
$$C \ and \ EF \ldots \ldots (75,840 - 38,090) \div 3,000 = 13,$$
$$EF \ ``\ AB \ldots \ldots \ldots 38,090 \div 3,000 = 13.$$

These conditions will be sufficiently near fulfilled if a pitch of 2¾ inches be taken for a distance of 6 feet from each end, and six inches for the remaining 5 feet between the stringers. The flange stresses do not require a six-inch pitch at the centre, but that value should not be exceeded, in order that the flanges may be properly bonded. The pitch in the cover plate on the upper flange should be 2¾ inches for a distance of 15 inches from each end, and over the remaining portion of its length that pitch may be doubled. In no case, however, should the pitch in a compression plate exceed about 16 times its thickness.

The end floor-beams are suspended from the pins by the plates shown riveted to the heavy end stiffeners, as at *AB*. These stiffeners transfer half the weight of the beam in addition to the 40,800 pounds at the stringer ends, to the suspension plates, or, $40,800 + 150 \times 8.5 = 42,100$ lbs. As these end stiffeners do not in this instance sustain this entire weight at any section, the latter cannot be analytically determined. They should be heavy, however, and will be taken as 5 × 4 36 lb. angles.

The number of rivets required between the end stiffeners and the web is 42,100 ÷ 3,000 = 14, but it is convenient to take 15 as shown.

The dimensions of the suspension plates cannot be fixed until the diameter of the pin is determined, and they will be found in a later Art.

The bill of material for the floor-beam with its weight will now be:

1	36 × $\frac{3}{8}$ Plate	17 ft. long	17 × 45	= 765
2	5 × 4 36 lb. angles	" " "	34 × 12	= 408
1	12 × $\frac{5}{16}$ cover plate	" " "	17 × 12.5	= 213
2	5 × 4 55 lb. angles	" " "	34 × 18$\frac{1}{3}$	= 623
4	3 × $\frac{5}{8}$ filling plates 2$\frac{1}{2}$	" "	9$\frac{1}{3}$ × 6$\frac{1}{4}$	= 58
4	3 × 2$\frac{1}{2}$ 16 lb. stiffening angles 3 ft. long	4 × 16	= 64
4	5 × $\frac{5}{8}$ end	plates 2$\frac{1}{3}$ " "	10.4 × 8$\frac{1}{3}$	= 87
4	5 × 4 36 lb. angles	3 " "	4 × 36	= 144
4	3 × 2$\frac{1}{2}$ 16 " "	1 " "	1$\frac{1}{3}$ × 16	= 21
	$\frac{3}{8}$ rivets			= 130

Total weight of beam . . . . 2,513 lbs.

The weight per lineal foot is then 2,513 ÷ 17 = 148 pounds, or less than that assumed, as it should be.

In the cases of large plate girders it is necessary to make splices in the web plate, a splice plate being used on each side of the web. The combined thickness of these splice plates should be at least 50 per cent. in excess of that of the plates spliced. The number of rivets on each side of the joint should be such that the total bearing resistance in the web, or the total shearing resistance of the rivets, shall at least equal the greatest possible transverse shear at the joint considered. At least two rows of rivets should be found on either side of the joint.

It is sometimes customary, if web plates have no splices, to take one-sixth of the web section as acting in either flange. If no rivet holes were punched for the stiffeners, this method would be allowable. But such rivet holes frequently take out considerable metal, and as the tension side of the plate only is affected, one-sixth of the remaining metal ceases to be a proper proportion. On the whole, therefore, it is better to neglect the bending resistance of the web, and allow it to balance, so far as it may, the effect of the rivet holes being out of the centre of gravity of the flange angles.

That depth of plate girder which will give the least weight depends entirely upon the manner and amount of the loading. With very heavy concentrated loads, it may be half the span; on the other hand, with very light loads it may be less than one-twentieth of the span. As all bending moments may be supposed to be caused by some uniform load, either fictitious or real, the following analytical discussion may be of some value as well as interest:

Economic Depth of Plate Girders with Uniform Flanges.

If a plate girder carries a uniform load, and is designed with flanges of uniform cross sectional area, the depth which will give the least weight of girder may easily be obtained.

Let l = span in feet.
" d = depth " "
" t = web thickness in inches.
" p = allowed working stress in lbs. per sq. in. for the flanges.
" p' = "   "   "   "   " end stiffeners.
" a = sectional area of each intermediate stiffener in *sq. ft.*
" n = number of stiffeners (intermediate).
" w = total load per lin. ft. of girder (in pounds).

The flange stress at centre will then be:

$$F = \frac{wl^2}{8d}.$$

The volume of the web plate in c. u. ft. will be:

$$\frac{ldt}{12}.$$

If one-sixth of the latter is taken to be concentrated in the flange, the volume of the two flanges in c. u. ft. will be:

$$\frac{2Fl}{144p} - 2\,\frac{1}{6}\,\frac{ldt}{12} = \frac{wl^3}{576pd} - \frac{ldt}{36}.$$

The volume of the end stiffeners will be:

$$\frac{wld}{144p'};$$

and that of the intermediate stiffeners:

$$nad.$$

The volume of the entire girder will then take the value:

$$V = \frac{wl^3}{576pd} + \frac{ldt}{18} + \frac{wld}{144p'} + nad \quad . \quad . \quad . \quad . \quad . \quad . \quad (5).$$

By taking the first derivative;

$$\frac{dV}{d(d)} = -\,\frac{wl^3}{576pd^2} + \frac{lt}{18} + \frac{wl}{144p^1} + na = 0.$$

Solving for d^2:

$$d^2 = \frac{wl^3}{576p\left(\dfrac{lt}{18} + \dfrac{wl}{144p^1} + na\right)}$$

$$.d = \frac{l}{2}\sqrt{\frac{wl}{p\left(8lt + \dfrac{wl}{p^1} + 144na\right)}} \quad . \quad . \quad . \quad (6).$$

If one-sixth of the web is not concentrated in each flange, $12lt$ will take the place of $8lt$ in Eq. (6).

If all stiffeners, both end and intermediate, are omitted, Eq. (6) will take form:

$$d = \frac{l}{4}\sqrt{\frac{w}{2pt}} \quad . \quad . \quad . \quad . \quad . \quad (7).$$

In reality p is seldom or never exactly the same for both

flanges, since it is the working stress in reference to the *gross section*. It will be sufficiently near, however, for all usual purposes to make it a mean of the two actual working stresses for the gross sections.

It should be borne in mind that local circumstances frequently compel a different depth from that given by Eq. (6). It will also be found that a considerable variation from that depth will cause a comparatively small variation in weight. Again, the difficulties of handling a deep girder, and the shop cost *per pound*, may, and usually does, make the economic depth a little less than that found by the aid of Eq. (6).

Art. 73.—Eye-Bars or Links.

An "eye-bar" or "link" is a tension member of a pin-connection bridge, fitted at each end with an eye for the insertion of a pin. Two views of an eye-bar are shown in Fig. 5, Pl. XII.; A is the body of the bar, D the neck, and C the eye. The head of an eye-bar is the enlarged portion in which the pin-hole is made. The eye-bar is one of the most important members of a pin-connection bridge, and the determination of the relative dimensions of the head has been the subject of much experimenting. A mathematical investigation, however, with the same object in view is a matter of considerable complexity, although an approximate solution of the problem may be obtained, and its agreement with the results of experiment is quite close.

Before taking a general view of the stresses which may arise in an eye-bar head, it must be premised that a difference of $\frac{1}{30}"$ to $\frac{1}{64}"$ between the diameter of the pin and that of the pin-hole is considered exceedingly good practice. Before the eye-bar is strained, therefore, there is a line of contact only between the pin and eye-bar head, but on account of the elasticity of the material, this line changes to a surface when the bar is under stress, and increases with the degree of stress to which the bar is subjected. This line and surface of contact is, of course, in the vicinity of K, Fig. 3, Pl. XII., *i.e.*, on that side of the pin toward the nearest end of the bar. The consequence of this is that, when the bar is strained, the portion about KA, Fig. 12, is subjected to direct compression and extension; that about BL, DH, and FM to direct tension and bending, while in the vicinity of CN (also CQ) there is a point of contra-flexure, and the stress in the direction of the circumference changes from compression to tension as H is approached from K.

It should have been said before that if w represents the width of an eye-bar, as shown, then its thickness, t, is generally included between the limits $\frac{1}{4}w$ and $\frac{1}{8}w$. These limits of the relative values of the quantities are seldom exceeded.

Fig. 2, Pl. XII., represents a method of laying down an eye-bar head which has been determined by a very extensive system of experiments given by a member of the British Institution of Civil Engineers, and one that has stood the test of long American experience; in short, there is probably no better method known. Let r represent the radius of the pin-hole, and w the width of the bar.

Then take $EN = 0.66w$. The curve $DRBK$ is a semicircle with a radius equal to $r + 0.66w$, with a centre, A, so taken on the centre line of the bar that $QB = 0.87w$. GF is a portion of the same curve, with A' as the centre $(A'C = AC)$; GH is any curve with a long radius joining GF gradually with the body of the bar. HG should be very gradual in order that there may be a large amount of metal in the vicinity of CG, for there the metal is subjected to flexure as well as direct tension. FD is a straight line parallel to the centre line of the bar.

Fig. 3 shows another method founded on the results of a mathematical investigation. Take r and w as before. Then $BC = AC = r + 0.87w$, $DH = \frac{2}{3}w = 0.66w$, $ED = EF = 2r + w$. DF is described with ED until $DCF = 45°$. BAB is described with BC until $BCA = 35°$. BN is drawn from L as a centre located in such a position as to cause that arc to be at the same time tangent to DN and AB. DN is a straight line drawn parallel to the axis of the bar. PF is any easy curve which will appear the best. The dotted lines in both Fig. 2 and Fig. 3 show the slope that should be given in order to clear a die.

The outline of the head is now usually formed of a portion of the circumference of a circle whose centre is the centre of the pin-hole. In such a case no dimension of the head should be less than the corresponding one determined by either of the methods just given.

Fig. 4 shows the head thickened in such a manner that the mean maximum intensity of pressure between pin and pin-hole shall not exceed a given amount, p. Let T represent the maximum intensity of tension in the body of the bar; then, as has been shown in discussing the pressure against the bodies of rivets:

$$wtT = 2rpt' \therefore t' = \frac{wtT}{2rp}.$$

Art. 74.—Size of Pins.

The exact analytical determination of the pin diameter in any particular case is, like many other matters, involving the elasticity of materials, an impossibility, although the problem in its simplest form was subject to a very able mathematical investigation by Charles Bender, C. E., in *Van Nostrand's Magazine* for October, 1873. One or two reasonable assumptions, which, in a great majority of cases, must be very nearly accurate, give the problem a very simple character. The first of these assumptions is that *the pressure applied to any pin has its centre at the centre of the surface of contact.* Fig 3 of Pl. XI. represents the half of a pin-connected joint, LL being the centre line, and by this assumption the centre of pressure between each of the eye-bars A, B, C, D, etc., and post bearing P, and the pin is located half-way between the faces of those members normal to the axis of the pin.

If, however, a pin is held by a compression member, such as an upper chord or post, then the centre of pressure in that member may be taken as *the centre of such a surface as will reduce the bearing intensity to its maximum limit.*

It is to be premised that the general considerations touching the distribution of pressure between rivets and plates given in Art. 70 hold equally true for pins. The greatest allowable bearing intensity between pins and eye-bars of wrought iron ranges from 10,000 to 12,500 pounds per square inch of the surface found by multiplying the diameter of pin by thickness of bar. The latter product is always considered the bearing surface.

If two bars only, such as A and D, act on each end of a pin it is clear that the centre line of the latter will be convex toward D. The result will be a movement of the centres of pressure of those bars toward each other ; so that the lower arm of A will be less than half the thickness of that member plus half that of B. The second assumption given above seems thus very reasonable, and may be extended to the case of a pair of eye-bars, only, at the end of a pin. When, on the other hand, a number of eye-bars of various sizes take hold of a pin, particularly if the bending moments have different directions at different sections of the pin, the axis of the latter may be essentially straight and the centres of pressure should be taken according to the first assumption. This is, in reality, the best practice in all cases, for if the centre of pressure departs from the axis of the bar, the latter will be subjected to a bending moment equal to the tension in the bar multiplied by the distance of the centre of pressure from its axis. Hence the necessity of so fixing the diameter of pin that it shall be as stiff as possible.

In Fig. 3 of Pl. XI., let a be the distance between the centres of eye-bars A and D ; a', that between D and B ; a'' that between B and E, etc., etc. These distances a, a', a'', etc., should always be taken as the thickness of the head plus one-eighth of an inch ; the latter amount representing about the proper clearance in the best work.

Then let T_a, T_d, T_b, etc., represent the total tensions in the bars A, D, B, etc. The bending moments about the centres of those bars will then be :

About centre of D ....$a\ T_a$
     "   "   " B ...$(a + a')\ T_a - T_d a'$.
     "   "   " E....$(a + a' + a'')\ T_a + T_b a'' - T_d(a' + a'')$.
Etc., etc., etc.,   etc.,   etc.

The rod R is a counter and does not usually act when the pin receives its greatest bending.

The preceding moments are all similarly formed and are about vertical axes until the centre of the post bearing, P, is reached. The tension T of the main tension brace T produces a moment about an axis normal to its own. Let it be supposed that the resultant moment of all the chord members A, B, C, D and E about the centre of P is right-handed looking vertically down, as shown by M' in Fig. 1.

Let M' represent that moment by any convenient scale. The moment of T by a^r, or Ta^v, will be right-handed looking upward; and let M_T represent that moment by the same scale as before. The latter line is drawn normal to the axis of the member T. The line M will now represent by the same scale the moment to which the pin is subjected at the centre of P, and its direction is that of the axis of the moment.

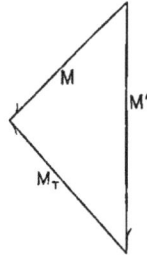

FIG. 1.

The greatest pin bending in the lower chord will usually take place with the greatest chord stresses, but the upper chord pins will receive their greatest moments by the greatest web stresses.

When a number of bars are coupled to the pin in such a joint as that shown in Fig. 3 of Pl. XI., it is usually necessary to test a number of sections in order to find the greatest moment; unless the bars are very nearly of the same size and placed alternately as shown when the greatest moment will be found at the centre of the pin.

It is frequently advisable, however, to employ different sized bars in order to reduce the bending moments; a small bar being placed at the end of the pin.

The same reduction of bending moments is brought about even more effectually by the arrangement of lower chord bars shown in Fig. 2.

In that figure it will be observed *that the lower chord eye-bars are so grouped on any one pin, that the stresses in them—for each half of the pin, form couples which have opposite signs* and, thus, to a great extent, or wholly, neutralize each other.

By varying the sizes or thicknesses of the bars and by resorting to the method of grouping shown in Fig. 2 (which

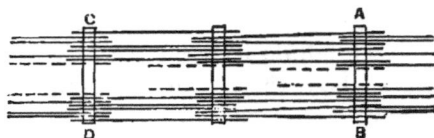

FIG. 2.

represents a portion of an actual lower chord) the bending moments in lower chord pins may easily be reduced to any desired extent in any case whatever.

It is evident that the resultant moment shown in Fig. 1 could be obtained by resolving the stress T into its vertical and horizontal components and combining their moments with those of the lower chord stresses, making the components of M vertical and horizontal instead of vertical and inclined.

The bending of pins is very much increased by thickened eye-bar heads, since the thickening increases the lever arm of the tensile stress in the eye-bar. The thickened eye is a most excellent thing for the bar, but necessitates an increased diameter of pin.

The preceding operations illustrate the general method of finding the bending moment to which a pin is subjected in all cases; the component moments are determined from the stresses in the individual truss members, and the resultant is then found by the moment triangle or polygon. The pin diameter is then readily found in the following manner.

If M is the external bending moment, I the moment of inertia of the normal section of the pin about its diameter D and K the intensity of stress in the fibres most remote from D, then it is known from the theory of flexure, since $I = \dfrac{\pi D^4}{64}$, that

$$M = \frac{K\pi D^3}{32} \therefore D = 2.2 \sqrt[3]{\frac{M}{K}} \quad \cdot \quad \cdot \quad \cdot \quad (1).$$

If K is known, Eq. (1) gives D at once after M is found by the general method exemplified by Fig. 1, or in any other manner.

For wrought-iron pins in the trusses of railway bridges, K is usually taken at 15,000 pounds. This value in Eq. (1) gives for wrought-iron pins:

$$D = 0.089 \sqrt[3]{M}. \quad \ldots \quad \ldots \quad \ldots \quad (2).$$

For steel pins, under similar conditions, K may be taken at 20,000 pounds, for which :

$$D = 0.081 \sqrt[3]{M}. \quad \ldots \quad \ldots \quad \ldots \quad (3).$$

There are numerous tables showing the bending moments of pins of all usual diameters with given values of K, so that in practice the computations expressed in Eqs. (1), (2) and (3) are seldom necessary. The value of M is determined for any particular case, after which, by the simple inspection of a table, the proper diameter may be chosen.

It is seen by Eq. (1) that the diameter of a pin varies directly as the cube root of M and inversely as the cube root of K.

It may sometimes happen that M_{T}, in Fig. 1, is so small that it may be neglected; in which case $M = M'$.

No pin should possess a diameter less than eight-tenths the width of the widest bar coupled to it.

When bending and bearing are properly provided for, a safe shearing resistance will be amply secured. If the apparent moment in the pin is sufficient to cause failure by flexure, it does not, by any means, follow that failure will actually take place; for the distortion of the pin beyond the elastic limit will relieve the outside eye-bars of a larger portion (in some cases perhaps all) of the stress in them. This result will produce a redistribution of stress in the eye-bars, by which some will be understrained and the others correspondingly overstrained. Thus, although the pin may not wholly fail, the safety of the joint will be sacrificed by the overstrained metal in the eye-bars.

Art. 75.—Camber.

Camber is the curve given to the chords of a bridge, caus-ing the centre to be higher than the ends, or rather it is the amount of rise of the centre above the ends. It is given to a truss so that the chords may not fall below a horizontal line when the load is applied. Fig. 8, Pl. XII., represents a truss with exaggerated camber. The actual amount varies from $\frac{1}{800}$th to $\frac{1}{1200}$th of the span.

Camber may be given to a truss either by lengthening the upper chord or shortening the lower one; the latter method is preferable because the upper chord is sometimes not hori-zontal, and different panel lengths would have to be shortened by different amounts.

On account of the unavoidable play at the joints of all work, the shortening of the lower chord, or lengthening of the upper, must be increased by about $\frac{1}{32}$d of an inch per panel in order to secure the desired camber.

The lower chord shortening is made uniformly throughout its length; that is, each panel length is shortened by a con-stant quantity. The true chords will, therefore, become arcs of circles of very large radii, and vertical posts will become radial.

By means of the equation of the circle, $y^2 = 2Rx - x^2$, R being the radius, the amount of shortening or lengthening of chord to produce a given camber may be determined if the play at the joints be omitted. In the equation above, y rep-resents the half span and x the camber desired, hence the radius

$$R = \frac{y^2 + x^2}{2x}.$$

This is the radius of the lower chord when cambered. Generally it will be near enough to put $R = \frac{y^2}{2x}$

The *angular length* of the lower chord will be

$$\alpha = 2 \sin^{-1} \frac{y}{R} = 2 \sin^{-1} \frac{2xy}{y^2 + x^2},$$

and the length in feet :

$$l = R\alpha.$$

The length of the upper chord will then be :

$$l' = (R + d)\,\alpha.$$

The difference in length of chords will be :

$$D = l' - l = d\alpha.$$

This is the amount by which the lower chord is to be shortened, or the upper lengthened, in order to produce the required camber, if no play or strains exist.

Since x is very small compared with y :

$$\alpha = 2\ sin^{-1}\frac{2xy}{y^2 + x^2} = \frac{4xy}{y^2 + x^2} = \frac{4x}{y}\ \text{(nearly)}.$$

If the span is l_1 :

$$\alpha = \frac{8\ x}{l_1};\ \text{and,}\ D = \frac{8dx}{l_1}.$$

If r is such a ratio that $d = rl_1$:

$$D = 8rx.$$

On account of the play at the joints, x should be taken a little larger than the camber desired.

Frequently r is about one-eighth, and for such a value :

$$D = x;$$

or, neglecting the play at the joints, *the difference in lengths of the chords should equal the camber.*

If the chords are to be horizontal under the greatest loads, while T and C represent the supposed uniform intensities of tension and compression in the lower and upper chords respectively, E and E' representing the coefficients of elasticity ;

$$\tfrac{1}{2}D = \frac{T}{E}l = \frac{C}{E'}l'.$$

This formula can only be approximate, for the chords are never exactly uniformly stressed, and the coefficient of elasticity is probably never the same throughout either chord.

Since d, the depth of truss, does not vary, these formulæ apply only to trusses of uniform depth.

A "through" truss has been supposed, but the same formulæ exactly apply to a deck bridge.

It is to be borne in mind that one-half the horizontal distance between the centres of end pins is to be taken for y in determining R. If this distance is assumed in designing the truss, then the panel length is to be found by dividing l or l' by the number of panels.

If the panel length is first assumed, and the camber produced by shortening or lengthening *it*, then this horizontal distance is essentially equal to the assumed chord length diminished or increased by $D = d\alpha$.

In order to hold the camber in a truss, the diagonals must be shortened, as shown in Fig. 9, Pl. XII. The diagonal which was *bd* before cambering, becomes *cd* afterward. *ad* and *bc* are supposed to be panels in the upper and lower chords respectively before putting in the camber; afterward *bc* becomes *cf*, while *ad* remains the same; the lower chord is supposed to be shortened. Let x be the amount of shortening of each panel of the lower chord $= 2bc = 2fc$; d, the depth of the truss; and p the original panel length equal to *ad*. Then

$$cd = \sqrt{dc^2 + cc^2} = \sqrt{d^2 + \left(p - \frac{x}{2}\right)^2}.$$

If the camber is produced by lengthening the upper chord, then *cf* is the original panel length, and *ad* the new one, and

$$cd = \sqrt{dc^2 + cc^2} = \sqrt{d^2 + \left(p + \frac{x}{2}\right)^2}.$$

In a triangular truss the diagonal *gc* Fig. 10, is changed to

$$gf = \sqrt{d^2 + \tfrac{1}{4}(p - x)^2}.$$

If the upper chord is lengthened, cg is the diagonal desired, and ff' the original panel length p. Hence,

$$cg = \sqrt{d^2 + \tfrac{1}{4}(p + x)^2}.$$

Each diagonal is to be shortened to the length cd.

In a draw-bridge each arm, in giving the camber, can be considered one span, but the whole amount of shortening in the lower chord of *one arm* must also be taken out of the upper chord at the centre. If this is not done, the ends will sink below their original positions.

Art. 76.—Economic Depth of Trusses with Parallel Chords.

The so-called economic depth of truss for a given span, is that depth which involves the least material or weight of metal in the bridge. This depth depends upon the intensity of moving load for each truss, the length of panel, the greatest allowable stresses, etc., etc. Various mathematical investigations have been made with a view to the determination of this depth of truss in terms of the length of span. But on account of the exceedingly intricate character of the problem, any feasible analysis must be based upon assumptions which simplify the analytical operations, but render the results only approximately true. These investigations, however, and the experience of American engineers, show that a depth varying from one-fifth to one-seventh the length of span will give the least weight of truss; the former for very heavy loads, as in two truss double track bridges, and the latter for light loads.

When the span becomes very long, *i. e.*, 400 to 500 feet, the depth of truss increases to an unusual height, and the cost of erection is correspondingly large. The depth is then frequently taken not larger than one-eighth the span, or even less.

Again, local conditions, such as the necessarily uniform depth (for the sake of appearance) of adjacent spans of varying length, sufficient depth of short spans for over-head bracing (very necessary for lateral stability), etc., in the majority

23

of cases exclude the use of the economic depth, even if it were exactly known.

It is to be borne in mind, also, that the lightest truss is not necessarily the cheapest. That bridge is the most economical which can be made ready for traffic for the least money.

Facility in working up details, and the least possible amount of time in the shop, are very important elements, indeed, in every design.

In fact, the lightest weight does not make the most economical bridge, for the reason that the shop cost per pound is greater than with a somewhat increased weight of metal. When it is borne in mind that a considerable variation may be made from the depth of least weight, without affecting that weight to any considerable extent (as actual computations show to be the case), it is easy to understand that the truly economic depth is materially less than that which gives precisely the least weight of material.

Long panels are an economic feature of any bridge possessing a system of floor-beams and stringers, as well as conducive to other points of merit. The resulting concentration of metal not only leads to less weight and rate of cost in the shop, but enhances, also, the stiffness and stability of the individual members.

For economy in weight, long panels require a greater depth than shorter panels.

This much may be said in regard to continuous trusses: On account of the existence of the points of contraflexure, they require considerably less depth than trusses that are not continuous, used on the same points of support. The depth of the latter, therefore, is a limit which should never be reached by the depth of the former.

Art. 77.—Fixed and Moving Loads.

Both fixed and moving loads depend upon the local circumstances of each case, and the former very much upon the character of the design. A depth from one-fifth to one-seventh the span will give a very light fixed weight, but a depth of one-twelfth the span will involve a considerable increase of weight, while the moving load remains the same.

The weight of a single-track railway floor, for the present (1885) existing moving loads, may be taken at about 400 pounds per foot.

The moving load, also, depends upon the length of span. If the span is very great, the probability of the whole bridge being covered with an excessively heavy moving load is very slight, if any exists at all. If the span is short, one or two locomotives may cover the whole bridge, thus causing the moving load, per foot, to be very great for the whole span.

Thus it is seen that the moving load, per foot, may *decrease* as the span *increases*.

The whole matter of moving loads for both highway and railway bridges is well illustrated by the following tables, taken from "A Bill to secure greater Safety for Public Travel over Bridges," introduced in the Sixty-second General Assembly of the State of Ohio, shortly after the Ashtabula disaster:

For City and Suburban Highway Bridges.

Span in feet.			Moving load per square foot.
o	to	30	110 pounds.
30	"	50	100 "
50	"	75	90 "
75	"	100	80 "
100	"	200	75 "
200	"	400	65 "

All other Highway Bridges.

Span in feet.			Moving load per square foot.
o	to	30	100 pounds.
30	"	50	90 "
50	"	75	80 "
75	"	100	75 "
100	"	200	60 "
200	"	400	50 "

Railway Bridges.

Span in feet.			Moving load per lineal foot of each track.
0	to	$7\frac{1}{2}$	9,000 pounds.
$7\frac{1}{2}$	"	10	7,500 "
10	"	$12\frac{1}{2}$	6,700 "
$12\frac{1}{2}$	"	15	6,000 "
15	"	20	5,000 "
20	"	30	4,300 "
30	"	40	3,700 "
40	"	50	3,300 "
50	"	75	3,200 "
75	"	100	3,100 "
100	"	150	3,000 "
150	"	200	2,900 "
200	"	300	2,800 "
300	"	400	2,700 "
400	"	500	2,500 "

Floor-beams and stringers are really bridges of short spans equal to their lengths, consequently they must be designed for the heavy loads belonging to those short spans. Fig. 1 shows the locomotive weight specified by

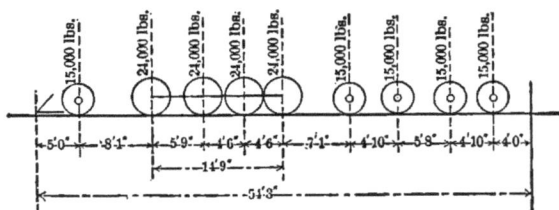

FIG. 1.

Mr. Theodore Cooper, C.E., in his "General Specifications for Iron Bridges and Viaducts," while Fig. 2 shows the heavy passenger locomotive used by Mr. Jas. M. Wilson, C.E., in his standard specifications for the Pennsylvania R. R.

These represent the heaviest engines of their type now in use. There is, however, a heavy decapod engine shown by Fig. 3, beginning to make its appearance.

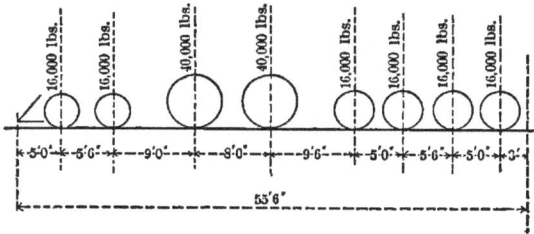

FIG. 2.

The moving load usually specified consists of two of any of these types of locomotives followed by a uniform train of 3,000 pounds per lineal foot. Occasionally the locomotive concentrations are followed by those of the train.

Besides the preceding heavy moving loads, there are in-

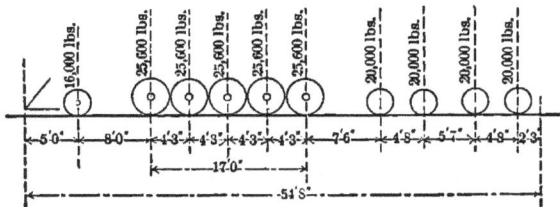

FIG. 3.

numerable lighter ones depending upon local circumstances of traffic.

The actual concentrations play a very important part in bridge computations. The old method of a uniform load, even with an engine excess, no longer fulfils the requirements of the best engineering practice, particularly in the treatment of short spans.

This is well illustrated by the following table which shows the uniform load per lineal foot which will produce the same chord stresses at the centre of the span as the actual concentrations shown in Fig. 1.

Span in feet.	Equiv. uniform load in lbs. per lin. ft.	Span in feet.	Equiv. uniform load in lbs. per lin. ft.
55	3,750	25	4,838
50	3,866	20	5,137
45	4,004	15	5,760
40	4,242	12	6,000
35	4,336	10	5,766
30	4,572	5	9,600

Above 55 feet the equivalent uniform load per lineal foot will slowly decrease until it reaches a value of about 3,200 lbs. for 100 feet and over, *i. e.*, supposing the moving load to consist of a train of such locomotives.

Art. 78.—Safety Factors and Working Stresses.

Although the subjects of safety factors and working stresses properly belong to the domain of the resistance of materials, they may here be touched upon in a general manner.

The fixed weight of a long span bridge is much greater, per foot, than that of a short span. Again, it has been seen in the preceding article that the moving load for a long span is much less than that for a short span. For both these reasons, the variations of stress in passing from a loaded to an unloaded condition are much greater in the material of a short span than that of a long one. Consequently, the material will be much more fatigued in a short span than in a long one.

Although the subject of the fatigue of metals is yet in an unsettled state, it is clearly established that these conditions of stress in short spans demand a larger safety factor, or smaller working stress, than those in the long spans.

Again, in any bridge or truss whatever, carrying a moving load, some parts are subject to a much greater variation of stress in the process of first being subject to, and then relieved of, loads than others.

Counter-braces may not be, and probably are not, strained at all by the fixed load ; but they take a proper working stress under the action of the moving load.

The condition of loading for greatest stress in any main web member, except those at the ends, is a partial covering of the span. But the fixed load is distributed over the whole span. Hence the variation of stress in the main web members will be greatest at the middle of the span, and least at the end. At the centre, however, the variation is much less than in the counter-braces.

The fatigue of the material, therefore, requires that the *greatest safety factors, or least working stresses, be found in the counter-braces; and that the working stresses in the main web members at the centre be greater than those in the counter-braces, but less than those in the main web members at the ends of the truss.*

The disposition of the moving load for the greatest chord stresses is, in all cases, essentially the same as that of the fixed load. Hence the variation of stress will be essentially the same throughout the chords, and the safety factor or working stress may be uniform throughout each chord ; the safety factor being the same as that in the end web members sustaining the same kind of stress.

If a structure is to carry a fixed load only, the safety factor may be three for wrought-iron and steel, and possibly as small for good qualities of cast-iron and timber. As a rule, however, cast-iron and timber require a larger safety factor than wrought-iron and steel. Local circumstances affect, to a great extent, working stress. If the risk (respecting life and property) attending failure is small, the safety factor may be small also. But if the risk is great, the safety factor must be correspondingly great.

In the truss members of long span bridges of wrought-iron and steel, the safety factors may vary from three and a half or four to five ; but in short spans of the same material, they should vary from about five to six or eight.

Good cast-iron should be found with safety factors varying from six to ten, while those for timber may vary from eight to twelve.

It is not to be supposed from these large safety factors that the determination of the stresses or the character of the various materials is so excessively uncertain. It is certainly true that there is some indetermination in these respects, but only a little in comparison with that connected with *the mode of application of the moving load.*

With a perfect condition of track, a rapidly moving train is supposed by many to approximate very closely to a suddenly applied load, although it is quite certain that it does not. For this reason some engineers have doubled the moving loads, in making their calculations, and then fixed the values of the safety factors as if all loads were gradually applied.

But no track is in perfect condition, and all rough places, or lack of continuity, such as rail joints more or less open, produce shocks which cause greater stress than any suddenly applied loads. The amounts of these last stresses are indeterminate, for the extent of their causes can scarcely be determined.

Again, Mr. J. W. Cloud, C. E., at the Philadelphia meeting of the American Institute of Mining Engineers, February, 1881, pointed out the existence of certain unrecognized stresses; such as those caused by the vertical component of the thrust of the connecting-rod of a locomotive, which alternates in direction twice in each revolution of the driving-wheels, thus producing a pulsating effect, as well as those which arise from the lack of balance of the driving-wheels in a vertical direction.

All these causes produce stresses which it is impossible to measure, and the safety factor must cover all uncertainties.

It is possible that a more highly perfected track and the production of more nearly uniform material in connection with an extended experience may justify the reduction of safety factors.

The following "Table of Tubular and Truss Bridges for Single and Double Track Railways, constructed of Iron and

Steel and having Spans exceeding 300 feet," gives the working stresses, loads, and other interesting data of some of the principal bridges of the world. It is taken (with the exception of No. 18) from the " Proceedings of the Institution of Civil Engineers" of Great Britain, Vol. LIV. The greater portion of it is there given in connection with a paper by Mr. T. C. Clarke, on " Long Span Bridges."

No.	Date of Erection.	Where Built.	Engineer.	Span in Feet between Points of Bearing.	Tons of Iron. (2000.00 lbs.)	Width between Centres of Trusses. F.In.	Panels Number.	Panels Length. F.In.	Height. F.In.	Total Dead Load of Iron and Timber in Track.	Live Load of Engines and Cars.	Tensile From Fixed Load only.	Tensile From Total Fixed and Moving Loads.	Compressive From Fixed Load only.	Compressive From Total Fixed and Moving Loads.	Test Load. Tons. (2000 lbs.)	Centre Deflection. In.	Dead Load of Iron, Timber, &c. Tons.	Deflection of Span from its own Weight on Removal of Scaffold'g. In.	Remarks.
1	1877	Susquehanna river; Havre de Grâce, U.S.A.	Phœnix Br. Co.	397	241	16.0	17	18.2	35.0	2150	2240	4900	10000	4400	9000	431	1.375	832		All rolled iron, except joint blocks. Pin connections; quadrangular truss.
2	1864	Ohio river; Steubenville, U. S. A.	J. H. Linville.	319	542	16.6	26	12.3	28.0	4000	3000	5700	10000	3100	6000					Top chords and posts cast-iron; rest rolled iron. Quadrangular truss.
3	1859	St. Lawrence river; Montreal, Canada.	R. Stephenson.	330	768	16.0	Ends Centre.		{22.0 30.0}	4800	2240	7680	11200	6060	8900				7.5	Tubular girder.
4	1870	Ohio river; Parkersburg and Bellaire, U. S. A.	J. H. Linville.	342	379	18.0	22	15.6	33.0	2500	3000	4545	10000	3636	8000					All rolled iron. Pin connections; quadrangular truss.
5	1862	Rhine river; Mayence.	Gerber.	345	402	15.1	13	25.4	{24.6 49.2}	4590	2890	7120	11600			504	0.27	805	2.25	Pauli system; all riveted work; lenticular girder.
6	1870	Ohio river; Louisville, U. S. A.	Albert Fink.	368	557	30.0	24	15.4	46.0	3668	2600	7000	Chords 12000 Diags. 10000	3500	6000	224	1.00			Top chords cast; rest, rolled iron. Pin connections; quadrangular truss.
7	1877	Kentucky river; Dixville, U. S. A.	C. Shaler Smith	375	476	18.0	20	18.9	37.6	2700	2037	5080	10000	4570	9000	371	1.62			All rolled iron. Pin connections; quadrangular truss.
8	1870	Ohio river; Louisville, U. S. A.	Albert Fink.	396	698	30.0	28	14.2	46.0	4168	2600	7400	Chords 12000 Diags. 10000	3700	6000	224	1.125			Top chords cast; rest rolled iron. Pin connections; quadrangular truss.

No.	Year	Location	Engineer															Remarks
9	1856	Vistula river; Dirschau.	Lentze.	397	939	21.8	Close Lattice.	28.6	6160	2128	7220	9720	7220	9720	337	2.13	1245.5	8.6	All rolled iron, one line of rails only taken. Close lattice with posts, two spans continuous.
10	1848	Conway	R. Stephenson.	400	1245.5	15.0	Tube.	25.6	6450	2240	9200	12400	4950	9000					Tubular girder.
11	1871	Ohio river; Cincinnati, U.S.A.	J. H. Linville.	415	930	19.0	20.9	20	41.6	5500	4500	5900	10000			185	0.13	444	0.90	Carriage way on each side of railway. Pin connections; quadrangular truss.
12	1861	Inn river; Passau.		320	366	15.0	14.0	23	30.0	2450	2680	6700	13700	5900	12000	430	1.20	1232	2.50	Riveted lattice. All rolled iron.
13	1859	Saltash.	I. K. Brunel.	455	1058	17.0	38.0	12	{30.0 / 00.0}	6500	2240	6650	8960							Lenticular truss.
14	1850	Menai Straits, Britannia; Wales.	Robert Stephenson.	460	1739	15.0	Tube.	30.0	7780	2240	10375	13336			278	0.63	1739	11.75	Tubular girder.
15	1868	Lch river; Kuilenberg, Holland.	G. Van Diesen.	492	2502	39.4*	13.1	38	{26.3 / 65.7}	{13200 / 3000 / 11800}		14560	8200	10000	666	1.32	3237	3.05	All rolled iron, riveted. Lattice, arched top.
16	1877	Ohio river; Cincinnati, U.S.A.	J. H. Linville.	515	1317	20.0	25.9	20	51.5	5400	1818	7479	10000	6600	9000	483	2.00	1384		All rolled iron. Pin connections; quadrangular truss; estimated camber, 4 in.; actual, 3⅜ in.

* This width is sufficient for two lines of rails, although only one has been laid.

SUPPLEMENTARY TABLE.

No.	Year	Location	Engineer														Remarks
1	1870	Zeglin river; Stettin.	Schröder.	302	503	27.3	Lattice.	17	19.8	{19.8 / 45.11}	3333	4024	10667	10667	10667			Double track.
2	1866	Mersey river; Runcorn.	W. Baker.	305	786	28.0			27.0	27.0	5667	3360	6618	10416	10667	4883	7795	382	1.25	All wrought iron. Estimated camber, 9 in.; actual, 9¼ in.
3	1872	Rhine river; Rheinhausen.	Hartwich.	317															
4	1876	Memel river; Tilsit.	Schwedler.	317	676.5	28.10	18	17.7	{17.0 / 39.1 / 24.7 / 46.2}	4700	5097	10667	10667					Double track.
5	1873	Vistula river; Thorn.	Schwedler.	319	678	37.9	18	{13.6 / 18.3}		4252	4830	10383	10383					" "
6	1869	Elb river; Hamburg.	Lohse.	325	674	27.6	26	12.4	{10.3 / 32.10 / 18.- / 37.0}	4144	4830	10667	10667					" "
7	1874	Elb river; Totschen.	Gerlich.	328	470	16.5	20	16.5		2864	2063	11378	11378	11378	396.5	1.53	Single track.
8	1872	Rhine river; Wesel.	Dreling.	334		27.5	28	12.4									Double "

No.	Date of Erection	Where Built	Engineer	Span in Feet between Points of Bearing	Tons of Iron (2000.00 lbs.)	Width between Centres of Trusses (F. In.)	Panels — Number	Panels — Length (F. In.)	Panels — Height (F. In.)	Total Dead Load of Iron and Timber in Track	Live Load of Engines and Cars	Tensile From Fixed Load only	Tensile From Total Fixed and Moving Loads	Compressive From Fixed Load only	Compressive From Total Fixed and Moving Loads	Test Load Tons (2000 lbs.)	Centre Deflection (In.)	Dead Load of Iron, Timber, &c. (Tons)	Deflection of Span from its own Weight on Removal of Scaffold'g. (In.)	Remarks
9	1875	Elb river; Hohnsdorf..	Grüttefien.	338	663	27.3	20	17.8	{ 23.7, 49.3 }	4292	4833	6000	9598	6000	9598	359.5	1.16	590		Double track.
10	1870	Maasriver; Crèvecoeur.		341	573	16.8	23	14.7	{ 23.0, 41.6 }	3455	2190									Single track.
11	1864	Old Rhine river; Griethausen.	Hartwich.	342	552	15.0	40	8.4	25.3	3413	2140		10383		10383	240	1.53			"
12	1870	Theiss river; Algyó.	Kőrösi.	342	506	16.5	20	17.0	34.2	2958	2132		11378		11378	349.5	1.38			"
13	1870	Rhine river; Mayence (New).	Pauli.	345	394	15.1	13	26.3	{ 0.0, 49.3 }	2455	2146	5689	11378	5689	11378					
14	1870	Rhine river; Düsseldorf.	Pichier.	347	738	28.0	27	12.4	{ 22.7, 44.5 }	4239	4272		10383		10383					Double track.
15	1878	Rhine river; Coblentz (New).	Hilf.	348																
16	1871	Hollandschdiep river; Moerdyk.	A Commission.	349	591	16.5	25	14.6	{ 19.8, 39.8 }	3212	2222	5042	8533	5042	8533	376	1.5	544		
17	1869	Waal river; Bommel.	G. Van Diesen.	408	963	17.2	27	14.11	{ 23.0, 42.7 }	4412	1915	6378	9241	6378	9954	394	1.32	914		
18	1880	Plattsmouth, Neb., Missouri river.	G. S. Morison.	400	422	22.0	16	25.00	50.00	2510	2000		{ 10000, 15000 }		15000 max.			502		Single track, Pratt except int. Steel posts

In No. 18, the 422 tons are for iron and steel. The writer is indebted to the kindness of Mr. Morison himself for the information in regard to this bridge.

Art. 79.—General Observations.

All abutting surfaces in bridges, or similar structures, should be very carefully machine finished.

Where pins bear against portions of the upper chord, as at *c*, *d*, and *e* of Fig. 5, Pl. III., the amount of bearing surface should be determined as for rivets, and sufficient area given by riveting on thickening plates, if necessary. A thickening plate is shown at the joints of the same figure. The number of rivets for the thickening plate is determined by the amount of pressure allowed on each one, as has already been shown.

If a finished piece is to fit into a finished cavity, however well the work may be done, there must be at least $\frac{1}{64}$ inch " play."

One end of a truss bridge, unless the span is very short, usually rests upon " expansion " rollers, from two to four inches in diameter. An approximate formula for the resistance of such rollers is given in the Appendix.

CHAPTER XI.

WIND STRESSES AND BRACED PIERS.

Art. 80.—Wind Pressure.

In a paper presented to the American Society of Civil Engineers (Transactions, Vol. X.), Mr. C. Schaler Smith gives some very valuable information in regard to wind pressure. The highest observed pressure which has come within his knowledge is 93 pounds per square foot. This pressure derailed a locomotive at East St. Louis, Mo., in 1871. In his own specification he says :—

"The portal, vertical, and horizontal bracing shall be proportioned for a wind pressure of 30 pounds per square foot on the surface of a train averaging 10 square feet per lineal foot, and on twice the vertical surface of one truss."

The wind pressure on a train is a moving load, and should be so considered, while the wind pressure on the trusses is a fixed load.

His experiments on the Rock Island draw-bridge showed that the wind pressure against the two trusses was over 1.8 times that on the exposed surface of one.

Again, quoting from his paper :—

" The Erie specifications are as follow :

Fixed load, roadway chord, 150 lbs. per lineal foot.
" " other " 150 " " " "
Moving " roadway " 300 " " " "
Iron in tension at 15,000 pounds.
" " compression, factor 4.

" The Pittsburg, Cincinnati and St. Louis Railway requires

300 pounds per foot for the train, and 30 pounds per square foot on one truss only.

"For the bridge over the Missouri, at Glasgow, 50 pounds per square foot on one truss, and 300 pounds per lineal foot of train were used.

"For the Eads bridge, at St. Louis, 50 pounds per square foot on the structure alone was the specified pressure.

"For the Kentucky River bridge the wind pressure was assumed at 31½ pounds per square foot on spans, train, and piers, and factor 4 was used in proportioning the bracing.

"The Portage bridge, New York, was built to resist 30 pounds per square foot on structure and train, and 50 pounds per square foot on the structure alone.

"The 520 feet span over the Ohio, at Cincinnati, was designed to withstand 50 pounds per square foot on structure alone, or 30 pounds per square foot on train and structure combined.

"A fully loaded passenger train, and the heaviest possible freight train, will leave the track at the respective pressures of 31¼ and 56½ pounds per square foot."

Engineers frequently specify 30 pounds per square foot of trusses and train combined, or 50 pounds per square foot of trusses alone.

300 pounds per linear foot of single track is also frequently used for moving wind pressure on train.

The following refers to the single track bridge at Plattsmouth, Neb., and is from the *Railroad Gazette*, 17th Dec., 1880: "The structure is also designed to resist a lateral wind pressure of 500 pounds per lineal foot on the floor, and 200 pounds per lineal foot on the top chord of the through spans and the bottom chord of the deck spans; these quantities are about equivalent to a wind pressure of 30 pounds per square foot on the bridge when covered by a train, and to 50 pounds per square foot on the empty bridge."

The following are a set of rules recommended for English practice, almost exactly in the words of the report :

Report of the Committee appointed to consider the Question of Wind Pressure on Railway Structures, by the Board of Trade of London, made on the 20th May, 1881.

The following rules were recommended :

(1). For railway bridges and viaducts, a maximum pressure of 56 pounds per square foot should be assumed for purposes of calculation.

(2). That when the bridge or viaduct is formed of close girders, and the tops of such girders are as high or higher than the tops of passing trains, the total wind pressure upon such bridge or viaduct should be ascertained by applying the full pressure of 56 pounds per square foot to the entire vertical surface of one main girder only. But if the top of a train passing over the bridge is higher than the tops of the main girders, the total wind pressure upon such bridge or viaduct should be ascertained by applying the full pressure of 56 pounds per square foot to the entire vertical surface, from the bottom of the main girders to the top of the train passing over the bridge.

(3). That when the bridge is of the lattice form, or of open construction, the wind pressure upon the outward or windward girder should be ascertained by applying the full pressure of 56 pounds per square foot, as if the girder were a close one, from the level of rails to the top of the train passing the bridge or viaduct, and by applying, in addition, the full pressure of 56 pounds per square foot to the ascertained vertical area of surface of the iron work of the same girder, situated below the level of the rails or above the top of a train passing over such bridge or viaduct. The wind pressure upon the inward or leeward girder or girders should be ascertained by applying a pressure per square foot to the ascertained vertical area of the surface of the iron work of one girder only, situated below the level of the rails, or above the top of a train passing over the said bridge or viaduct, according to the following scale :

(*a*). If the surface area of the open spaces does not exceed $\frac{2}{3}$ of the whole area included within the outline of the girder, the pressure should be taken at 28 pounds per square foot.

(*b*). If the surface area of the open spaces lie between $\frac{2}{3}$ and $\frac{3}{4}$ of the whole area included within the outline of the girder, the pressure should be taken at 42 pounds per square foot.

(*c*). If the surface area of the open spaces be greater than $\frac{3}{4}$ of the whole area included within the outline of the girder, the pressure should be taken at 56 pounds per square foot.

(4). That the pressure upon arches and piers of bridges and viaducts should be ascertained, as nearly as possible, in conformity with the rules above stated.

(5). That in order to insure a proper margin of safety for bridges and viaducts, in respect of the strains caused by wind pressure, they should be made of sufficient strength to withstand a strain of 4 times the amount due to the pressure calculated by the foregoing rules. And that for cases where the tendency of the wind to overthrow structures is counterbalanced by gravity alone, a safety factor of 2 will be sufficient.

JOHN HAWKSHAW.
W. G. ARMSTRONG.
W. H. BARLOW.
G. G. STOKES.
W. YOLLAND.

The evidence before us does not enable us to judge of the lateral extent of the extreme high pressures occasionally recorded by anemometers, and we think it desirable that experiments should be made to determine this question. If the lateral extent of exceptionally heavy gusts should prove to be very small, it would become a question whether some relaxation might not be permitted in the requirements of this report.

W. G. ARMSTRONG.
G. G. STOKES."

Up to the date of the above report, the highest pressure per square foot ever recorded at Glasgow was 47 pounds;

while the highest ever recorded at Bidston, near Liverpool, was 90 pounds per square foot. At another time, at Bidston, 80 pounds per square foot was recorded.

The above committee also found that, if P is the maximum pressure per square foot, V the maximum run of wind in miles per hour, both these quantities being observed by anemometers, the following equation very nearly held true :

$$P = \frac{V^2}{100}.$$

Art. 81.—Sway Bracing.

The construction of the upper and lower sway bracing of a truss must, so far as the jar and oscillation of a moving road are concerned, be a matter of judgment ; but the stresses due to the action of the wind may be determined with sufficient accuracy.

Although a moving train will partially shelter one truss, it seems no more than prudent, with the ordinary open style of American bridge, to consider the action of the wind as existing constantly, during the passing of a train, over the whole of the projection of each truss in the bridge on a plane normal to the direction of the wind. If this is considered excessive, however, for low trusses, that portion of the windward truss sheltered by the train may be omitted.

Let Fig. 1 and Fig. 2 represent a single concellation railway truss bridge, with vertical and diagonal bracing, and let the wind be supposed to blow in the direction shown by the arrow, which is normal to the planes of the trusses. Primed letters belong to the truss $DC'N'O,'$ but all are not shown.

As the truss is a " through " one, all wind pressure against the floor system will act in the lower chord.

With the wind pressure between thirty and forty pounds per square foot, the following loads may be taken at the various panel points :

At *C', G', K', L', M', N', C, G, K, L, M, N*....... 0.35 tons.
" *D* and *O'*.................................. ... 0.18 "
" Intermediate points......................... 0.35 "
" *B* and *O*....... .... 0.18 "
" Intermediate points......................... 3.01 "

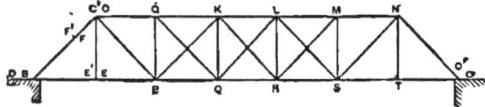

FIG. 1.

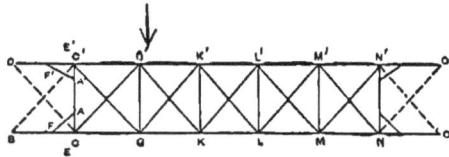

FIG. 2.

The amount 3.01 tons involves the pressure against the train, which is taken at 300 pounds per foot of track. The panel length is fourteen feet, hence the panel train load is $14 \times 300 = 4,200$ pounds $= 2.10$ tons. The wind pressure against the floor system is assumed to be 0.56 ton per panel, while the panel pressure against each truss is 0.35 ton. The sum of the three quantities is 3.01 tons.

The panel train loads (2.10 tons) constitute a continuous, moving load; the wind pressure against the trusses and floor system, however, forms a *fixed load.*

The following are the truss dimensions, including the lengths of the braces *AF* and *A'F'* :

Panel length $= 14.00$ feet.   Height of truss $= 16.00$ feet.
Width, *BD* $= 14.00$ "   *BC* $= 21.26$ "
       CF $= 7.00$ "   *AC* $= 4.04$ "

$$FA = 8.08 \text{ feet.}$$

Normal from C on $FA = CF \times sin\ 30° = 3.5$ feet.

Let H represent half the total wind pressure concentrated in the two upper chords; this will be resisted (if the bridge is not blown bodily off the abutments or piers) by an equal force of friction developed at the feet B and D, or O and O' of the end posts. Let H' and H'' be the forces developed at D and B, respectively. These horizontal forces will tend to overturn the trusses in a vertical plane normal to the axis of the bridge. A vertically upward reaction, V, will be developed at B, and an equal downward one (a portion of the weight of the truss $DC'N'O'$) at C'. Considering the left end of the truss, the following three conditional equations of equilibrium must be fulfilled :

$$H' + H'' + H = 0 \quad . \quad . \quad . \quad . \quad . \quad . \quad (1).$$

$$V + V' = 0 \quad . \quad . \quad . \quad . \quad . \quad . \quad (2).$$

$$(H' + H'') \times 16 + V \times 14 = 0 \quad . \quad . \quad . \quad . \quad . \quad . \quad (3).$$

The vertical force acting at C' is represented by V'.

These three equations are not sufficient for the determination of the four quantities H', H'', V, and V'. The forces H' and H'' are therefore indeterminate in magnitude, except in this respect, their sum must be equal and opposite to H.

With the form of portal bracing shown in Fig. 1, it will be assumed in this article that $H'' = 0$ and $H' = -H$. Other and better forms of portal, together with other assumptions in regard to the horizontal reactions H' and H'', will be given in succeeding articles.

From the data already given :

$$H = 6 \times 0.35 = 2.10 \text{ tons.}$$

Hence by Eqs. (3) and (2) ;

$$V = 2.40 \text{ tons} = -V'.$$

At the point E' let there be supposed to act two forces

equal, opposite, and parallel to H and H'; these two forces will balance each other. Instead of the two forces H and H', there may then be taken two couples, $M' = H' \times 14$, and $M = H \times 16$.

In the same manner at E let two forces equal, opposite, and parallel to V and V' be supposed to act. Then, instead of V and V', there will exist two couples, $M'' = V \times 14$, and $M''' = V' \times 14$.

The couples whose moments are M and M''' balance each other, as is shown by Eq. (3). The couples whose moments are M' and M'' have axes at right angles, consequently their resultant will be :

$$M_1 = \sqrt{M'^2 + M''^2} = 14 \sqrt{(2.1)^2 + (2.4)^2} = 44.66 \text{ ft. tons.}$$

The plane in which M_1 acts, contains the chords BO and $C'N'$, and the direction of the couple is such that it causes compression in $C'N'$ and tension in BO. Consequently for those stresses

$$(BO) = - (C'N') = 44.66 \div 21.26 = + 2.1 \text{ tons.}$$

As a check ; $(BO) = + V \times 14 \div 16 = + 2.1$ tons.

The following stresses in the members of the portal of the bridge may now be written :

$$(A'F') = H \times 21.26 \div 3.5 = + 12.8 \text{ tons.}$$
$$(A'C') = - 12.8 \times sin\ 30° = - 6.4 \text{ ``}$$
$$(F'C') = - 12.8 \times cos\ 30° = - 11.1 \text{ ``}$$

The greatest bending moment in DC' exists at F', and is :

$$M_2 = 2.1 \times (21.26 - 7.00) = 29.95 \text{ ft. tons.}$$

The stress in BC is :

$$(BC) = - \frac{21.26}{14} \times H = - \frac{21.26}{16} \times V = - 3.19 \text{ tons.}$$

The compressive stress in DC', due to the vertical loading, is relieved by the same amount.

The greatest bending moment in CC' exists at A', and has for its value :

$$M_3 = (BC) \times (14.00 - 4.04) = 31.77 \text{ ft. tons.}$$

With the wind in the direction taken, the brace AF must be supposed not to act at all. *Both moments M_2 and M_3 produce bending in the plane of the portal.*

The end post (DC'), always of uniform cross section, must be able to resist with a proper safety factor, at F', the bending moment M_2. The sway brace CC' must be able to resist the moment M_3 at *both* the points A and A'.

The ordinary truss stresses in the sway bracing remain to be found.

In both upper and lower sway bracing the inclined members are tension ones only, while those normal to the planes of the trusses (in the direction of the wind) sustain compression only.

In the upper chord the truss $C'GMN'$ has the two points of support C' and N'. The following trigonometric quantities will be required :

$$\text{Angle } G'KK' = 45° \qquad tan\ 45° = 1.$$

$$sec\ 45° = 1.414.$$

The upper web stresses are the following :

$$
\begin{aligned}
(KK') = & = -0.35 \text{ tons.} \\
(G'K) = +\ 2 \times 0.35 \times sec\ 45° & = +1.00 \text{ "} \\
(G'G) = -\ 3 \times 0.35 & = -1.05 \text{ "} \\
(GC') = +\ 4 \times 0.35 \times sec\ 45° & = +2.00 \text{ "} \\
(C'C) = & = +0.35 \text{ "}
\end{aligned}
$$

The resultant upper chord stresses are the following :

$$
\begin{aligned}
(C'G) = -\ 4 \times 0.35 \times tan\ 45° & = -1.40 \text{ tons.} \\
(G'M) = -\ 2 \text{ " } \text{ " } \text{ " } -1.40 & = -2.10 \text{ "}
\end{aligned}
$$

$$(GK) = + 4 \times 0.35 \times \tan 45° \qquad = + \; 1.40 \text{ tons.}$$
$$(KL) = + 2 \text{ `` `` ``} + 1.40 = + 2.10 \text{ ``}$$

The lower resultant web stresses are the following, remembering that the train* pressure is a moving load, and that D and O' are the supporting points for the lower sway truss:

$$(Q'R) = + \; 6 \times 0.30 \times \sec 45° \qquad\qquad = + \; 2.56 \text{ tons.}$$
$$(QQ') = - \; 6 \times 0.30 \qquad\qquad\qquad - 0.35 = - 2.15 \text{ ``}$$
$$(P'Q) = + 10 \times 0.30 \times \sec 45° + 2 \times 0.63 \times \sec 45° = + \; 6.04 \text{ ``}$$
$$(PP') = - 10 \times 0.30 \qquad\quad - 3 \times 0.63 + \; 0.28 = - 4.61 \text{ ``}$$
$$(E'P) = + 15 \times 0.30 \times \sec 45° + 4 \times 0.63 \times \sec 45° = + \; 9.96 \text{ ``}$$
$$(EE') = - 15 \times 0.30 \qquad\quad - 5 \times 0.63 + \; 0.28 = - 7.37 \text{ ``}$$
$$(DE) = + 21 \times 0.30 \times \sec 45° + 6 \times 0.63 \times \sec 45° = + 14.31 \text{ ``}$$

The $+ 0.28$ ton, which is a release, is due to the fact that the *half* panel wind pressure against the floor system is added to 0.35 ton, and taken *once* too many times in each of the struts.

The quantity 0.30 will be at once recognized as $2.10 \div 7$. The counters $S'T$ and $R'S$ are not required to resist wind stresses, but should never be omitted, in order that the general stiffness of the bridge may be increased; their cross sections may be the same as that of $Q'R$.

The lower resultant chord stresses are the following:

$$(DE') = - (6 \times 0.63 + 3 \times 2.10) \quad \tan 45° \quad = - 10.08 \text{ tons.}$$
$$(E'P') = - (4 \times 0.63 + 2 \times 2.10) \quad \text{`` } - 10.08 = - 16.80 \text{ ``}$$
$$(P'S') = - (2 \times 0.63 + \qquad 2.10) \quad \text{`` } - 16.80 = - 20.16 \text{ ``}$$

$$(EP') = - (DE') + 2.10 = + 12.18 \text{ tons.}$$
$$(PQ) = - (E'P') + 2.10 = + 18.90 \text{ ``}$$
$$(QR) = - (P'S') + 2.10 = + 22.26 \text{ ``}$$

Although not a part of the lower chord of the truss under consideration, BE sustains the stress:

$$(BE) = + 2.10 \text{ tons.}$$

---

* The train is taken as passing from right to left.

If the wind blows in the opposite direction to that assumed, the chord stresses which have been determined for $C'N'$ will be found in CN, and *vice versa.* Precisely corresponding changes are to be made in the lower chords. The stresses in the sway struts would not be changed. That diagonal in each panel which is not stressed in the preceding instance, would sustain a tensile stress exactly equal to that already found in the other diagonal.

It is therefore necessary to make calculations for but one direction of the wind.

So far as equilibrium is concerned, in the preceding investigation, there might be taken $H'' = -H$ and $H' = 0$. In such a case BC would be subjected to a bending moment at F equal to $-M_2$; and the bending moment in CC', at A, would be $-M_3$; while the stresses in FA, AC, and CF would be respectively $-(FA')$, $-(A'C')$, and $-(C'F')$. For these reasons all parts of the portal should be built to sustain the stresses and moments which have been found when affected by opposite signs.

It should be remembered that the parts EC, $E'C'$, and CC' are subjected to combined direct stresses and bendings to the respective amounts that have been found.

Those portions of the lower sway struts EE', PP', etc., extending from the windward rail to the lower chord BO (with the direction of the wind first assumed), are each subjected to a compressive stress, in addition to those already found, nearly equal to an amount to be determined in the following manner : Let N represent the number of panels in the sway truss, and n the number of any strut, from the farther end of the truss, counting the end itself zero, *i. e.,* for PP', n will be 5. In the case taken $N = 7$. The amount desired will then be *the panel train wind load multiplied by* $\frac{n}{N}$ *added to the panel wind pressure against the floor system;* or in the example,

$$2.10 \times \frac{n}{7} + 0.56 = 0.30 \times n + 0.56.$$

This compression in the struts arises from the fact that the wind pressure against train and floor system is not applied at panel points, but *on the struts between their ends*, and that a panel load of the former must be added to the ordinary strut stress which exists with the head of the train at the strut considered.

This involves, however, a very small error on the side of safety, since the pressure is divided between the two rails. Considering both directions of the wind, it will be seen that these struts are subjected to this amount of compression from end to end, in addition to the regular truss stresses.

All the preceding wind stresses are to be combined with those due to the vertical loading, wherever they act in the same piece.

If the portals are vertical, the stresses (BO) and $(C'N')$, due to the moment M_1, will be zero; also the span of the upper sway truss will be equal to that of the lower. No other changes will occur.

If the bridge is a deck one, when possible, the ends of the chords should be secured directly to the piers or abutments, as no bending will then take place in the end posts. If this is not possible, the calculations will be precisely the same as those already indicated, with possible changes of signs in some of the stresses in the end posts or braces. In deck bridges, however, the wind pressure against floor system and train will be found in the upper chord.

The method of treatment which has been exemplified is, therefore, perfectly general and sufficient for all cases.

In deck bridges, tension sway braces (contained in planes normal to the trusses) are introduced, extending from either chord of one truss to the diagonally opposite one in the adjacent truss. So far as pure equilibrium is concerned, when horizontal sway trusses are present, these are superfluous; but they are very efficient in their influence on lateral stability. The actual stresses, in these members, in any given case, are indeterminate, but their greatest possible values are easily fixed. Let the total wind pressure exerted at a pair of opposite panel points in the two upper chords be represented by

P', and let α represent the angle between a horizontal line and the tension brace in question. Then the greatest possible stress which is required will be : $T' = P'\ \sec\ \alpha$.

This assumes that *all* the wind pressure is carried to the lower chords and resisted by the lower sway truss.

In the case of vertical end posts, where the upper chord ends are not secured directly to piers or abutments, the stresses in the end lateral diagonals become perfectly determinate. In fact, in such a case, the braces AF, $A'F'$, Figs. 1 and 2, become the diagonals in question. Let P represent half the *total* wind pressure in the two upper chords. The tensile stress in either one of these diagonals (if α retains its preceding signification) will then be ;

$$T = P_1\ \sec\ \alpha.$$

Let P represent the total wind pressure against the bridge, and P'' the total wind pressure against the train when it covers the whole span. Then let W and W' represent the total weight of bridge and train respectively ; also let f be the coefficient of friction between the foot of end post and the supporting surface underneath. In order that neither truss shall possibly be moved bodily by the wind, with the bridge empty or covered by a train, there must exist the following relations :

$$P < f\left(\frac{W}{2} - 2V\right)\ ;\ \text{or, } P + P'' < f\left(\frac{W + W'}{2} - 2V_1\right).$$

In the cases of through bridges, or of deck bridges with upper chords *not* secured to abutments, V_1 is to be found by applying the general form of Eq. (3) to *both bridges and train.*

If the ends of the chords are secured to the piers or abutments, the resistances of these fastenings will take the place of the frictional resistances.

One other effect of the wind pressure against the train remains to be noticed. The normal action of this pressure will not permit the train's weight to be distributed between the two chords which carry it, according to the law of the lever. Let p'' represent a panel wind pressure against the train, and let h represent the height of its centre of action above the points of support. Also let b represent the horizontal distance between centres of trusses; then

$$t = \frac{p''h}{b} \quad . \quad . \quad . \quad . \quad . \quad (4),$$

will be the amount of load which is transferred from the windward to the leeward truss. In other words, the panel leeward load will exceed the panel windward one by $2t$. If, therefore, w' is a panel moving load, the action of the wind will cause all moving load truss stresses to be increased by an amount found by multiplying the stresses, determined without regard to the wind, by $\frac{t}{w^1}$. Also, if s is the distance (normal) between two adjacent parallel stringers, the increase of load on one, and decrease of that on the other, will be:

$$t_1 = \frac{p''h}{s} \quad . \quad . \quad . \quad . \quad . \quad (5).$$

Eq. (5) gives the variation of load on the floor beam also. Without essential error, h may be measured downward from the centre of the body of the car.

Specifications sometimes require calculations to be made with an unloaded bridge. In such a case the methods are precisely the same as the preceding, with the train wind pressure omitted.

Art. 82.—Transverse Bracing for Transferring Wind Stresses from One Chord to Another—Concentrated Reaction.

In the preceding Article it has been supposed that the wind pressure is resisted by sway trusses in the horizontal planes of both upper and lower chord. It may sometimes be desirable to transfer *all* the wind pressure to the lower chord, or to the upper.

The section of a through truss bridge, in which it is desired to carry all the wind pressure to the lower chord, is

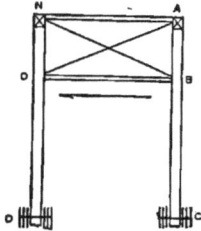

FIG. 1.

represented in Fig. 1. *AC* and *ND* are posts directly opposite to each other in the two trusses. *AN* and *OB* are lateral *struts*, while *AO* and *BN* are lateral ties. Let the wind be supposed to blow from right to left, as shown by the arrow. According to the principles of the preceding Article, in consequence of its direction the wind will relieve the truss *AC* of a part of the weight which it carries, and add the same amount to that carried by the truss *DN*.

If the direction of the wind were reversed, the truss *DN* would be relieved, and *AC* would receive the increase of loading.

Let this relief (or increase) of truss load, per panel, be denoted by *w ; it will act as though hung from B.*

The following notation, also, will be used :

$$AB = d = ON. \qquad BC = a = OD.$$

$$DC = AN = b.$$

$F =$ total wind pressure, per panel (for one truss), on

$$\tfrac{1}{2}(AB + BC).$$

$F' =$ total wind pressure, per panel (for one truss), on

$$\tfrac{1}{2}AB.$$

With the assumed direction of the wind, the tie *AO* will not be stressed. As usual, the plus sign (+) will indicate tension, while the minus (−) sign will indicate compression.

In this article the total horizontal reaction, equal to $2(F + F')$, will be taken as concentrated at *D*.

There will then result:

Relief in truss $AC = w = 2\left(\dfrac{F'\,(a + d) + Fa}{b}\right).$. (1).

Compression in $AN = \quad -(AN) = -F'.$ . . (2).

Tension in $BN \quad = \quad +(BN) = + w \sec ABN.$ (3).

Compression in $BO = \quad -(BO) = -(F + w \tan ABN)$

$$= \quad -\left(F + 2F' + [F + F']\frac{2a}{d}\right)$$

$$= F - 2(F + F')\left(\frac{a + d}{d}\right). \quad (4).$$

Compression in $ND = \quad -(ND) = -w.$ . . (5).

The horizontal force $2(F + F')$ acts toward C, at D, producing a bending in DN which has its greatest moment M at O. Hence:

$$M = 2(F + F')a. \quad . \quad . \quad . \quad . \quad . \quad (6).$$

If K is the greatest intensity of compressive stress (due to flexure) in the cross section of the post DN, at O, d_1 the greatest distance of any compressed fibre from the neutral axis of the cross section, and I the moment of inertia of the section about an axis passing through its centre of gravity and lying in the plane of the truss; then, by the well-known formula—

$$K = \frac{d_1\,M}{I}. \quad . \quad . \quad . \quad . \quad (7).$$

At O there will then exist the intensity of compression:

$-\left(\dfrac{w}{q} + K\right)$; in which q is the area of cross section of the column. The intensity of compression $-\left(\dfrac{w}{q}+K\right)$ is in addition to the regular truss stresses arising from vertical and wind loads.

If, for lack of head room, a flanged beam only is used, as shown in Fig. 2, instead of the lateral bracing of Fig. 1, then that beam will be subjected to combined compression and bending. Let $+F$ represent the total wind pressure, per panel, *for both trusses*, on $\frac{1}{2}AB$, and let w represent the release of weight in AB and increase in CD.

Also let

$AB = a$, and $BD = b$. $\frac{1}{2}F$ is to be taken as applied at A and w, at the same point. Equal and opposite forces are also to be supposed

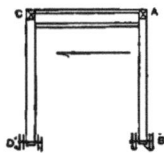

FIG. 2.

to act at D. The moment

$M = Fa = wb$, exists at C and gives:

$$w = \frac{Fa}{b}.$$

With the direction of wind shown by the arrow, the bending caused by w will increase uniformly from nothing at A to

$$M = wb$$

at C. The bending moment, therefore, to be resisted by this beam AC, *and by the joints between it and the chords A and C,* is :

$$M = Fa = wb. \quad . \quad . \quad . \quad . \quad (8).$$

The direct compression in AC is :

$$(AC) = -\tfrac{1}{2}F. \quad . \quad . \quad . \quad . \quad (9).$$

Hence, if q is the area of cross section of the beam, and if K, I, and d_1 retain the same general signification as in Eq. (7), the greatest intensity of compression in the beam (at its ends) will be :

$$-\left(\frac{\frac{1}{2}F}{q}+\frac{d_1 M}{I}\right). \quad . \quad . \quad . \quad . \quad . \quad (10).$$

The direct compression in CD is

$$(CD) = -w. \quad . \quad . \quad . \quad . \quad . \quad (11).$$

The bending moment in CD at C is:

$$M = Fa. \quad . \quad . \quad . \quad . \quad . \quad (12).$$

The greatest compressive intensity is found at once by Eq. (10), after writing w for $\frac{1}{2}F$, and giving to the remaining notation its general signification.

The two preceding cases are those of through trusses. In the case of a deck truss the lateral bracing is of much more simple character; it is shown in Fig. 3. At C and A are the two lower chords. CA is a lateral strut, while BC and DA are lateral ties. No parts are subjected to bending.

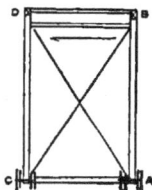

FIG. 3.

If F is the panel wind pressure (for both trusses) acting along AC, there will result:

$$(CA) = -\tfrac{1}{2}F. \quad . \quad . \quad . \quad . \quad . \quad (13).$$

$$(BA) = -w. \quad . \quad . \quad . \quad . \quad . \quad (14).$$

$$(BC) = +\sqrt{F^2 + w^2}. \quad . \quad . \quad . \quad (15).$$

The horizontal component of (BC) is equal to F, and acts at B. Thus all wind pressure is carried to the upper chord.

The compression (BA) is in addition to the regular truss stresses induced by the vertical and wind loads.

By these methods all the wind pressure may be carried to either chord. The truss stresses of the sway truss in the horizontal plane of that chord have already been found in the preceding Article, or rather, the methods for finding them and the effect of w on the stresses in the vertical trusses have there been completely given.

The wind has been taken in one direction only; with the other direction, opposite but symmetrically located parts would be stressed by the amounts found.

Art. 83.—Transverse Bracing with Distributed Reactions.

In the preceding articles it has been assumed that the horizontal reactions of the wind pressure were concentrated at the extremity (top or bottom, as the case may be) of one post in the tranverse panel considered. This assumption, however, may not be admitted; or some other may be substituted in its place.

Let Fig. 1 represent a transverse panel, with the wind blowing in the direction shown by the arrow.

As before, the following notation will be used:

$$AB = ON = d. \qquad BC = OD = a.$$
$$DC = AN = b.$$

$F =$ total wind pressure, per panel, for one truss, on $\frac{1}{2}(AB+BC)$.
$F' =$ " " " " " " " " $\frac{1}{2}AB$.
$w =$ relief of load in truss AC.

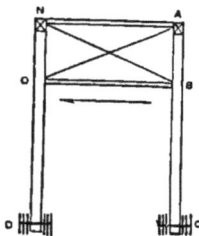

Fig. 1.

Instead of concentrating the entire horizontal reaction at D, if n is a quantity less than unity, there will be assumed:

Horizontal reaction at $D = 2n(F + F')$.
" " " $C = 2(1 - n)(F + F')$.

The wind pressures on $\frac{1}{2}BC = \frac{1}{2}OD$ act directly at C and D in the horizontal sway truss, and, consequently, will be omitted from consideration.

As in the preceding article:

$$w = \frac{2(F'(a + d) + Fa)}{b}. \qquad \qquad (1).$$

Taking moments about B:

$$(AN) = -\left[F' + 2(1 - n)(F + F')\frac{a}{d}\right]. \quad \cdot \quad \cdot \quad (2).$$

Taking moments about N:

$$(OB) = -\left[\frac{2n(F + F')(a + d)}{d} - F)\right]. \quad \cdot \quad \cdot \quad \cdot \quad (3).$$

Taking moments about the intersection of AN and OB at the distance infinity (∞) from the figure:

$$(BN)\infty \cos ABN = +[w\infty + 2n(F + F')a] ;$$

$$\therefore \ (BN) = + w \sec ABN = \frac{2(F'(a + d) + Fa)}{b} \sec ABN. (4).$$

The stress in BN, therefore, remains the same whatever may be the assumptions in regard to the horizontal reactions.

The bending moment at O, about an axis lying in the plane of the vertical truss, will be:

$$M = 2n(F + F')a. \quad \cdot \quad \cdot \quad \cdot \quad (5).$$

Since the windward truss is always relieved of a part of its weight the bending moment $2(1 - n)(F + F')a$, at B, will seldom or never be needed.

The value of M, from Eq. (5), put in Eq. (7) of the preceding Article, and in the expression following that equation, will enable the greatest compressive intensity in the post to be found.

If the transverse panel, Fig. 1, represents the portal of a bridge, the distances AB and BC, or d and a, represent *inclined distances in the plane of the portal. F' will (or may) then include, also, the reaction of the horizontal sway truss in the plane of AN*, while F will include the reaction of the horizontal sway truss in the plane of OB, if there is such a sway truss.

25

If $n = \frac{1}{2}$, as is sometimes assumed:

$$(AN) = -\left[F' + (F + F')\frac{a}{d}\right]. \quad \cdots \quad (6).$$

$$(OB) = -\left[F' + (F + F')\frac{a}{d}\right] = (AN). \quad \cdots \quad (7).$$

If $n = 1$ in the formulæ of this Article, those of the cor-responding cases in the preceding Article at once follow.

In Fig. 2 let the notation be as follows:

$$AB = CD = a. \qquad BD = CA = b.$$

Total wind pressure for *both trusses*, per panel, along

$$AC = F.$$

Horizontal reaction at $D = nF.$ . . . . . (8).
" " " $B = (1 - n) F.$ . . . (9).

If Fig. 2 represents a portal, F will or may include the reaction of a horizontal sway truss.

The bending moment on both DC and CA, at C, also on the joint at the same point, is:

$$M_1 = nFa. \quad \cdots \quad (10).$$

This is the greatest bending in DC and CA of that kind which produces *compression* in the lower flange of the beam CA.

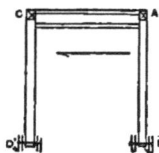

FIG. 2.

The relief of panel load in the truss AB and increase of that in CD is;

$$w = \frac{Fa}{b} \quad \cdots \quad (10).$$

Let x represent any variable portion of CA; then the bending moment at any point of CA is:

$$M = n\,Fa - wx = n\,Fa - \frac{Fa}{b}\,x. \quad . \quad . \quad (11).$$

For the point or joint A, x becomes equal to b, while the expression for the moment is:

$$M'_1 = -(1 - n)\,Fa. \quad . \quad . \quad . \quad (12).$$

This is the greatest bending of the kind opposite to M_1, in CA. It is also the greatest bending in AB. The connections at C must resist the moment M_1, while those at A must resist M'_1.

> The direct compression in CA is $\frac{1}{2}F$.
>   "   " " CD " w.

$$\text{If } n = \tfrac{1}{2}; \quad M_1 = -M'_1 = \tfrac{1}{2}Fa. \quad . \quad . \quad . \quad (13).$$

These various bending moments, substituted in Eq. (10) of the preceding Article for M, will enable the greatest intensities of stress in the members CA, CD, and AB to be at once found.

Art. 84.—Stresses in Braced Piers.

The general treatment of stresses in braced piers may be exemplified by that of a single "bent" represented by a skeleton diagram in Fig. 1, in which the horizonal web members are compressive ones. The plane of the "bent" is vertical and normal to the centre line of the truss whose end rests upon it; or if the track is curved, this plane is normal to it. The bent shown in Fig. 1 may be considered one of a pair, in parallel planes, which, being braced together, compose the complete braced pier. The dotted rectangle $AMNB$ represents a skeleton section of the truss supported by the piers, the upper chords of which rest upon the top of the pier at A and B. A skeleton section of the train is also shown.

The direction of the wind is supposed to be shown by the arrows a, normal to the track at the top of the pier. If the trusses are loaded with a train, the wind pressure against them and the train will be carried to the top of the piers in the manner shown in Art. 81. The wind will also act against the pier itself.

Let the train be supposed to cover the whole of the two spans adjacent to the top of the bent (in all ordinary cases one of these spans will be the distance between two adjacent bents); then let H represent *half* the total pressure against trusses, and P_1'' half that on the train covering the two spans.

The height of the centre of action of P_1'' above AB, Fig. 1, is h. Also let $b = AB$. The pressure P_1'' will *decrease* the train reaction at A and *increase* that at B by the amount:

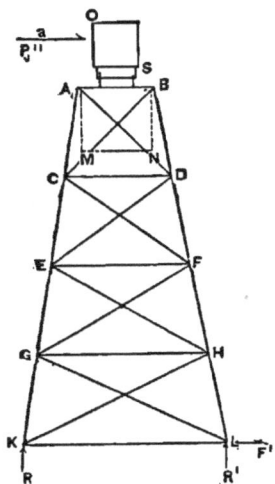

FIG. 1.

$$V_1 = \frac{P_1''h}{b}. \qquad \ldots \ldots \ldots \quad (1)$$

Let h' represent the vertical distance of the centre of action of H from the horizontal line AB.

The wind pressure on the truss $AMNB$ will cause an increase of truss reaction at A, and an equal decrease of that at B, which will be denoted by V, and its value will be:

$$V = \frac{Hh'}{b},$$

consequently if

$$t' = V_1 - V = \frac{P_1''h}{b} - \frac{Hh'}{b} ;$$

the total horizontal force to be taken as acting at A, and with the wind, will be $(H + P_1'' - 2t' \, tan \, \alpha)$ added to the wind pressure acting directly at A. In Fig. 2, *cd* represents this force, laid down to any desired scale. The small segments measured to the right of d represent the panel wind pressures against the pier at the points C, E, G, and K, while those shown on the left of c represent the panel pressures at D, B, F, H, and L. The panel pressures at A, B, K, and L are half those at the other points.

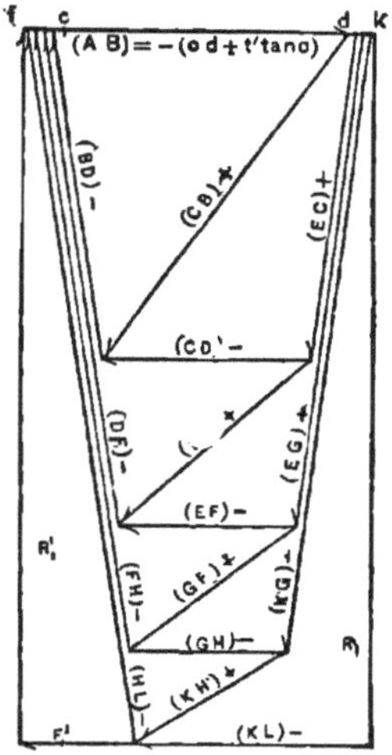

Let W represent the total weight of adjacent trusses and moving load resting at the top of the pier.

Let W_1 represent the panel weight of the pier itself resting at the points C, E, G, D, F, H: $\frac{1}{2} W_1$ will be taken as applied at the points A, B, K, and L; then the resultant reactions at A and B, with the wind blowing, will be, respectively,

FIG. 2.

$$\frac{W}{2} - t' \text{ and } \frac{W}{2} + t'. \quad . \quad . \quad . \quad . \quad (2).$$

It has been implicitly supposed that two equal and opposite forces, equal in magnitude and parallel to P_1'', act along AB. One of these forms, with P_1'' itself, the couple $P_1'' h$; the other is the wind pressure which, combined with the half panel pressure at A, and $(H - 2t' \tan \alpha)$, is represented by cd in Fig. 2.

The quantity t' is the force of a couple whose lever arm is b. One force t' is therefore supposed to act at A, and the other at B. V_1 will be considered larger than V; hence t' will act upward at A and downward at B. If α is the angle between AK or BL and a vertical line, the t' at B will cause a compression in AB equal to $t' \tan \alpha$, while the t' at A will pull to the left by the same amount. Consequently the force $2t' \tan \alpha$ will act on the point A and toward the left.

In the diagrams and in the equations which follow, positive and negative signs indicate tensile and compressive stresses, respectively.

The stresses due to vertical loads at A and B, and th? other panel points, will be the following :

$$(AC)'' = -\left(\frac{W}{2} - t' + \frac{W_{\prime}}{2}\right) \sec \alpha.$$

$$(CE)'' = -\left(\text{``} \quad \text{``} + \frac{3W_{\prime}}{2}\right) \quad \text{``}$$

$$(EG)'' = -\left(\text{``} \quad \text{``} + \frac{5W_{\prime}}{2}\right) \quad \text{``}$$

$$(GK)'' = -\left(\text{``} \quad \text{``} + \frac{7W_{\prime}}{2}\right) \quad \text{``}$$

$$(BD)'' = -\left(\frac{W}{2} + t' + \frac{W_{\prime}}{2}\right) \quad \text{``}$$

$$(DF)'' = -\left(\text{``} \quad \text{``} + \frac{3W_{\prime}}{2}\right) \quad \text{``}$$

$$(FH)'' = -\left(\text{``} \quad \text{``} + \frac{5W_{\prime}}{2}\right) \quad \text{``}$$

$$(HL)'' = -\left(\text{``} \quad \text{``} + \frac{7W_{\prime}}{2}\right) \quad \text{``}$$

$$(AB)'' = -\left(\frac{W}{2} + \frac{W_{\prime}}{2}\right) \tan \alpha.$$

$$(CD)'' = -W_{\prime} \tan \alpha.$$

$$(EF)'' = -\text{``} \quad \text{``}$$

$$(GH)'' = -\text{``} \quad \text{``}$$

$$(KL)'' = +\left(\frac{W}{2} - t' + \frac{7W_{\prime}}{2}\right) \tan \alpha.$$

The difference between the horizontal component in HL and $(KL)''$ is $2t' \tan \alpha$, and it acts towards the right.

The stresses caused by the horizontal wind pressure acting through A, B, C, D, etc., are shown in Fig. 2, as has already ocen noticed. The diagonals sloping similarly to AD are assumed not to be stressed. The diagram explains itself.

The resultant stresses, finally, are to be found by combining the results of the diagram in Fig. 2 with those expressed by the equations already written. They are the following:

$$(\overline{AC}) = -\left(\frac{W}{2} - t' + \frac{W_i}{2}\right) \sec \alpha.$$

$$(\overline{CE}) = -\left(\quad\text{``}\quad\text{``} + \frac{3\,W_i}{2}\right)\quad\text{``}\quad + (CE).$$

$$(\overline{EG}) = -\left(\quad\text{``}\quad\text{``} + \frac{5\,W_i}{2}\right)\quad\text{``}\quad + (EG).$$

$$(\overline{GK}) = -\left(\quad\text{``}\quad\text{``} + \frac{7\,W_i}{2}\right)\quad\text{``}\quad + (GK).$$

$$(\overline{BD}) = -\left(\frac{W}{2} + t' + \frac{W_i}{2}\right) \sec \alpha - (BD).$$

$$(\overline{DF}) = -\left(\quad\text{``}\quad\text{``} + \frac{3\,W_i}{2}\right)\quad\text{``}\quad - (DF).$$

$$(\overline{FH}) = -\left(\quad\text{``}\quad\text{``} + \frac{5\,W_i}{2}\right)\quad\text{``}\quad - (FH).$$

$$(\overline{HL}) = -\left(\quad\text{``}\quad\text{``} + \frac{7\,W_i}{2}\right)\quad\text{``}\quad - (HL).$$

$$(\overline{AB}) = -\tfrac{1}{2}(W + W_i)\tan \alpha - (AB).$$

$$(\overline{CD}) = -W_i \tan \alpha - (CD).$$

$$(\overline{EF}) = -\text{``}\quad\text{``}\quad - (EF).$$

$$(\overline{GH}) = -\text{``}\quad\text{``}\quad - (GH).$$

$$(\overline{KL}) = +\left(\frac{W}{2} - t' + \frac{7\,W_i}{2}\right)\tan \alpha - (KL).$$

It is not necessary to reproduce the stresses in the oblique web members, since they can be scaled directly from Fig. 2.

All the stresses may be checked by the method of moments in the usual manner, and such checks should always be applied.

The two reactions R and R' are the following:

$$R = \tfrac{1}{4}(W - 2t' + 8\,W_i) + R_1.$$

$$R' = \tfrac{1}{4}(W + 2t' + 8\,W_i) + R_1'.$$

It is to be remembered that R_1' is to be taken as *positive* in these expressions; also that $R_1' = - R_1$, as shown in Fig. 2.

The lateral force F_1 to be resisted at the foot of the bent by friction or some special device, is the total wind pressure against the train, truss, and bent.

If f' is the coefficient of friction at K and L, Fig. 1, the lateral resistance of friction offered at K is $f'R$, and that at L, $f'R'$. It is supposed that both the reactions R and R' are *upward*, also that both coefficients of friction are the same.

The expression for (KL) has been written on the assumption that all frictional resistance is exerted at L. Strictly, however, the stress in KL may be taken as:

$$(\overline{KL})_1 = (\overline{KL}) - f'R;$$

always supposing, numerically, $(\overline{KL}) > f'R$.

The circumstances of particular cases frequently require calculations to be made with the structure free of moving load, as well as covered with it. In such a case it is only necessary to put for W, in the preceding operations, the weight of trusses only.

The wind has been taken in but one direction only, though the pier is to be designed for both directions, since it is only necessary in the resultant stresses to change the letters B, D, F, H, L, to A, C, E, G, K, and *vice versa*.

If MN, Fig. 1, should coincide with AB, or if the truss should rest upon the top of the pier, it would only be necessary to take $t' = I_1'' + V$, remembering that h is the distance (vertical) from the centre of P_1'' to the top of the pier.

It should be stated that $2t'\ tan\ \alpha$ may be treated as a single force acting toward the left and along AB, Fig. 1. It will then give rise to the diagram in Fig. 3, which shows all the stresses produced by its action. In that Fig. ad represents $2t'\ tan\ \alpha$. In such a treatment of the question, cd,

Fig. 2, would represent $H + P_1''$ added to the half panel pressure at A.   The resultant stresses would then be found by combining the results of the two diagrams with those of the equations.   All the results of these two methods will not agree ; the latter will give the greatest.   This ambiguity cannot be avoided, for it results from the fact that the pier cannot be so divided as to sever these members only.

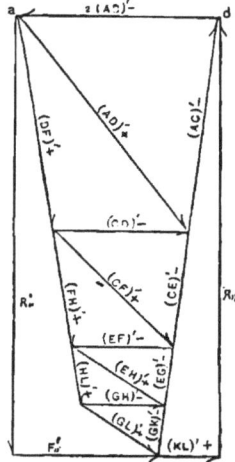

FIG. 3.

Mr. J. A. Powers, C. E., has called the attention of the writer to the fact that the web members of a braced pier carrying a double track railway, similar to that shown in Fig. 4, will receive their greatest stresses with the windward track only loaded.

The vertical member GH may be supposed to carry its proper proportion of the load which rests on each track.   This supposition, however, does not affect the statement made above.

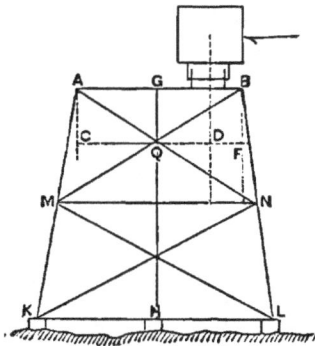

FIG 4.

Let the wind have the direction shown by the arrow, and let W, as before, represent the fixed and moving weight resting on GB, while w' is that part of W which is carried to B.   If GH acts ·

$$w' = \frac{QD}{QF} W.$$

If GH does not act :

$$w' = \frac{CD}{CF} W.$$

In the latter case, the beam AB will carry $w_1' = \dfrac{DF}{CF} W$ to A.

If the angle $FBN = CAM = \alpha$, the force

$$h = w' \ tan \ \alpha - w_1' \ tan \ \alpha$$

will act along AB as an unbalanced horizontal one.   If GH acts, $w_1' \ tan \ \alpha$ becomes equal to zero.

Then in the preceding investigation, there is to be put, $(H + h)$ for H, while w' is to be taken as acting vertically down at B, and w_1' or o (as the case may be) at A.   The preceding methods and diagrams remain exactly the same as before.

In the formulæ, however, w_1' or o is to be put for the $\dfrac{W}{2}$ at A, Fig. 1, and w' for that at B in the same figure.   Nothing else is changed.

If W rests on AG and GB at the same time, a horizontal force equal and opposite to h is developed at A, Fig. 4. Hence h will be balanced and disappear.

If W rests on AG alone, with the direction of the wind remaining the same, h will change its direction, thus giving much smaller web stresses than those existing with W on GB alone.

If, for any reason, the load on a single track pier does not rest over its centre, h will have a definite value, and the above considerations must govern the determination of the web stresses.   This condition may exist if it becomes necessary to place braced piers under a single track railway curve.

The stresses caused by the traction, or pull, of the locomotive, in the members of a braced pier, are simple in character and easily determined.

In such a case, the pier is simply a cantilever with the traction, or pull, as a single force acting at its extremity. The traction acts along the line of the rails, and the length of the cantilever is the height of the pier.   The stresses thus determined are to be combined with those already found.

Art. 85.—Complete Design of a Railway Bridge.

The main sections and details of this design shown on Pls. XI. and XII. are based on the following general specifications.

The span length, depth of truss, panel division, moving load, fixed loads, and stresses resulting from the preceding, shall be as determined in Art. 11.

The clear width between trusses shall be 14 feet.

A wind load of 150 pounds per lin. ft. of span shall be taken for the upper chord, and the same amount for the lower, and shall be treated as a fixed load in each chord. In addition to this fixed load, 300 pounds per lin. ft. of span shall be taken as a moving wind load for the lower chord, since the train passes along the latter.

The greatest allowed tensile stresses under the preceding loads shall be :

For lower chord eye-bars	10,000 lbs. per sq. in.
" main brace eye-bars nearest end of span	10,000 " " " "
" first counter-brace	7,333 " " " "
" vertical adjacent to end of span	7,333 " " " "
Other tension braces to be proportioned according to location between	7,333 and 10,000 " " " "
For plate hangers on floor-beams (net section)	8,000 " " " "
" bottom flanges of floor-beams and stringers (net section)	8,000 " " " "
" lateral braces . . . . , ...	12,000 " " " "

The greatest allowed compressive stresses shall be :

For upper chord and end posts:

Flat ends.   Pin ends.

$$p = \frac{7,800}{1 + \dfrac{l^2}{50,000r^2}} \quad \cdots \quad p = \frac{7,800}{1 + \dfrac{l^2}{30,000r^2}} \quad \cdots \quad (1).$$

For intermediate posts at centre of span, a reduction of 20 per cent. shall be made from the preceding values, and all other intermediate post stresses shall be proportioned according to location, between end and centre values.

In the preceding column formulæ, "p" is pounds per square inch ; "l," length ; and "r" the radius of gyration of normal section in direction of failure, and in the same unit as "l."

For lateral compression braces the above values may be increased 20 per cent.

For top flanges of stringers and floor-beams the greatest compressive stress shall be 7,000 pounds per square inch of gross section.

The greatest mean bearing intensity of pressure between pins and pin-holes or rivets and rivet-holes, shall be 12,000 pounds per square inch. In stringers and floor-beams and their connections with each other or with the trusses, this value shall be reduced 25 per cent.

The greatest shearing intensity in rivets or pins shall be 7,500 pounds per square inch, and in stringers and floor-beams and their connections with each other and the trusses, this amount shall be reduced 25 per cent.

The greatest bending stress in the extreme fibres of wrought-iron pins shall be 15,000 pounds per square inch, and the centres of pressure shall be taken at the centres of the bearing surfaces.

No cast-iron whatever shall be permitted in any part of the structure, and all parts shall be accessible for inspection and painting.

The unsupported width of any plate in compression shall not exceed thirty times its thickness.

The pitch of rivets in compression members shall not exceed sixteen times the thickness of the thinnest plate through which the rivets pass.

An initial stress of 5,000 pounds shall be added to that produced by the vertical loading in all adjustable tension members.

These meagre specifications are sufficient for the design when it is premised that all details of construction, such as eye-bar heads, connections, etc., shall be consistent with the best engineering practice.

Although it is customary to add the total stress of adjustment to that caused by the vertical loading, in adjustable tension members, the practice is not strictly correct. Just what part of the initial stress should be added is not exactly determinate, but it is certainly not the whole. For this reason the apparently small value of 5,000 pounds has been taken.

As the widths of the compression members depend, to some extent, on the thickness of the tension members, and as the design of the latter is of the greatest simplicity, it conduces to the greatest convenience to begin with them. All eye-bars should be as thin as considerations of resistance will permit, as pin bending will then be reduced to a minimum.

By taking the stresses from Fig. 1 of Pl. II., and subjecting them to the preceding specifications, the following sections are obtained :

Brace.	*Total stress.*	*Allowed stress.*	*Sections.*
2	45,553 lbs.	7,333 lbs. per sq. in.	$2 - 4'' \times \frac{3}{4}''$ Bars.
3	167,800 "	10,000 " " " "	$2 - 6 \times 1\frac{3}{8}$ "
5	119,817 "	9,340 " " " "	$2 - 6 \times 1\frac{1}{4}$ "
7	77,445 "	8,670 " " " "	$2 - 5 \times \frac{7}{8}$ "
9	47,654 "	8,000 " " " "	$2 - 2''\frac{1}{4}$ 0 "
10	21,560 "	7,333 " " " "	$2 - 1\frac{5}{8}$ " "

Lower chord.

1 = 2	131,520 "	10,000 " " " "	$2 - 5'' \times 1''\frac{3}{8}$ "
3	225,424 "	10,000 " " " "	$4 - 5'' \times 1\frac{1}{8}$. "
4	289,238 "	10,000 " " " "	$6 - 5'' \times 1$ "
5	322,220 "	10,000 " " " "	$6 - 5 \times 1\frac{1}{16}$ "

There should be as little diversity in widths of bars as possible, but varying thicknesses within standard limits are easily produced in the mill. Again, a small number of large bars is cheaper to produce than a large number of small bars, on account of the smaller number of pieces. Hence the aim should be to produce large pieces, though not too heavy to be handled conveniently in the shop.

Before designing the pins the intermediate post sections

should be determined. These posts are secured to pins at each end, and although they are constrained, by some extent in the vertical plane of the pin axis, it is only slightly so, and they should be considered columns, with pin ends in all directions. They will each be built of two channels laced in the usual manner. The depth of the channel is an important matter, but the length of no column in a truss should exceed forty times its least diameter, and in the present case the depth will be taken at ten inches. The channels will be placed as shown in the figure with a clear separation of ten inches.

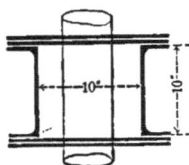

FIG. 1.

Pin plates will be riveted to the flanges of the channels at each end, through which the pin will pass, leaving the axis of the latter parallel to the channel webs and normal to the planes of the trusses. The least radius of the post section will be parallel to the pin axis and will be the same as that of one channel about an axis normal to its web, or about 3.9 inches. The length of the post between pin centres is 27 feet, or 324 inches. But in the plane normal to the truss the column is shortened six feet by the transverse bracing as shown in Fig. 16 of Pl. XII. Hence, in the plane of the pin axis $l \div r = 252 \div 3.9 = 65$; and in the plane normal to the preceding $r = 5.8 \therefore l \div r = 324 \div 5.8 = 56$. As the post is considered with pin ends in all directions the first value of $l \div r$ will be used.

Eq. (1) then gives for a post at the end $p = 6,840$; and for one at the centre $0.8 \times 6,840 = 5,472$. Now since $(6,840 - 5,472) \div 3 = 456$;

For vertical brace 4.. $p = 6,840 - 456 = 6,384$ lbs. per sq. in.
"   "   " 6.. $p = 6,384 - 456 = 5,928$ "   "   "   "
"   "   " 8.. $p = 5,928 - 456 = 5,472$ "   "   "   "

The initial stresses in the counters intersecting at the top of vertical brace 8 increase the stresses in that member 8,000 pounds. The preceding quantities then give the following results:

Total stress.	Allowed stress.	Member.
Vertical 4....99,819 pounds....6,384 pounds....2 — 10″ 79 lb. channels.		
" 6....66,193 " ....5,928 " ....2 — 10″ 56 " "		
" 8....42,611 " ....5,472 " ....2 — 10″ 48 " "		

The last sectional area shows a material excess over that required, but a 48-lb. channel is about the lightest rolled, and this excess is usually found in the centre posts of trusses.

The lacing on these posts will be $2\frac{1}{4} \times \frac{5}{16}$ placed at an angle of about 60° with the post axis.

In order to determine the lower chord pin bending some

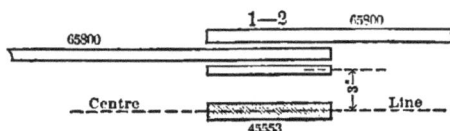

FIG. 2.

diameter of pin must be assumed, for the moment at the centre of the pin will depend partially on the thickness of the pin plates, which bear against the pins. A diameter of $4\frac{3}{8}$ inches will be taken; hence each inch in length of the pin will take $4,375 \times 12,000 = 52,500$ pounds. It will be seen hereafter that the floor-beams will be riveted into the posts below the pins in such a manner that the pin plates will not only carry the column pressures to the pin, but the floor-beam loads also. Hence, in determining the thicknesses of bearing areas in the pins, these two loads and the pressures

FIG. 3.

due to initial stresses in the counter rods must be added. The vertical brace 8 is the only post subject to the initial stresses in the counters and the vertical component of each counter at its top is 4,000 pounds, or 8,000 for the two.

The total lower chord load has already been found to be 45,-553 pounds in the case of brace 2. This maximum lower chord panel load will not usually occur with the greatest post stress, but all possible cases are covered by combining the two. The bearing thickness at each side of each post is thus found to be:

For brace 4.......(99,819 + 45,553)   ÷ 52,500 × 2 = 1.4 inches.
"   "   6.......(66,193 + 45,553)   ÷ 52,500 "   = 1.06 "
"   "   8.......(34,611 + 45,553 + 8,000) ÷ 52,500 "   = 0.84 "

By regarding the principles affecting pin bending as developed in Art. 74, it will be found that the arrangements of

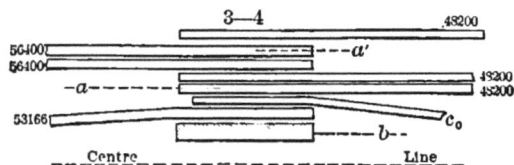

FIG. 4.

lower chord eye-bars and braces shown in Figs. 2, 3, 4 and 5 will reduce the lower chord pin bending to the least amounts possible.

The values of these least pin moments for the principal

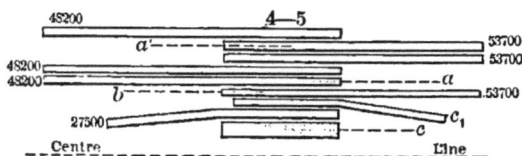

FIG. 5.

sections are found to be as follows, and they can be verified by remembering the thicknesses of the eye-bars (already determined) and the fact that a play of one-eighth of an inch is allowed between each contiguous pair of heads.

Joint 1—2.

The section of plate hanger at the end of the floor-beam is shown shaded, and the distance between its centre and that of either of vertical braces 2 is 3 inches. Taking moments about the centre of the plate-hanger:

Moment about Vert. axis............65,800 × 1.5 = 98,700 in. lbs.
  "   "   Hor.   " ....22,800 × 3.0 = 68,400 "   "
Hence resultant moment = $\sqrt{(98,700)^2 + (68,400)^2}$ = 120,000 " " . (2).

Joint 2—3.

The bearing area of post 4 on the pin is shown shaded. In the remaining cases it will be assumed that the adjacent lower chord panel stresses take their greatest values together, in accordance with which assumption the eye-bars will be stressed for this joint as shown in Fig. 3. The force 78,400 pounds is the corresponding stress in one eye-bar of brace 3. The tangent of the inclination of the latter to a horizontal line is 1.32, hence the vertical component of brace 3 (one eye-bar) is (112,800 − 65,800) 1.32 = 62,040 pounds.

The moments about vertical axes are:

About a.......65,800 × 2.625 − 56,400 × 1.25 = 102,225 in. lbs. . (3).
  " b.......65,800 × 4.0   − 112,800 × 2.0 = 37,600 " "

The moment about a horizontal axis through c is:

$$62,040 \times 1.5 = 93,060 \text{ in. lbs.}$$

The vertical moment about c is the same as that about b, hence the resultant moment about c is:

$$\sqrt{(37,600)^2 + (93,060)^2} = 100,360 \text{ in. lbs.} \qquad . \quad . \quad . \qquad (4).$$

Joint 3—4.

By the preceding method the resultant moment of the inclined brace was resolved into vertical and horizontal components; in this and the following cases, however, the resultant moment itself will be taken. As before, the post bearing is shaded in Fig. 4. The head of the counter rod c_0 is assumed to be one inch thick. The secant of the inclination of the inclined bar to a horizontal line is 1.66. Hence the inclined stress is (3 × 48.200 − 2 × 56,400) 1.66 = 53,166 pounds. The vertical moments are then as follows:

26

About a' . . . 48,200 × 1.2 $\qquad$ = 57,840 in. lbs.
" $\quad a$ . . (112,800 − 96,400) × 2.94 $\quad$ = 48,216 " "
" $\quad b$ . . . . 48,200 × 3.5 − 16,400 × 6.44 = 63,100 " "

The inclined moment about b is;

$$53,166 \times 2.31 = 122,800 \text{ in. lbs.}$$

In Fig. 6 ab is normal to the inclined brace, and represents 122,800 inch-pounds by scale, while bc is vertical, and represents 63,100 inch-pounds.

Hence the resultant moment about b is represented by ac, and has the value:

$$R = 97,500 \text{ inch pounds} \quad . \quad . \quad . \quad . \quad . \quad (5).$$

Joint 4—5.

The thickness of the head of the centre rod c is 1.5 inches; and the inclined eye-bar stress of brace 7 is

$$(3 \times 53,700 - 3 \times 48,200) \ 1.66 = 27,500 \text{ pounds.}$$

The vertical moments are as follows:

About a' 48,200 × 1,155 $\qquad\qquad\qquad\qquad$ = $\quad$ 55,671 in. lbs. (6).
" $\quad a$ 2 (53.700 − 48,200) × 2.875 $\qquad\qquad$ = $\quad$ 31,625 " "
" $\quad b$ 2 (53.700 − 48,200) × 4.03 − 48,200 × 1.155 = − 11,340 " "
" $\quad d$ 3 (53,700 − 48,200) × 3.7 − 11,340 $\qquad$ = $\quad$ 49,700 " "

The inclined moment about d is

$$27,500 \times 2.6 = 71,500 \text{ in. lbs.}$$

Hence the resultant moment about d is, by Fig. 7:

$$R = 57,000 \text{ inch-pounds} \quad . \quad . \quad . \quad . \quad (7).$$

In addition to the preceding, the moments of the greatest vertical components of the inclined eye-bar stresses about the centres of the post bearings should be examined. It is here unnecessary to go into all these in detail. The greatest occurs at joint 3—4. The vertical component of the maximum stress is $(119,817 \div 2)\ 0.8 = 47,930$ lbs. Hence the moment in question is:

$$47,930 \times 2.31 = 110,700 \text{ in. lbs.} \quad . \quad . \quad . \quad (8).$$

The preceding results show that while it is quite unnecessary to take moments at the centre of all bearings, a thorough examination of the lower chord joints must be made in order to find the greatest moments.

Eq. (2) gives the greatest resultant moment of 120,000 inch-pounds. Hence a wrought-iron pin 4,375 inches in diameter will be sufficient to meet the requirements of the specifications. But since the bars in braces 3 and 5 are 6 inches wide, and since the eye-bar heads are no thicker than the bodies of bars, the requirement of 12,000 pounds per square inch bearing pressure against pins cannot be met by a less diameter than 5 inches in the case of brace 3. For a reason that will appear hereafter, the diameter of the large pins will be taken at $5\frac{3}{8}$ inches.

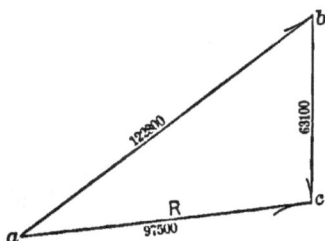

FIG. 6.

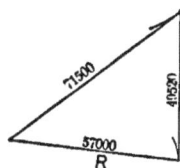

FIG. 7.

FIG. 8.

In the cases of braces 7 and 9 a much smaller pin may be used, and while it is not economy in the shop to have a large number of pin diameters, two, or even three, are not too many for a span of this length. The cosine of the inclination of brace 7 to a horizontal line is 0.61, hence the horizontal component of the greatest stress in one of its eye-bars is $(77,445 \div 2) \times 0.61 = 23,620$ pounds. It will be seen hereafter that the side plates of the upper chord panels 3 and 4 will be 0.5 inch thick, and since $4\frac{3}{16} \times 12,000 \div 2 = 25,000$, it appears that with a pin diameter of $4\frac{3}{16}$ inches no thickening plates at pin-holes D, E, F, and M will be needed. It will be found that the thickness of bearing plates at the top of brace 6 must be $\frac{11}{16}$ inch, and the clearance at each side of eye-bar head (between $\frac{1}{2}$-inch side plate of chord and pin plate of post) will be about $\frac{3}{16}$ inch. The vertical component of eye-bar stress (for brace 7) will be $(77,445 \div 2) \times 0.8 = 30,980$ pounds. Hence:

Pin moment at centre of eye-bar head = $23,620 \times 0.875 = 20,670$ in. lbs.
Vertical   "   "   post pin plate = $23,620 \times 1,875 = 44,290$ " "
Inclined   "   "   "   = $38,720 \times 1.00 = 38,720$ " "

The resultant of the last two moments is shown by Fig. 8 to be:

$$R = 37,000 \text{ inch-pounds} \quad . \quad . \quad . \quad . \quad . \quad (9).$$

The moment of the greatest vertical component in brace 7 at the bottom of post 8 is:

$$30,980 \times 2.6 = 80,550 \text{ inch-pounds}. \quad . \quad (10).$$

Eqs. (6), (7), (9), and (10) show moments far below the allowed resisting capacity of a $4\frac{3}{16}$ wrought-iron pin, *i. e.*, 108,000 inch pounds.

Hence, *at C, L, K, and J* $5\frac{3}{8}$ *inch pins will be used, while at D, E, and I,* $4\frac{3}{16}$ *pins will be taken.*

Before determining the diameters of the pins in the inclined end post, it will be necessary to fix the sections of that member, and it

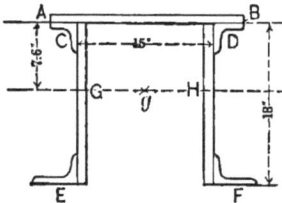

FIG. 9.

will be convenient to find those of the upper chord at the same time. As each upper chord panel is a beam of considerable span, carrying its own weight, the depth should not be small, and it will be taken at eighteen (18) inches. The radius of gyration of the normal section about a horizontal axis through its centre of gravity must first be found.

As the upper chord stress in panels 3 and 4 is about 322,000 pounds, the area of that panel section will not be far from 44 square inches. Fig. 9 represents a trial section of that area.

AB is a   21 × $\frac{9}{16}$ inch cover plate.
C and D are   3 × 3   "   21 lb. angles.
E " F "   5 × 3   "   50   "
G " H "   18 × $\frac{1}{2}$ "   side plates.

The 5-inch legs of E and F are horizontal. The centres of gravity of C and D are 0.9 inch from lower surface of AB, and those of E and F are 0.9 inch from lower surface of the horizontal 5-inch legs. Static moments about a horizontal line through the centre of gravity of the section of AB give:

3 × 3 angles . . . 2 × 2.1 × 1.2 =   5.04
5 × 3   "   . . . 2 × 5.0 × 17.4 = 174.00
Side plates   . . .   18 × 9.3 = 167.4
                                ‾‾‾‾‾‾‾
Total 346.44

Hence, 346.44 ÷ 44 = 7.9 inches; or *the centre of gravity g of the entire section is 7.6 inches from the lower surface of AB.* In such computations some dimensions are taken a little full because adjacent surfaces do not have mathematical contact.

The elements of the moment of inertia of the section about the horizontal axis GH through g take the values:

Cover plate . . . . . . 11.81 × $\overline{7.9}^2$ = 737.06

3 × 3 angles   . . . . . $\begin{cases} 2 & × 2.0 = & 4.0 \\ 4.2 & × \overline{6.7}^2 = & 188.54 \end{cases}$

5 × 3   "   . . . . . $\begin{cases} 2 & × 4.5 = & 9.0 \\ 10.0 & × \overline{9.5}^2 = & 902.5 \end{cases}$

Side plates . . . . $\begin{cases} 18 × \overline{18}^2 ÷ 12 & = 486.0 \\ 18 × \overline{1.4}^2 & = 35.28 \end{cases}$
                                          ‾‾‾‾‾‾‾
Moment of inertia 2,362.38

The moment of inertia of AB about a horizontal axis through its own centre of gravity is so small that it has been neglected. The 3 × 3 and 5 × 3 angles each have a moment of inertia of 2 about a horizontal axis through the centre of gravity of each respective section. The least radius of gyration about a horizontal axis for the entire section then becomes:

$$\sqrt{2{,}362.38 \div 44} = 7.33 \text{ inches.}$$

The panel length is 20.55 ft. Hence

$$l \div r = 20.55 \times 12 \div 7.33 = 33.7.$$

The preceding value in Eqs. (1) gives:

Flat ends.	*Pin ends.*
$p = 7{,}620$ *lbs. per sq. in.*	$p = 7{,}510$ *lbs. per sq. in.*

For one pin and one flat end . . . . $p = (7{,}620 + 7{,}510) \div 2$
$$= 7{,}565 \text{ lbs. per sq. in.}$$

The upper chord will be continuous after the bridge is erected, but the extremities will be hinged at the upper ends of the inclined end posts in the manner shown in Fig. 4 of Pl. XI. Hence upper chord panel 1 will have one pin end and one flat end; all other panels will be flat end columns. The upper chord sections will now be as follows:

Upper chord 1.
Required area = 225,424 ÷ 7,565 = 30.0 sq. ins.

1 — 21 × $\frac{7}{16}$ inch cover plate.	9.2 "	"
2 — 3 × 3 " 18 lb. angles	3.6 "	"
2 — 5 × 3 " 30 " "	6.0 "	"
2 — 18 × $\frac{3}{8}$ " side plates.	13.5 "	"

Total . . . . 32.3 "   "

Upper chord 2.
Required area = 289,238 ÷ 7,620 = 38.0 sq. ins.

1 — 21 × $\frac{7}{16}$ inch cover plate .	9.2 "	"
2 — 3 × 3 " 18 lb. angles.	3.6 "	"
2 — 5 × 3 " 36 " "	7.2 "	"
2 — 18 × $\frac{1}{2}$ " side plates . .	18.0 "	"

Total . . . . 38.0 "   "

Upper chord 3 *and* 4.

Required area = 322,220 ÷ 7,620 = 42.3 sq. ins.

    1 — 21 × $\frac{7}{16}$ inch cover plate = 9.2 " "

    2 — 3 × 3 " 18 lb. angles 3.6 " "

    2 — 5 × 3 " 47 " " 9.4 " "

    2 — 18 × $\frac{9}{16}$ " side plates. 20.25 " "

                Total............42.45 " "

The end post bears on pins at top and bottom; hence it is a pin-ended column. It is about 408 inches long, and it will be most convenient to take its depth identical with that of the upper chord, or 18 inches. The radius of gyration may then be taken, as before, at 7.33 inches. Hence, $l \div r$ = 408 ÷ 7.33 = 55.7. The second formula of Eq. (1) then gives:

$$p = 7,070 \ lbs. \ per \ sq. \ in.$$

Inclined end post.

Required area = 217,750 ÷ 7,070 = 30.8 sq. ins.

    1 — 21 × $\frac{7}{16}$ cover plate .... 9.2 " "

    2 — 3 × 3 18 lb. angles .. 3.6 " "

    2 — 5 × 3 30 " " .. 6.0 " "

    2 — 18 × $\frac{3}{8}$ side plates .... 13.5 " "

                Total............. 32.3 " "

All these actual areas agree sufficiently near in character and amount with the trial section to make re-computations of the radius of gyration quite unnecessary. A very little experience makes such a result possible in all ordinary cases. A very close but approximate rule for all box or semi-closed sections like those just considered, is to take the radius of gyration at four-tenths (0.4) the depth of the side plates. In the present case it would make r = 0.4 × 18 = 7.2 inches, while the exact value is 7.33 inches.

In building a section such as these, the angles C and D, Fig. 9, should be made as light as possible, in order that the cover plate AB may be, to a considerable extent, balanced by the heavy angles E, F. In this manner the centre of gravity, *g*, of the section may be brought down sufficiently near to the mid depth to give all the space needed inside the chord for the eye-bar heads, if the pin axis should be made to pass through *g*, at the same time there is gained the incidental but important advantage of an increased moment of inertia. If the chord were subject to no bending from its own weight, the axis of every pin should pass through the centre of gravity of the section. It has been shown in Art. 69 that this flexure cannot be satisfactorily neutralized by the direct stress, particularly if the chord is continuous, as in the present case. It is best, therefore, to reduce the bending stresses by making the chord depth as great as possible. For these reasons it was taken at eighteen (18) inches. If the panels were non-continuous the greatest stress per sq. in. in the exterior fibres of panels 3 and 4 would be :

$$K = \frac{Md}{I} = \frac{88,800 \times 10.4}{2,362} = 390 \text{ lbs.}$$

Now, when it is remembered that the chord is continuous it is evident that flexure may be neglected in the sections found. This point, however, should always receive careful attention.

In the present case, *the axes of pins in the upper chord and end posts will be placed eight (8) inches from the line A B*, thus allowing a small counter moment from the direct stress due to a lever arm of 0.4 inch.

A compression member with the same degree of end constraint in all directions ought to have equal capacity for resistance in all directions. If the radius of gyration be taken in different directions about *g*, Fig. 9, for the different sections as formed, it will be found that this condition is fulfilled.

Finally, the unsupported width of any plate in compression, measured transversely between rivet heads, should not be more than about thirty times the thickness. An examination of the sections will show that this condition also has been fulfilled.

The details about the pin bearings at the upper and lower ends of the inclined end post may now be considered.

Fig. 4 of Pl. XI., shows the detail at the upper end of the inclined end post. The end panel of the upper chord does not rest its extremity immediately against the upper end of brace 1, but they are separated along the line *ab* by about the distance of $\frac{3}{8}$ inch, and each bears directly against the end pin. *The diameter of the latter is taken by trial at* 5$\frac{1}{8}$ *inches.* By the specifications the bearing value of this pin against a one-inch plate is 5$\frac{1}{8}$ × 12,000 = 61,500 pounds. Hence for plates $\frac{1}{4}$, $\frac{5}{16}$, $\frac{3}{8}$, $\frac{1}{2}$ and $\frac{9}{16}$ inch the bearing values will be as follows:

$$61,500 \times \tfrac{1}{4} = 15,375 \text{ pounds.}$$
$$\text{``} \quad \times \tfrac{5}{16} = 19,220 \quad \text{``}$$
$$\text{``} \quad \times \tfrac{3}{8} = 23,060 \quad \text{``}$$
$$\text{``} \quad \times \tfrac{1}{2} = 30,750 \quad \text{``}$$
$$\text{``} \quad \times \tfrac{9}{16} = 34,596 \quad \text{``}$$

The arrangement of thickening plates for upper chord 1 is clearly shown by Fig. 4; there is a half-inch plate inside and next the $\frac{3}{8}$ths web and a $\frac{3}{8}$ jaw plate inside the half-inch thickener. Against the web outside is a $\frac{9}{16}$ inch thickener. The total bearing thickness is then $\frac{3}{8} + \frac{1}{2} + \frac{3}{8} + \frac{9}{16} = 1\frac{13}{16}$ inches. Hence 61,500 × $1\frac{13}{16}$ = 111,500 pounds. The half of the stress in upper chord 1 is 112,712 pounds and the two quantities are sufficiently near in amount. All rivets about the joint are $\frac{3}{4}$ inch in diameter. The shearing resistance of one rivet at 7,500 pounds per sq. in. is 3,300 pounds, while the bearing values against the various plates are:

$$\tfrac{3}{4} \text{ rivet against } \tfrac{5}{16} \text{ plate} = 2,800 \text{ pounds.}$$
$$\text{`` `` ``} \quad \tfrac{3}{8} \text{ ``} = 3,400 \quad \text{``}$$
$$\text{`` `` ``} \quad \tfrac{1}{2} \text{ ``} = 4,500 \quad \text{``}$$

The total bearing pressure against the jaw and thickening plates is 23,060 + 30,750 + 34,596 = 88,406 pounds, and there are 21 rivets through those plates, as shown in Fig. 4. Applying the bearing and shearing values given above to the number and distribution of rivets located in that figure, it will be seen that there is a little excess of both those resistances.

The same figure shows the number and distribution of both rivets and thickening plates at the upper end of the inclined end post. There is a $\frac{3}{4}$ inch jaw plate outside, then a half-inch thickener and a $\frac{5}{16}$ plate between that and the web. There is also a quarter-inch thickener inside. The amount of bearing thickness is thus the same as for the upper chord. An examination of the number and location of the rivets will show that there is again a little excess in the bearing and shearing resistances. *The pitch of rivets in the immediate vicinity of the joint is three (3) inches, in all other parts of the upper chord and end posts it will be six (6) inches.*

The preceding arrangement is for one side of the chord and end post, since both sides, of course, are alike. As Fig. 4 shows, the pin passes *through* the jaw plates only. The four (4) jaw plates hold the post and upper chord securely together in case of any derailment or other accident tending to knock the end post out of place. They are further reinforced by the light $\frac{3}{8}$ inch cover plate shown at *a*. The latter also performs an important office in transferring upper lateral loads to the end posts. One portion of it, or both, must, of course, be riveted in the field. It will be observed that each jaw plate has a "play" or clearance of $\frac{1}{4}$ inch, to provide for imperfections of workmanship and secure ready erection. The figure shows what rivets must be countersunk, both outside and inside.

The eye-bars of brace 3 lie adjacent to the interiors of the upper chord and end post, while those of brace 2 are inside of the first. Assuming that the greatest stresses in those braces occur together (which is a small error on the side of safety), the end pin will be subjected to the bending moments shown in Fig. 10. The component moments are as follows:

$$\text{For brace 3.} \quad \frac{167,800}{2} \times 1.71 = 143,470 \text{ in. lbs.}$$

$$\text{"} \quad \text{"} \quad 2. \quad \frac{45,553}{2} \times 2.90 = 66,053 \text{ " "}$$

$$\text{"} \quad \text{"} \quad 1. \quad \frac{217,750}{2} \times 0.62 = 67,500 \text{ " "}$$

The latter moment arises from the fact that the upper chord and end-post bearings have their centres separated by 0.62 inch.

In the figure, *bc* is normal to brace 3, and *ac* is horizontal, while *ad* is normal to brace 1. Hence *db* is the resultant moment of 220,000 inch pounds. A pin 5⅜ inch in diameter will a little more than supply the required resistance with $K = 15,000$, as Eq. (1) of Art. 74 demonstrates, or as may more

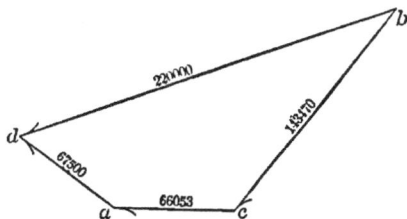

FIG. 10.

simply be found by reference to any reliable table of pin moments. The thickening plates and rivets just found will now be a very little excessive, but they will be retained.

These large rounds frequently vary in standard sizes by quarter-inches, and a 5¼ inch diameter may be turned to 5⅜ with little waste.

Fig. 7 of Pl. XI. shows the lower end of the inclined end post with the number and location of the ¾ rivets and thickening plates. The operation of designing them is precisely similar to those already employed, and they will not now be repeated. An examination of the plates and rivets in connection with the preceding values, will show that the shearing and bearing resistances of both the rivets, plates, and 5⅜ inch (assumed) pin required by the specifications are secured. The line *ab* is 2¼ inches below the centre of the pin-hole, giving about 2 inches of solid metal below the pin.

Figs. 8, 9 and 10, of Pl. XI., show two elevations and a plan of the pedestal at the lower extremity of each end post. The centre of the pin-hole is taken six (6) inches above the bottom ¾ inch plate. The figures show with perfect clearness the arrangement of the various parts. The 4 and 19 inch spaces give ample clearance for the sides of the end post which enter them.

The vertical component of the end-post stress is 172,820

pounds. The total bearing thickness under each half of the pin is $\frac{5}{8} + \frac{5}{8} + \frac{3}{8} = 1\frac{5}{8}$ inches. But $64{,}500 \times 1\frac{5}{8} = 104{,}610$; or greater than $172{,}820 \div 2 = 86{,}410$. Hence, ample bearing surface at 12,000 pounds per square inch is secured. Now since the half of the end-post bearing is at the centre of each of the 4 inch spaces, it may at first sight appear as if either equal bearing areas ought to be found each side of those spaces, or as if all ought to be on one side. But it is better to mass the metal as much as possible; at the same time the weight should be distributed somewhat on the $\frac{3}{4}$ inch plate. The arrangement shown accomplishes these results and gives a little excess of bearing area. The 5 by 3 inch angles are but 15 inches long, while the $\frac{3}{4}$ inch bottom plate is 24 by 36 inches.

The bearing thickness at a is 1.25 inches; hence the upward pressure at that surface is $64{,}500 \times 1.25 = 80{,}625$ pounds.

The 4 inch space gives about $1\frac{3}{16}$ inch total clearance for the side of the end-post and the eye-bar ($1\frac{3}{8}$ inches thick) of lower chord panel 1, or $\frac{1}{4}$ inch each for the three clearance spaces thus formed.

The pin moment about the centre of the end-post bearing is:

$$80{,}625 \times 1.875 \text{ inches} = 151{,}171 \textit{ inch lbs.} \quad . \quad . \quad . \quad (11).$$

Again, taking moments about the centre of the angle bearing b, there are two moments with horizontal axes but with opposite signs formed by the upward pressure at a, and the half vertical component in the inclined end post, thus:

$$+ 80{,}625 \times 4.81 = + 387{,}806 \textit{ in. lbs.}$$
$$- 86{,}413 \times 3.00 = - 259{,}230 \text{ " "}$$
$$\textit{Resultant} = + 128{,}576 \text{ " "}$$

The stress in the $5 \times 1\frac{3}{8}$ inch eye-bar of lower chord 1 has the following moment about a vertical axis passing through the centre of b:

$$65{,}760 \times 1.14 = 74{,}970 \textit{ in. lbs.}$$

Hence the resultant moment about the centre of b is:

$$\sqrt{(128,576)^2 + (74,970)^2} = 150,000 \ in. \ lbs. \quad . \quad . \quad . \quad (12).$$

As the moment (11) is greater than (12) and far less than the resisting capacity of the assumed $5\frac{3}{8}$ inch pin, *the latter diameter will be retained.* A smaller pin would give sufficient bending resistance, but would necessitate additional metal in the thickening plates, and would increase the variety in pins and pin-holes.

It is frequently desirable to hang one pair of eye-bars (either braces 2 or 3) *outside* of chord and end post at the upper end of the latter. In such a case the angle flanges at b, Fig. 4, Pl. XI., would be cut away, and more rivets would need to be countersunk about the pin-hole on the outside of the outer jaw plate.

In the present instance, however, the pin necessary at the upper chord end is but little different from those required by braces 3 and 5. Hence, for the sake of uniformity, the pins at B, C, A, L, K and J, will be given a diameter of $5\frac{3}{8}$ inches, while the others are $4\frac{3}{16}$ inches.

The pin-plates at the upper and lower end of the intermediate posts will now be found.

It will be assumed that the maximum post stress and the greatest panel load occur together. This is not possible, but it is difficult to determine the exact maximum load on the lower pin-plate, and the assumption involves a safe error. It will farther be assumed that the greatest panel load for the intermediate posts is the same as the greatest load on brace 2. This also involves a slight safe error.

In consequence of these assumptions and the additional fact that the smaller part of the load in each lower pin-plate is the panel moving load, no addition for impact will be made in fixing the thickness of the pin-plates.

The manner of supporting the ends of the floor-beams is clearly shown in Figs. 14, 15 and 16, Pl. XI. They are built into the posts below the pin. The channels are continued 30 inches below the centres of the pin-holes, and a 4 by 4 inch

36 pound angle is riveted to each as shown at a a Fig. 14. The end stiffeners of the floor-beam (Fig, 2, Pl. XI.) are brought against these latter and riveted fast to them in erection. The number of rivets required for this connection will be found later on. In order to freely admit the end stiffeners of the floor-beam, the 10 inch channels of the post will be separated 10 inches.

Vertical Brace 4.

The top pin plates will carry 99,819 lbs.
" bottom " " " " 99,816 + 45,553 = 145,372 lbs.

Since $5\frac{3}{8} \times 12,000 = 64,500$ pounds, the total thickness of bottom pin plates will be $145,372 \div 64,500 = 2.25$ inches ; and since the shearing resistance of one $\frac{3}{4}$ inch rivet is 3,300 pounds, the total number of rivets in the channel flanges will be $145,372 \div 3,300 = 44$. The lower part of Fig. 14, Pl. XI. shows the required arrangement of pin plates and rivets There are three $\frac{3}{8}$ inch outside pin plates on each side of the post. The rivets about the pin-hole on the outside will be countersunk in order that the eye-bar of brace 3 may lie close against the post.

The total thickness of pin plates at the top of the post will be $99,819 \div 64,500 = 1\frac{9}{16}$ inches, and the total number of rivets, $99,819 \div 3,300 = 30$. The upper part of Fig. 14, Pl. XI. shows the required arrangement of pin plates and rivets. As the total number of the latter must be divided by 4, 32 rivets are used. There is one $\frac{3}{8}$ inch outside pin plate and one $\frac{7}{16}$ inch inside plate riveted to the former. The object of placing the latter inside is to keep the upper chord as narrow as possible.

Vertical Brace 6.

The top pin plates will carry 66,193 lbs.
" bottom " " " " 66,193 + 45,553 = 111,746 lbs.

The total thickness of bottom pin plate will be $111,746 \div 64,500 = 1\frac{7}{8}$ inches, and that of the upper $66,193 \div 50,250$

$= 1\frac{5}{10}$ inches, since the upper pin is $4\frac{3}{16}$ inches in diameter and $4\frac{3}{16} \times 12,000 = 50,250$ pounds. The total numbers of $\frac{3}{4}$ rivets below and above, respectively, are $111,746 \div 3,300 = 34$ and $66,193 \div 3,300 = 20$. Fig. 15, Pl. XI., shows the required pin plates and rivets. At the bottom there is a $\frac{3}{8}$ inch plate next to the channels and a half-inch plate outside. At the top there is a $\frac{3}{8}$ inch plate outside and a $\frac{5}{16}$ inch plate inside, as shown.

Vertical Brace 8.

Four adjustable ties meet the upper extremity of this post, and it has already been shown that each tie adds 4,000 pounds to the post stress. Hence, the

top pin plates will carry $34,611 \times 16,000 \qquad = 50,611$ lbs.
bottom   "   "   " $34,611 + 16,000 + 45,553 = 96,164$ "

The total thickness of bottom pin plates will be $96,164 \div 50,250 = 1\frac{7}{8}$ inches; and that of the top plates $50,611 \div 50,250 = 1$ inch. The total numbers of rivets required are $96,164 \div 3,300 = 29$ and $50,611 \div 3,300 = 16$. Fig. 16, Pl. XI., clearly shows the arrangement of plates and rivets. There are more rivets shown at the top than is necessary for bearing or shearing alone, for the reason that the notch a must be cut out of one channel to let the counterbraces 9 take hold of the pin *inside* the post, as there is not room enough outside.

Upper Chord Joints and Thickening Plates.

It will readily be seen that the pin-hole at the joint point between upper chord 1 and 2 is the only one needing a thickening plate. The greatest tension in an eye-bar of brace 5 is $119,817 \div 2 = 59,909$ pounds, and the sine of its inclination to a vertical line is 0.61. Hence, its horizontal component is $59,909 \times 0.61 = 36,544$ pounds. The thickness of the side plates of upper 2 is $\frac{1}{2}$ inch; hence, $\frac{1}{2} \times 12,000 \times 5\frac{3}{8} = 32,250$ pounds. A little over 4,000 pounds, then, is all that need be resisted by a thickening plate. This might safely be neglected, but the $\frac{5}{16}$ joint plate shown by Fig. 13, Pl. XI., will be extended to cover the pin-hole.

Precisely the same operation shows that no thickening plates are needed at the other pin-holes.

There will be joints in the upper chord as near as possible to, and on the left of the pin-holes at *C*, *D* and *E* of Fig. 1, Pl. II., and at corresponding points in the other half of the truss.

These joints are formed as shown at Fig. 13, Pl. XI. At *C* the joint will be 12 inches from the centre of the pin-hole. It is formed by riveting top and bottom and side plates to the chords, as shown. All the joints are formed precisely like this, except that in the other cases the $\frac{5}{16}$ plate between the angles extends each side of the outer one, as shown on the left only.

Latticing and Batten Plates.

The dimensions of latticing and batten plates are matters of judgment and experience. Evidently no segment of a column between lattice points ought to be less in resistance per square unit of section than the column as a whole, but experiments are yet lacking to give quantitative results. Single latticing with centre lines making angles of 60° with the axis of the member will be used here.

On the under side of the upper chord and end post the lattice bars will be 4 inches by $\frac{3}{8}$, and each end will be held by

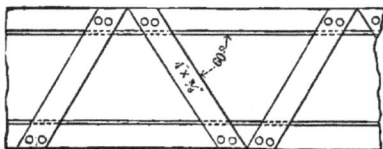

FIG. 11.

two rivets. Fig. 11 shows this latticing. It will weigh about eight (8) pounds per lineal foot of member. The two battens (one at each end) on the under side of the end post and those at the ends of the upper chord (four in all) will be 21 inches by 21 inches by $\frac{3}{8}$ inch. All other battens (one on that side of each vertical post opposite to the chord joint) will be $21 \times 15 \times \frac{3}{8}$ inches. The bottom plate of each joint forms, of course, a batten.

On the intermediate posts, the latticing will be single and

60° as before, but the lattice bars will be 2 inches by $\frac{5}{16}$ inch. This latticing (both sides) will weigh about 9 pounds per lineal foot of post.

Floor-beam Supports.

The method of suspending the floor-beam from the pin at the lower extremity of brace 2 is shown in Fig. 2, Pl. XI. Two plates riveted to the end stiffeners of the beam take the $5\frac{3}{8}$ pin with its centre line six inches above the upper flange. The greatest moving load carried by the beam end has been already found to be 34,600 pounds. One-third of this will be added for impact, and as the fixed load is 8,000 pounds, the total load to be resisted by the plate hangers becomes :

$$\frac{4}{3} \times 34{,}600 + 8{,}000 = 54{,}130 \text{ pounds.}$$

The greatest allowable load per square inch in these hangers is 8,000 pounds; hence the required net area is 54,130 ÷ 8,000 = 6.8 square inches. These plates will be taken 12 inches wide. By deduction of the pin-hole the available net width becomes 12 − 5.375 = 6.625 inches. One plate 12 × $\frac{9}{16}$ and another 12 × $\frac{1}{2}$ inch gives the required area. Rivets $\frac{7}{8}$ inch in diameter will hold these plates to the end stiffeners. The shearing resistance in this case is less than the bearing, and the former for one rivet at 7,500 pounds per square inch, is 4,500 pounds. Hence the required number of rivets is 54,130 ÷ 4,500 = 12 rivets. In consequence of the deflection of the beam some of the upper ones will be subjected to slight tension. Hence 16 rivets are shown. The figure shows the number and distribution of rivets and plates. The vertical pitch of rivets is 3 inches.

The manner of attaching the floor-beams to the lower extremities of the intermediate posts is shown by Fig. 14, Pl. XI. *a* and *a* are 4 × 4 inch 36 pound angles 27 inches long riveted to the inner surfaces of the 10-inch channels, as shown.

The vertical centre lines of the rivet rows in the channels

27

are coincident with the central lines of the latter, thus insur-
ing an equal division of the floor-beam load between the
pin-plates.   The number and distribution of the $\frac{3}{8}$ inch rivets
in the angles *a, a* will, of course, be the same as those in the
plate hangers of Fig. 2, Pl. XI.

Both these methods of supporting floor-beams insure a
central application to the pin, and the latter insures additional stiffeners to the floor system and entire structure.

The two lines of $\frac{3}{8}$ inch rivets take less than a square inch
of section from the two 10 inch 50 pound channels. Hence the
remaining net section of our nine square inches is more than
sufficient to carry the total floor-beam load at all points.

Upper Lateral System.

A wind pressure of 150 pounds per lineal foot will be taken
as acting in the horizontal plane of the upper chord.   The
panel wind load will then be 20.55 × 150 = 3,083 pounds.
Fig. 12 shows a half plan of the upper lateral system.   The
diagonals are ties, and the other members are struts.

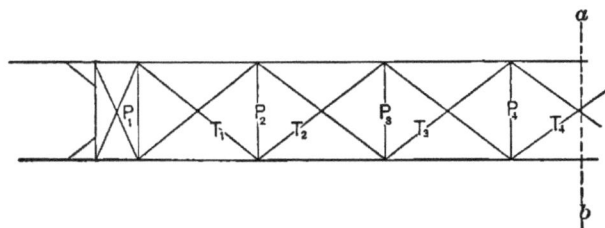

FIG. 12.

The secant of the inclination of T_1 to a horizontal line
normal to the axis of the bridge is 1.57, and 3,083 × 1.57 =
4,848 pounds.

The wind load in the upper chord is a fixed one.   The line
ab is the centre of the span.

The following stresses may now be written, remembering
that as the tension diagonals will each be adjustable, 5,000
pounds must be added for initial stress:

$$T_4 = \quad - \quad + 5{,}000 = 5{,}000 \text{ pounds.}$$
$$T_3 = 4{,}848 + \text{'} \quad \text{"} \quad = 9{,}848 \quad \text{"}$$
$$T_2 = 9{,}696 + \quad \text{"} \quad = 14{,}696 \quad \text{"}$$
$$T_1 = 14{,}544 + \quad \text{"} \quad = 19{,}544 \quad \text{"}$$

$$P_4 = 3{,}083 + 3{,}200 = 6{,}283 \text{ pounds}$$
$$P_3 = 6{,}166 + \quad \text{"} \quad = 9{,}366 \quad \text{"}$$
$$P_2 = 9{,}249 + \quad \text{"} \quad = 12{,}449 \quad \text{"}$$
$$P_1 = 12{,}332 + \quad \text{"} \quad = 15{,}532 \quad \text{"}$$

The greatest allowable stresses in the lateral systems may be taken as follows :

For tension . . 14,000 pounds per sq. in.

" compression . . $$\dfrac{9{,}300}{1 + \dfrac{l^2}{30{,}000 \, r^2}} \quad \text{"} \quad \text{"} \qquad (13).$$

The latter formula is for flat-end members of angle iron, as those will be used for lateral compression members. It will be observed that it gives less value than the formula (1) for columns of the box type, like those used in the trusses. Under these stresses the tension members become :

$$T_4 \quad . \quad . \quad . \quad . \quad . \quad . \quad . \quad 1 - 1\tfrac{1}{8} \text{ O.}$$
$$T_3 \quad . \quad . \quad . \quad . \quad . \quad . \quad . \quad 1 - 1\tfrac{1}{8} \text{ O.}$$
$$T_2 \quad . \quad . \quad . \quad . \quad . \quad . \quad 1 - 1\tfrac{1}{4} \text{ O.}$$
$$T_1 \quad . \quad . \quad . \quad . \quad . \quad . \quad 1 - 1\tfrac{3}{8} \text{ O.}$$

It is not advisable to have any tension member less in sectional area than 1 square inch. Hence, T_3 and T_4 are a little larger than the stresses require.

All the struts except P_1 will be formed of $3 - 3 \times 3$ inch angles. A section of this strut is shown at Fig. 11. Pl. XI. The two angles c lie on the upper chord and are riveted to it, as shown. These two angles are designed to carry all the stress of the strut. The only office of the angle a is to keep the strut stiff in a vertical plane; it takes hold of the lower flange of the chord with two rivets, as shown at c, Fig. 4, Pl.

XI. The strut is thus of the same depth as the chord, and takes hold of both those members in such a manner as to give them great rigidity.   At each end of the strut there is a 20 × 12 × $\frac{5}{8}$ inch plate, and there is a set of single 60° lacing, 2 × $\frac{5}{16}$ inch.   The radius of gyration of the section of the two top angles about a vertical line midway between the two, is 1.3 inches.   The length of the strut is about 192 inches. Hence Eq. (13) gives 5,300 pounds as the greatest allowable stress per square inch.   As P_2 requires only the lightest angle that ought to be used, all these struts will be made alike.

$$\left.\begin{array}{l} P_4 \quad . \quad . \quad . \quad . \\ P_3 \quad . \quad . \quad . \quad . \\ P_2 \quad . \quad . \quad . \quad . \end{array}\right\} 3 - 3'' \times 2\tfrac{1}{2}'' \text{ 15 pound angles.}$$

As the detail for these struts would give some trouble at the end of the upper chord, P_1 will be a single 6″ × 4″ angle with the 6″ leg horizontal, as shown at *a*, Fig. 4, Pl. XI.

The tie T_1 is attached to the upper chord by the detail shown at *b*, Fig. 11, Pl. XI.   A piece of 6″ by 4″, 60 pound angle, 12 inches long, with the 6 inch leg lying on the cover plate of the chord, carries two pieces of 3″ by 3″, 21 pound angles about 5½ inches long, and with edges parallel to the axis of T_1.   One end of each of the latter angles rests squarely against the vertical 4-inch leg of the 6″ by 4″ angle.   Six three-quarter inch rivets are then passed through the angles in the manner shown.   Each such rivet will resist about 4,000 pounds in single shear.   The tie T_1 passes through the 4-inch leg of the heavy angle (between the 3-inch angles), and carries a nut at its end, which gives the requisite adjustment.

T_2 and all the other lateral ties are held by the same detail, except that 4 rivets only are needed, as shown at *b′*.

Transverse Bracing.

A skeleton sketch of the intermediate transverse bracing for the vertical plane of any two opposite posts is shown in Fig. 16, Pl. XII.   If the upper lateral system is designed to carry the whole wind load to the ends of the upper chord, as has been supposed, the duty of the intermediate transverse bracing is entirely indeterminate. As the sections must be determined in some manner, however, the method of Art. 82, Fig. 1, will be applied.

The slight analytical superabundance of stability thus secured is no more than is required by a rapidly moving load.

F' of Art. 82 will here be taken as 3,083 pounds and $F = 0$. Also, $b = 17$ ft. and $a = 6$ ft. It will here be assumed that all the wind load is applied in the windward truss. This is the usual assumption in practice, although the conditions taken in Art. 82 are *exactly* true. Eq. (1) of that Art. then gives:

$$w = (3,083 \times 27) \div 17 = 4,932 \text{ pounds.}$$

As the tangent of the inclination of T to a vertical line is 2.833 and the secant, 3.0, the stresses are:

$T = 4,932 \times 3 + 5,000 = \quad 19,796$ pounds. $1\frac{3}{8}$ o.
$P = 4,932 \times 2,833 + 5,000 = 18,974 \quad$ " $\quad 2 - 3'' \times 3''$ 24 lb. angles.

The sizes are based on the same allowed working stresses as for the upper laterals. The strut P is shown at Fig. 19, Pl. XII. The two angles are held $1\frac{3}{4}$ inches apart by separators, and present a horizontal upper surface. The ends are secured to a batten plate in proper position on the post and in the manner shown. The separation of the angles permits the tie T to pass between them and through the batten plate and take a nut inside the post. The washer a is formed from a piece of 3 by 2 inch angle, one-half inch thick, with the 2 inch leg sheared off until the proper angle is formed. The 3 by 2 inch angle b forms a check to keep the angle washer a in place.

The length of the strut P is 192 inches, while the radius of gyration of a 3 by 3 inch angle about an axis through its centre of gravity and parallel to one leg is 0.92 inch. Hence the allowable stress per sq. in. by Eq. (13), is 4,000 pounds.

Both ends of T are held by precisely the same detail. At the upper end of the post, however, the pin-plates take the place of the battens at the intermediate points. The rivets securing the batten plate to the post are seen to give an excess of resistance.

Portal Bracing.

Fig. 17, Pl. XII., shows a skeleton sketch of the portal bracing. The sketch is taken in the plane of the portal. The strut P is placed 7 ft. 6 in. from the top of the post, while the length of the latter is 33.8 ft.

The computations are made precisely as in connection with Fig. 16, Pl. XII. The force acting at the upper extremity of the end post is 12,332 pounds. The tangent of the angle between T and the end post is 2.27, while the secant is 2.48. Hence, if $w = 12,332 \times 33.8 \div 17 = 24,664$, then :

$T = 24,664 \times 2.48 = 61,170$ pounds. . 1 — 6″ × 4″ 60 lb. angle.
$P = 24,664 \times 2.27 = 55,990$ "   . . 1 — 6″ × 6″ 75 "   "

As shown by the preceding results, this bracing is composed entirely of angles. This is done in order to secure the utmost stiffness or rigidity in the portal.

Fig. 12, Pl. XI., shows the method of securing the ends of the members T and P. The upper extremity of T is shown with the six-inch leg of the angle lying on the end post at a. In order that the proper number of three-quarter inch rivets may be brought into play, a $\frac{3}{8}$ inch plate lies underneath the angle, as shown. The method of securing the lower end of T, and each end of P is clearly shown at b. Three-quarter inch rivets are used for all these connections. At the intersection of the two Ts, one is cut and a firm joint is made by a centre plate a half-inch thick, aided by angle lugs.

The length of the strut P is about 192 inches, and its radius of gyration, 1.9 inches. Hence Eq. (13) gives the working stress at 7,000 pounds per square inch. The tension allowed in this angle bracing is 12,000 pounds per square inch.

An ornamental wrought-iron bracket may be placed in the angle between P and the end post.

Lower Lateral Bracing.

The lower lateral bracing is shown in skeleton plan by Fig. 6, Pl. XII. It is designed to resist a uniform fixed wind load of 150 pounds per lineal foot in addition to a uniform moving wind load of 300 pounds per lineal foot. The fixed panel load, therefore, will be $20.55 \times 150 = 3,083$ pounds, and the moving panel load, $20.55 \times 300 = 6,166$ pounds. The secant of the angle between T_1 and P_1 is 1.57. The line *ab* is the centre line of the span. Remembering that there are nine panels in the lower lateral system, and that the greatest allowable tension is 14,000 pounds, the following stresses and sizes may at once be written from the preceding data:

T_1 stress......$58,164 + 5,000 = 63,164$ *lbs.* $1 - 2''\frac{3}{8}$ *round.*
T_2 " ......$44,700 + $ " $= 49,700$ " $1 - 2''\frac{1}{8}$ "
T_3 " ......$32,313 + $ " $= 37,313$ " $1 - 1\frac{7}{8}$ "
T_4 " ......$21,000 + $ " $= 26,000$ " $1 - 1\frac{5}{8}$ "
T_5 " ......$10,770 + $ " $= 15,770$ " $1 - 1\frac{1}{4}$ "

The initial stress is included in the total by the addition of 5,000 pounds, as has been done before.

The floor-beams form the struts in the lower lateral system, except in the cases of the end struts P_1, hence, only the latter need be provided. The friction on the wall-plate will evidently relieve P_1 of some of its stress, but it is uncertain to what extent; hence, initial tension only will be neglected. Consequently:

P_1 stress......$41,620$ *lbs*......$1 - 6'' \times 4''$ 50 *lb. angle.*

The 6 inch leg of this angle is horizontal, and *it is riveted to the top flanges of the stringers where it crosses the latter.* By this means the stringer ends are held rigidly in position, and the general stiffness is increased.

The rods of the lower lateral system will necessarily pass through the webs of the stringers, and will be secured to the webs of the floor-beams (as closely as possible to their upper flanges) by precisely the detail shown in Fig. 19, Pl. XII.

The web-plate now takes the place of the batten in that figure.

The lower lateral T_1 takes hold of the pedestal by the clevis and 3 inch pin bolt shown in Fig. 10, Pl. XI.

A $9'' \times 7'' \times \frac{3}{4}''$ plate is riveted above and below the $\frac{3}{4}$ inch base plate of the pedestal in order to give the proper bearing area. The method of securing the end of P_1 to the pedestal is shown by the same figure with perfect clearness.

Expansion Rollers.

A set of expansion rollers under one end of each truss must be provided. A diameter of $2\frac{11}{16}$ inches will be assumed.

According to Appendix II., the resistance per lineal inch of a roller is:

$$\frac{4}{3} R \sqrt{2w^3 \frac{E + E'}{E E'}} = \frac{4}{3} R \sqrt{\frac{4w^3}{E}} \text{ (for } E = E'\text{)}.$$

Since all metal is wrought-iron $E = E'$. The greatest allowable intensity of pressure on the roller will be taken at 12,000 pounds per square inch, or $w = 12,000$. Also $R = 1.47$ and $E = 26,000,000$. Hence the allowable load per lineal inch, by the above formula, is 1,050 pounds. The maximum vertical component of the end-post stress is 172,820 pounds. Hence the number of lineal inches of roller bearing required is $172,820 \div 1,050 = 164$ inches. The set of rollers shown by Fig. 11, Pl. XII., gives very closely the required amount. The clear space between each adjacent pair of rollers is $\frac{3}{8}$ inch. The ends of each roller are turned down to $\frac{3}{4}$ inch and pass through a $2\frac{1}{2} \times \frac{3}{4}$ inch wrought-iron strap on the outside of which each roller end takes a nut. The rollers are thus held rigidly in their proper relative positions. A $1 \times \frac{1}{4}$ inch collar is turned down at the centre of each roller to take the $1 \times \frac{3}{16}$ inch shoulder which is shown in the wall plate, and by means of which all lateral motion of the rollers is prevented.

Wall Plates.

The mean pressure per square inch on the total surface of

the wall plate should not exceed 200 pounds. The total area of wall-plate surface shown in Fig. 12, Pl. XII., is $30 \times 30 = 900$; hence the total allowable weight is $900 \times 200 = 180,000$ pounds. The maximum total vertical pressure of 172,820 pounds is thus provided for.

The plan of the wall plate at the roller end is shown in the figure; the upper elevation also belongs to it. At the fixed end either the pedestal or the masonry must be sufficiently high to fill the roller space. Both those alterations, however, are unadvisable for obvious reasons. It is better to fill the roller space with the wall plate. The lower elevation of Fig. 12, Pl. XII., shows the arrangement to be adopted. The same $\frac{3}{4}$ inch thick wall plate is to be taken, $4 - 3\frac{1}{2} \times 3\frac{1}{2} \times \frac{5}{8}$ inch angles then run in the direction of the rollers across the entire plate. At right angles to these 4 lines of $3 \times 3 \times \frac{1}{2}$ inch angles are placed. The latter are cut to fit in between the former and will, of course, require filling strips underneath. The top of this gridiron arrangement is then planed off until the proper height of wall plate is reached. The rectangular spaces thus formed are then filled with Portland cement rammed hard and flush with the planed upper surfaces of the angles. A solid wall plate is thus formed with the interior surfaces completely protected against corrosion.

At diagonally opposite corners are seen the holes for the $1\frac{3}{8}$ inch anchor bolts.

Wind Pressure on Chords and End Posts.

The effect of the wind load on both upper and lower chords has been shown in detail in Art. 81; the principles there established remain to be applied here. The chord stresses in the lateral trusses are the same in kind as those produced by the vertical loading in one lower chord and one upper chord. Just to what extent these wind stresses may be allowed to exist without necessitating any increased chord section is a matter of experience only; but as the greatest wind stresses and those due to the vertical loading so rarely combine in most localities that with the working stresses specified in this case *the wind load may be allowed to ·reach ⅜ths the value of the greatest vertical loading without requiring any increase in chord section,* in all localities not ordinarily subject to cyclones or tornadoes.

In the present instance the total fixed and moving vertical load is equivalent to about 2,040 pounds per lineal foot of each truss. The total wind load in the lower chord is 450 pounds per lineal foot, and its depth of truss is only 17 feet. If reduced to the same truss depth as that for the vertical load, *i. e.,* 27 feet, it would be $450 \times \dfrac{27}{17} = 720$ pounds.

Again, the overturning effect of the wind on the train (discussed at the close of Art. 81) throws on the leeward truss the additional weight of $\dfrac{300 \times 8}{17} = 140$ pounds per lineal foot. It is assumed that the centre of wind pressure on the train is 8 feet above the end supports of the floor-beam. The total wind effect on the loading of the leeward truss is then $720 + 140 = 860$ pounds per lineal foot, or a little in excess of four-tenths the vertical loading. As three-eighths the vertical loading is 765 pounds per lineal foot, the chord sections should be increased for 95 pounds per lineal foot. As the increase in area, however, would be but one-sixth of an inch for one chord, and as the transverse bracing will slightly relieve the leeward truss, no change will be made.

If the lower chord needs no revision, the upper need not be considered.

The total pressure of wind against the upper extremities of the end posts, and, hence, against their lower extremities also, has already been seen to be 12,332 pounds. If this is assumed to be equally divided between the end-post feet, each of the latter will carry 6,166 pounds. Each end of P (in the portal, Fig. 17, Pl. XII.) is 26.3 feet from the end-post foot. Hence at the former point the end post suffers the bending moment $6,166 \times 12 \times 26.3 = 1,945,990$ *in. lbs.* The moment of inertia of the end-post section about the neutral axis normal to the cover plate is 1,886, and since the half total width of the chord is 12.5 inches, the stress per square inch in the extreme fibres of the 5 × 3 angles is

$$K' = 1{,}945{,}990 \times 12.5 \div 1{,}886 = 12{,}904 \; pounds \; per \; sq. \; in.$$

The direct post stress will be about 7,200 more, or a total of 20,104 *pounds per sq. in.*, whereas 15,000 should not be exceeded.

If $K =$ limit of compressive bending stress per square inch, which must not be exceeded; $M =$ bending moment in inch pounds; I the moment of inertia of the total post section about a neutral axis normal to the cover-plate; $2d =$ total width of end post; $P =$ total direct stress of compression, due to vertical loading and overturning action of the wind against the train (t of Art. 81 is its panel value), and $A =$ total area of section; then the sectional area must be increased until the following equation holds true:

$$K = \frac{Md}{I} + \frac{P}{A} \quad . \quad . \quad . \quad . \quad . \quad . \quad (14).$$

It has been seen above that the wind effect in the leeward truss is 140 lbs. per lin. ft., or 20.55 × 140 = 11,508 lbs. per panel. Hence, 4 × 11,508 × 1.26 = 14,500 lbs. is that part of P due to the wind; or:

$$P = 14{,}500 + 217{,}750 = 232{,}250 \text{ lbs.}$$
Also, $2d = 25$; or $d = 12.5$ inches;
And, $M = 1{,}945{,}990 \; in. \; lbs.$

If $2 - 3'' \times 3''$ 25 *lb.* angles are riveted on the outside of each side plate of each post in the manner shown in Fig. 13, the centres of gravity of those angles will be 8.3 inches from the neutral axis about which I is taken, and the moment of inertia of each angle section about a parallel axis through its own centre of gravity is 2.25; hence:

$$I = 1{,}886 + 4 \times 2.5 \times (8.3)^2 + 4 \times 2.25 = 2{,}585.$$

Finally:
$$A = 32.3 + 4 \times 2.5 = 42.3 \text{ sq. ins.}$$

These quantities placed in Eq. (14) give:

$$K = 9,412 + 5,500 = 14,912 \; lbs. \text{ per } sq. \; in.$$

which shows that the desired section is obtained.

Fig. 13 shows an elevation of parts of the end post, which is supposed to be' intersected by the portal strut at *ab*. *cc'*, and *dd'* are the 3″ × 3″ 25 *lb*. angles. *c* is 9 ft. below *ab* and *c'* 2 ft. 9 in. above it. *dd'* is half the length of *cc'*. Below *c* and above *c'*, the preceding figures show that no increase of section is needed.

If much increase is needed, unequal legged angles with

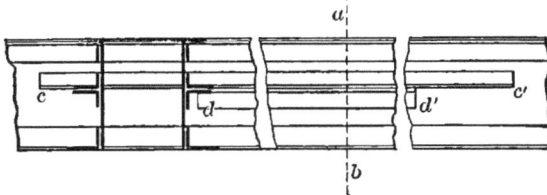

FIG. 13.

the larger legs normal to the side plates can be most advantageously used.

It will ordinarily be sufficiently accurate to increase the section in the ratio of $\dfrac{K'}{K}$.

These computations show what an important factor the wind load may be in a country subject to tornadoes and cyclones. In such exposed localities the chord-wind stresses should not be allowed to exceed 25 per cent. of those due to the vertical loading without providing correspondingly increased sections.

The wind effect on the stringers mentioned in Art. 81 is such a small per centage of the vertical load that it need not be considered.

Conclusion.

It is not necessary here to produce in detail the complete list of weights of all the parts, although this must invariably be done in practice. If the estimated weight comes out greater than the assumed, a revision of the computations must be made with a sufficiently increased fixed weight to exceed, at least by a little, the estimated weight. If, on the other hand, the estimated weight is considerably less than the assumed, the latter may be reduced in a recomputation, in order to reach a proper degree of economy.

APPENDIX I.

THE THEOREM OF THREE MOMENTS.

ART. I.—The object of this theorem is the determination of the relation existing between the bending moments which are found in any continuous beam at any three adjacent points of support. In the most general case to which the theorem applies, the section of the beam is supposed to be variable, the points of support are not supposed to be in the same level, and at any point, or all points, of support there may be constraint applied to the beam, external to the load which it is to carry; or, what is equivalent to the last condition, the beam may not be straight at any point of support before flexure takes place.

Before establishing the theorem itself, some preliminary matters must receive attention.

In Fig. 1, let ABC represent the centre line of any bent beam; AF, a vertical line through A; CF, a horizontal line

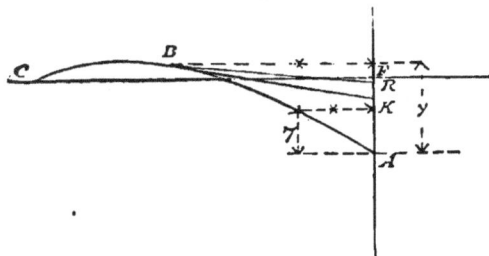

FIG I.

through C, while A is the section of the beam at which the deflection (vertical or horizontal) in reference to C, the bending moment, the shearing stress, etc., are to be determined.

As shown in the figure, let x be the horizontal co-ordinate measured from A, and y the vertical one measured from the same point; then let z be the horizontal distance from the same point to the point of application of any external vertical force P. To complete the notation, let D be the deflection desired; M_1, the moment of the external forces about A; S, the shear at A; P', the strain (extension or compression) per unit of length of a fibre parallel to the neutral surface and situated at a normal distance of unity from it; I, the general expression of the moment of inertia of a normal cross-section of the beam, taken in reference to the neutral axis of that section; E the coefficient of elasticity for the material of the beam; and M the moment of the external forces for any section, as B.

Again, let $\mathit{\Delta}$ be an indefinitely small portion of any normal cross-section of the beam, and let y' be an ordinate normal to the neutral axis of the same section. By the "common theory" of flexure, the intensity of stress at the distance y' from the neutral surface is $(y'P'E)$. Consequently the stress developed in the portion $\mathit{\Delta}$, of the section, is $EP'y'\mathit{\Delta}$, and the resisting moment of that stress is $EP'y'^2\mathit{\Delta}$.

The resisting moment of the whole section will therefore be found by taking the sum of all such moments for its whole area.

Hence:

$$M = EP' \Sigma y^2 \mathit{\Delta} = EP'I.$$

Hence, also,

$$P' = \frac{M}{EI}.$$

If n represents an indefinitely short portion of the neutral surface, the strain for such a length of fibre at unit's distance from that surface will be nP'.

If the beam were originally straight and horizontal, n would be equal to dx.

P' being supposed small, the effect of the strain nP' at any section, B, is to cause the end K, of the tangent BK, to move vertically through the distance $nP'x$.

If BK and BR (taken equal) are the positions of the tangents before and after flexure, $nP'x$ will be the vertical distance between K and R.

By precisely the same kinematical principle, the expression $nP'y$ will be the horizontal movement of A in reference to B.

Let $\Sigma nP'x$ and $\Sigma nP'y$ represent summations extending from A to C, then will those expressions be the vertical and horizontal deflections, respectively, of A in reference to C. It is evident that these operations are perfectly general, and that x and y may be taken in any direction whatever.

The following general, but, strictly, approximate equations, relating to the subject of flexure, may now be written:

$$S = \Sigma P \qquad \ldots \ldots \ldots \ldots \quad (1).$$

$$M_1 = \Sigma Pz \qquad \ldots \ldots \ldots \ldots \quad (2).$$

$$P' = \frac{M}{EI} \qquad \ldots \ldots \ldots \ldots \quad (3).$$

$$\Sigma nP' = \Sigma n \frac{M}{EI} \qquad \ldots \ldots \ldots \quad (4).$$

$$D = \Sigma nP'x = \Sigma \frac{nMx}{EI} \qquad \ldots \ldots \quad (5).$$

$$D_h = \Sigma nP'y = \Sigma \frac{nMy}{EI} \qquad \ldots \ldots \quad (6).$$

D_h represents horizontal deflection.

ART. 2.—Some elementary but general considerations in reference to that portion of a continuous beam included between two adjacent points of support must next be noticed.

If a beam is simply supported at each end, the reactions are found by dividing the applied loads according to the simple principle of the lever. If, however, either or both ends are not simply supported, the reaction, in general, is greater at one end and less at the other, than would be found

by the law of the lever; a portion of the reaction at one end
is, as it were, transferred to the other.   This transference can
only be accomplished by the application of a couple to the
beam, the forces of the couple being applied at the two adjacent points of support ; the span, consequently, will be the
lever arm of the couple.   The existence of equilibrium requires the application to the beam of an equal and opposite
couple.   It is only necessary, however, to consider, in connection with the span AB, the one shown in Fig. 2.   Further,
from what has immediately preceded, it appears that the
force of this couple is equal to the difference between the
actual reaction at either point of support and that found by
the law of the lever.   The bending caused by this couple will
evidently be of an opposite kind to that existing in a beam
simply supported at each end.

These results are represented graphically in Fig. 2. A and
B are points of support, and AB is the beam; AR and BR'
are the reactions according to the law of the lever; $RF =$

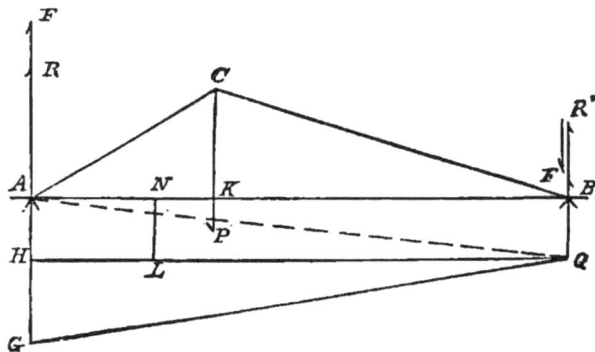

FIG. 2.

$R'F$ is the force of the applied couple; consequently $AF =$
$AR + RF$ and $BF = BR' - (R'F = RF)$ are the reactions
after the couple is applied.   As is well known, lines parallel
to CK, drawn in the triangle ACB, represent the bending moments at the various sections of the beam, when the reactions are AR and BR'.   Finally, vertical lines parallel to AG,

in the triangle *QHG*, will represent the bending moments caused by the force *R'F*.

In the general case there may also be applied to the beam two equal and opposite couples, having axes passing through *A* and *B* respectively. The effect of such couples will be nothing so far as the reactions are concerned, but they will cause uniform bending between *A* and *B*. This uniform or constant moment may be represented by vertical lines drawn parallel to *AH* or *LN* (equal to each other) between the lines *AB* and *HQ*. The resultant moments to which the various sections of the beam are subjected will then be represented by the *algebraic* sum of the three vertical ordinates included between the lines *ACB* and *GQ*. Let that resultant be called *M*.

Let the moment *GA* be called M_a, and the moment $BQ = LN = HA$, M_b. Also designate the moment caused by the load *P*, shown by lines parallel to *CK* in *ACB*, by M_1. Then let *x* be any horizontal distance measured from *A* towards *B*; *l* the horizontal distance *AB*; and *z* the distance of the point of application, *K*, of the force *P* from *A*. With this notation there can be at once written:

$$M = M_a\left(\frac{l-x}{l}\right) + M_b\left(\frac{x}{l}\right) + M_1 . \quad . \quad . \quad (7).$$

Eq. (7) is simply the general form of Eq. (2).

It is to be noticed that Fig. 2 does not show all the moments M_a, M_b and M_1 to be of the same sign, but, for convenience, they are so written in Eq. (7).

ART. 3.—The formula which represents the theorem of three moments can now be written without difficulty. The method to be followed involves the improvements added by Prof. H. T. Eddy, and is the same as that given by him in the "American Journal of Mathematics," Vol. I., No. 1.

Fig. 3 shows a portion of a continuous beam, including two spans and three points of support. The deflections will be supposed measured from the horizontal line *NQ*. The spans are represented by l_a and l_c; the vertical distances of

NQ from the points of support by c_a, c_b and c_c; the moments at the same points by M_a, M_b and M_c, while the letters S and R represent shears and reactions respectively.

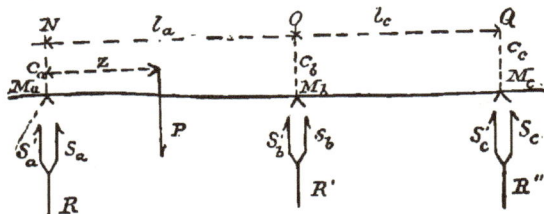

FIG. 3.

In order to make the case general, it will be supposed that the beam is curved in a vertical plane, and has an elbow at b, before flexure, and that, at that point of support, the tangent of its inclination to a horizontal line, toward the span l_a is t, while t' represents the tangent on the other side of the same point of support; also let d and d' be the vertical distances, before bending takes place, of the points a and c, respectively, below the tangents at the point b.

A portion of the difference between c_a and c_b is due to the original inclination, whose tangent is t, and the original lack of straightness, and is not caused by the bending; that portion which is due to the bending, however, is, remembering Eq. (5):

$$D = c_a - c_b - l_a t - d = \sum_b^a \frac{Mxn}{EI}$$

By the aid of Eq. (7) this equation may be written:

$$E(c_a - c_b - l_a t - d) = \sum_b^a \left[\left\{ M_a\left(\frac{l-x}{l}\right) + M_b\left(\frac{x}{l}\right) + M_1 \right\} \frac{xn}{I} \right] \quad . \quad . \quad (8).$$

In this equation, it is to be remembered, both x and z (involved in M_1) are measured from support a toward support b. Now let a similar equation be written for the span l_c, in which the variables x and z will be measured from c toward b. There will then result:

$$E\left(c_c - c_b - l_c t' - d'\right) = \Sigma_b^c \left[\left\{ M_c\left(\frac{l-x}{l}\right) + M_b\left(\frac{x}{l}\right) + M_1 \right\} \frac{xn}{I} \right] \quad \cdots \quad (9).$$

When the general sign of summation is displaced by the integral sign, n becomes the differential of the axis of the beam, or ds. But ds may be represented by udx, u being such a function of x as becomes unity if the axis of the beam is originally straight and parallel to the axis of x. The Eqs. (8) and (9) may then be reduced to simpler forms by the following methods:

In Eq. (8) put

$$\Sigma_b^a \left(\frac{l-x}{l}\right)\frac{xn}{I} = \frac{1}{l_a}\int_b^a \frac{u\left(l_a - x\right)xdx}{I} =$$

$$\frac{x_a}{l_a}\int_b^a \frac{u\left(l_a - x\right)dx}{I} \quad \cdots \quad (10).$$

Also,

$$\frac{x_a}{l_a}\int_b^a \frac{u\left(l_a - x\right)dx}{I} = \frac{i_a x_a}{l_a}\int_b^a u\left(l_a - x\right)dx \quad \cdots \quad (11).$$

Also,

$$\frac{i_a x_a}{l_a}\int_b^a u\left(l_a - x\right)dx = \frac{i_a x_a u_a}{l_a}\int_b^a \left(l_a - x\right)dx =$$

$$\frac{i_a x_a u_a l_a}{2} \quad \cdots \quad (12).$$

In the same manner:

$$\Sigma_b^a \frac{x^2 n}{l_a I} = \frac{1}{l_a}\int_b^a \frac{u x^2 dx}{I} = \frac{x_a'}{l_a}\int_b^a \frac{u x dx}{I} \quad \cdots \quad (13).$$

Also,

$$\frac{x_a'}{l_a}\int_b^a \frac{u x dx}{I} = \frac{i_a' x_a'}{l_a}\int_b^a u x dx \quad \cdots \quad (14).$$

And,

$$\frac{i_a' x_a'}{l_a} \int_b^a u x\, dx = \frac{i_a' x_a' u_a'}{l_a} \int_b^a x\, dx = \frac{i_a' x_a' u_a' l_a}{2} \quad . \quad . \quad (15).$$

Again, in the same manner:

$$\Sigma_b^a \frac{M_1 x n}{I} = i_{1a} u_{1a} \Sigma M_1 x\, \varDelta x \quad . \quad . \quad . \quad (16).$$

Using Eqs. (10) to (16), Eq. (8) may be written:

$$E(c_c - c_b - l_a t - d) = \frac{l_a}{2}(M_a u_a i_a x_a + M_b u_a' i_a' x_a') +$$
$$u_{1a} i_{1a} \Sigma_b^a M x\, \varDelta x \quad . \quad . \quad (17).$$

Proceeding in precisely the same manner with the span l_c, Eq. (9) becomes:

$$E(c_c - c_b - l_c t' - d') = \frac{l_c}{2}(M_c u_c i_c x_c + M_b u_c' i_c' x_c') +$$
$$u_{1c} i_{1c} \Sigma_b^c M_1 x\, \varDelta x \quad . \quad . \quad (18).$$

The quantities x_a and x_c are to be determined by applying Eq. (10) to the span indicated by the subscript; while u_a, i_a, u_c and i_c are to be determined by using Eqs. (11) and (12) in the same way. Similar observations apply to u_a', i_a', x_a', u_c', i_c' and x_c', taken in connection with Eqs. (13), (14) and (15).

If I is not a continuous function of x, the various integrations of Eqs. (10), (11), (13), and (14) must give place to summations (Σ) taken between the proper limits.

Dividing Eqs. (17) and (18) by l_a and l_c, respectively, and adding the results:

$$E\left(\frac{c_a - c_b}{l_a} + \frac{c_c - c_b}{l_c} - T - \frac{d}{l_a} - \frac{d'}{l_c}\right) = \frac{u_a i_a}{l_a} \Sigma_b^a M_1 x \varDelta x +$$
$$\frac{u_c i_c}{l_c} \Sigma_b^c M_1 x \varDelta x + \tfrac{1}{2}(M_a u_a i_a x_a + M_b u_a' i_a' x_a' +$$
$$M_c u_c i_c x_c + M_b u_c' i_c' x_c') \quad . \quad . \quad . \quad (19).$$

in which $T = t + t'$.

Eq. (19) is the most general form of the theorem of three moments if E, the coefficient of elasticity, is a constant quantity. Indeed, that equation expresses, as it stands, the "theorem" for a variable coefficient of elasticity if (ie) be written instead of i; e representing a quantity determined in a manner exactly similar to that used in connection with the quantity i.

In the ordinary case of an engineer's experience, $T = 0$, $d = d' = 0$, $I = constant$, $u = u_a = u_c = etc. = c' = secant\ of\ the$ $inclination\ for\ which\ t = -t'\ is\ the\ tangent$; consequently

$$i_a = i_a' = i_c = i_c' = i_{,a} = i_{,c} = \frac{1}{I}.$$

From Eq. (10),

$$x_a = \frac{2l_a}{6}, \qquad x_c = \frac{2l_c}{6};$$

From Eq. (13),

$$x_a' = \frac{4l_a}{6}, \qquad x_c' = \frac{4l_c}{6}.$$

The summation $\Sigma M_1 x \Delta x$ can be readily made by referring to Fig. 2.

The moment represented by CK in that figure is,

$$P\left(\frac{l-z}{l}\right) \cdot z;$$

consequently the moment at any point between A and K, due to P, is,

$$M_1 = P\left(\frac{l-z}{l}\right) \cdot z \cdot \frac{x}{z} = P\left(\frac{l-z}{l}\right)x.$$

Between K and B,

$$M_1' = \left(\frac{l-x}{l-z}\right) \cdot CK = P\frac{z}{l}(l-x).$$

Using these quantities for the span l_a:

$$\Sigma_b^a M_1 x \varDelta x = \int_0^{z} M_1 x dx + \int_z^{la} M_1' x dx = \tfrac{1}{6} P (l_a^2 - z^2) z.$$

For the span l_c, the subscript a is to be changed to c.

Introducing all these quantities, Eq. (19) becomes, after providing for any number of weights, P:

$$\frac{6EI}{c'} \left(\frac{c_a - c_b}{l_a} + \frac{c_c - c_b}{l_c} \right) = M_a l_a + 2M_b (l_a + l_c) + M_c l_c +$$

$$\frac{1}{l_a} \overset{a}{\Sigma} P(l_a^2 - z^2)z + \frac{1}{l_c} \overset{c}{\Sigma} P(l_c^2 - z^2)z . \quad . \quad . \quad (20).$$

Eq. (20), with c' equal to unity, is the form in which the theorem of three moments is usually given ; with c' equal to unity or not, *it applies only to a beam which is straight before flexure*, since $T = t + t' = 0 = d = d'$.

If such a beam rests on the supports a, b, and c, before bending takes place, $\dfrac{c_a - c_b}{l_a} = -\dfrac{c_c - c_b}{l_c}$, and the first member of Eq. (20) becomes zero.

If, in the general case to which Eq. (19) applies, the deflections c_a, c_b, and c_c belong to the beam in a position of no bending, the first member of that equation disappears, since it is the sum of the deflections *due to bending only*, for the spans l_a and l_c, divided by those spans, and each of those quantities is zero by the equation immediately preceding, Eq. (8). Also, if the beam or truss belonging to each span is straight between the points of support (*such points being supposed in the same level or not*), $u_a = u_a' = u_{1a} = constant$, and $u_c = u_c' = u_{1c} = another\ constant$. If, finally, I be again taken as constant, x_a and x_c, as well as $\Sigma M_1 x \varDelta x$, will have the values found above.

From these considerations it at once follows that the second member of Eq. (20), put equal to zero, expresses the theorem of three moments for a beam or truss straight between points of support, when those points are not in the same level, but when they belong to a configuration of no bending in the beam. Such an equation, however, does not belong to a beam not straight between points of support.

The shear at either end of any span, as l_a, is next to be found, and it can be at once written by referring to the observations made in connection with Fig. 2. It was there seen that the reaction found by the simple law of the lever is to be increased or decreased for the continuous beam, by an amount found by dividing the difference of the moments at the extremities of any span by the span itself. Referring, therefore, to Fig. 3, for the shears S, there may at once be written :

$$S_a = \overset{a}{\Sigma} P \frac{l_a - z}{l_a} - \frac{M_a - M_b}{l_a} \quad . \quad . \quad . \quad (21).$$

$$S_b' = \overset{a}{\Sigma} P \frac{z}{l_a} + \frac{M_a - M_b}{l_a} \quad . \quad . \quad . \quad . \quad (22).$$

$$S_b = \overset{c}{\Sigma} P \frac{z}{l_c} + \frac{M_c - M_b}{l_c} \quad . \quad . \quad . \quad . \quad (23).$$

$$S_c' = \overset{c}{\Sigma} P \frac{l_c - z}{l_c} - \frac{M_c - M_b}{l_c} \quad . \quad . \quad . \quad (24).$$

The negative sign is put before the fraction $\dfrac{M_a - M_b}{l_a}$, in Eq. (21), because in Fig. 2 the moments M_a and M_b are represented opposite in sign to that caused by P, while in Eq. (7) the three moments are given the same sign, as has already been noticed.

Eqs. (21) to (24) are so written as to make an upward reaction positive, and they may, perhaps, be more simply found by taking moments about either end of a span. For example, taking moments about the right end of l_a:

$$S_a l_a - \overset{a}{\Sigma} P(l_a - z) + M_a = M_b.$$

From this, Eq. (21) at once results. Again, moments about the left end of the same span give :

$$S_b' l_a - \overset{a}{\Sigma} Pz + M_b = M_a.$$

This equation gives Eq. (22), and the same process will give the others.

If the loading over the different spans is of uniform intensity, then, in general, $P = wdz$; w being the intensity. Consequently:

$$\Sigma P(l^2 - z^2)\,z = \int_0^l w\,(l^2 - z^2)\,zdz = w\,\frac{l^4}{4}.$$

In all equations, therefore, for $\frac{1}{l_a} \overset{a}{\Sigma} P(l_a^2 - z^2)\,z$ there is to be placed the term $w_a\frac{l_a^3}{4}$; and for $\frac{1}{l_c} \overset{c}{\Sigma} P(l_c^2 - z^2)\,z$, the term $w_c\frac{l_c^3}{4}$. The letters a and c mean, of course, that reference is made to the spans l_a and l_c.

From Fig. 3, there may at once be written:

$$R \ = S_a' + S_a \ . \quad . \quad . \quad . \quad . \quad . \quad (25).$$

$$R' \ = S_b' + S_b \ . \quad . \quad . \quad . \quad . \quad . \quad (26).$$

$$R'' = S_c' + S_c \ . \quad . \quad . \quad . \quad . \quad . \quad (27).$$

$$\text{etc.} = \text{etc.} + \text{etc.}$$

APPENDIX II.

AN approximate expression for the resistance of a roller may easily be written, and although the approximation may be considered a loose one, it furnishes an excellent basis for an accurate empirical formula.

The following investigation contains the improvements by Prof. J. B. Johnson and Prof. H. T. Eddy on the method originally given by the author.

The roller will be assumed to be composed of indefinitely thin vertical slices parallel to its axis. It will also be assumed that the layers or slices act independently of each other.

Let E' be the coefficient of elasticity of the metal over the roller.

Let E be the coefficient of elasticity of the metal of the roller.

Let R be the radius of the roller and R' the thickness of the metal above it.

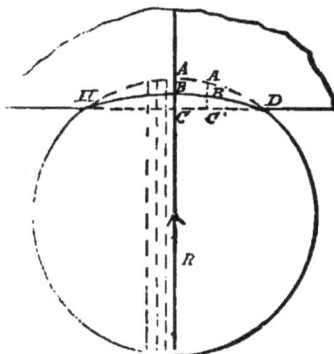

FIG. 1.

Let w = intensity of pressure at A.

  " p =   "   "   any other point.

  " P = total weight which the roller sustains per unit of length.

  " x be measured horizontally from A as the origin.

  " d = AC.

  " e = DC.

From Fig. 1 :

$$AB = \frac{wR}{E}; \; A'B' = \frac{pR}{E}.$$

$$BC = \frac{wR'}{E'}; \; C'B' = \frac{pR'}{E'}.$$

$$\therefore d = AC = AB + BC = w\left(\frac{R}{E} + \frac{R'}{E'}\right); \quad \cdot \quad \cdot \quad \cdot \quad (1).$$

And

$$A'C' = A'B' + B'C' = p\left(\frac{R}{E} + \frac{R'}{E'}\right). \quad \cdot \quad \cdot \quad \cdot \quad (2).$$

Dividing Eq. (2) by Eq. (1):

$$p = A'C'\frac{w}{d}.$$

But

$$P = \int_{-e}^{+e} p\,dx = \frac{w}{d}\int_{-e}^{+e} A'C'dx.$$

If the curve DAH be assumed to be a parabola, as may be done without essential error, there will result:

$$\int_{e}^{+e} A'C'dx = \frac{4}{3}ed.$$

Hence:

$$P = \frac{4}{3}we \quad \cdot \quad \cdot \quad \cdot \quad \cdot \quad \cdot \quad \cdot \quad (3).$$

But :

$$e = \sqrt{2Rd - d^2} = \sqrt{2Rd} \text{ nearly.}$$

By inserting the value of d from Eq. (1) in the value of e, just determined, then placing the result in Eq. (7):

$$P = \frac{4}{3}\sqrt{2w^3R\left(\frac{R}{E} + \frac{R'}{E'}\right)} \quad \cdot \quad \cdot \quad \cdot \quad \cdot \quad (4).$$

If $R = R'$:

$$P = \frac{4}{3} R \sqrt{2w^3 \frac{E + E'}{EE'}} \quad \cdots \quad \cdots \quad (5).$$

The preceding expressions are for one unit of length. If the length of the roller is l, its total resistance is

$$P' = Pl = \frac{4}{3} l \sqrt{2w^3 R \left(\frac{R}{E} + \frac{R'}{E'}\right)} \quad \cdots \quad (6).$$

Or if $R = R'$:

$$P' = \frac{4}{3} Rl \sqrt{2w^3 \frac{E + E'}{EE'}} \quad \cdots \quad \cdots \quad (7).$$

In ordinary bridge practice Eq. (7) is sufficiently near for all cases.

A simple expression for conical rollers may be obtained by using Eqs. (4) or (5).

As shown in Fig. 2, let z be the distance, parallel to the axis, of any section from the apex of the cone; then consider

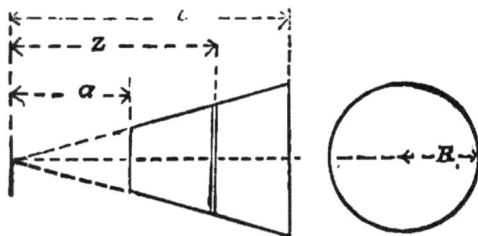

FIG. 2.

a portion of the conical roller whose length is dz. Let R_1 be the radius of the base. The radius of the section under consideration will then be

$$R = \frac{z}{l} R_1;$$

and the weight it will sustain, if $R_1 = R'$;

$$dP' = \frac{R_1}{l} \sqrt{2w_3 \frac{E + E'}{EE'}} \cdot zdz.$$

Hence :

$$P' = \int_a^l dP' = \frac{l^2 - a^2}{2l} R_1 \sqrt{2w^3 \frac{E + E'}{EE'}} \quad . \quad . \quad . \quad (8).$$

Eqs. (6), (7), and (8) give ultimate resistances if w is the ultimate intensity of resistance for the roller.

It is to be observed that the main assumptions on which the investigation is based lead to an error on the side of safety.

If for wrought iron, $w = 12,000$ pounds per square inch, and $E = E' = 28,000,000$ pounds, Eq. (5) gives :

$$P = \frac{8}{3} R \sqrt{\frac{w^3}{E}} = 664 \, R.$$

APPENDIX III.

THE SCHWEDLER TRUSS.

THE general principle applied in Chapter III. to bowstring trusses, enables the characteristics of the Schwedler truss to be very simply shown. In fact, that truss is a special bowstring, having the least possible number of diagonal braces under the conditions assumed.

Fig. 1 represents the elevation of such a truss, and the problem involved is the determination of such depths, near

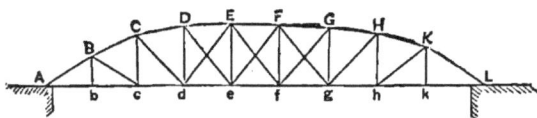

FIG. 1.

the ends, that one diagonal only will be needed in each panel; it being premised that the inclined web members or diagonals are to sustain tension only.

Let W = total (upper and lower chord) panel fixed load.

" R' = half the fixed load (or weight) of the bridge.

" w = panel moving load.

" l = length of span.

" d = any vertical brace or truss depth, as Cc.

" d_1 = vertical brace or truss depth, as Dd, adjacent to d and toward centre.

" p = panel length.

" α = inclination of any diagonal, as Cd, to the horizontal lower chord, i. e., $Cdc = \alpha$.

447

Let x = distance from A to the intersection of the prolongation of any upper chord panel, in the left half of the truss, with the prolongation to the left of the lower chord.

" y = the normal distance from that point of intersection to the prolongation of the diagonal immediately under the upper chord panel prolonged.

Let the moving load pass on the bridge from A toward L, and let n be the number of panel-moving loads from A.

The reaction at A, for any position of the moving load will be :

$$R = R^1 + nw\left(1 - \frac{(n+1)p}{2l}\right). \quad . \quad . \quad . \quad (1).$$

Then let the truss be imagined divided through the panel immediately in front of the train.

If moments be taken about the point of intersection denoted by x, and if T represents the tension in the diagonal just in front of the train, whose lever arm is y, there will result :

$$Ty = Rx - n(W + w)\left(x + \frac{(n+1)p}{2}\right). \quad . \quad (2).$$

Eq. (2) is so written, it is important to notice, that if the second member is greater than zero, or positive, T will be tension. Hence, if T is tension :

$$Ty \geq O.$$

But, $y = (x + (n+1)p) \sin \alpha. \quad . \quad . \quad . \quad (3).$

Also, from similar triangles ;

$$\frac{d_1 - d}{p} = \frac{d_1}{x + (n+1)p} \quad \therefore x + (n+1)p = \frac{pd_1}{d_1 - d}. \quad . \quad (4).$$

By the aid of Eq. (3) :

$$T = \frac{-R(n + 1)p + n(W + w)\dfrac{(n + 1)p}{2}}{(x + (n + 1)p)\,sin\,\alpha} + \frac{R - n(W + w)}{sin\,\alpha} \geq 0.$$

Hence, by the aid of Eq. (4):

$$x + (n + 1)p \geq \frac{\left(R - \dfrac{n(W + w)}{2}\right)(n + 1)p}{R - n(W + w)} \leq \frac{pd_1}{d_1 - d}.$$

Using the last two members of this inequality:

$$\frac{d}{d_1} \geq 1 - \frac{R - n(W + w)}{\left(R - \dfrac{n(W + w)}{2}\right)(n+1)} \quad \cdots \quad (5).$$

$$\text{Or}\,;\, d \geq d_1\left(1 - \frac{R - n(W + w)}{\left(R - \dfrac{n(W + w)}{2}\right)(n + 1)}\right). \quad \cdots \quad (6).$$

The second member of (6) is the least value of the depth *d* which can exist without inducing compression in the diagonal under consideration. This diagonal is the one immediately in front of the train, and the principles given in Chapter III. show that if this position does not induce compression, no other will.

Inequality (6) shows that if *R* is greater than $n(W + w)$, or $R > n(W + w)$, *d* will always be less than d_1. If $R = n(W + w)$, then $d = d_1$.

Again, if $R < n(W + w)$, then will *d* be greater than d_1, if tension is to be found in the diagonal. But it is not admissible to make $d > d_1$; *hence, when n becomes so great that R is less than $n(W + w)$, or $R < n(W + w)$, T will be compression, and the diagonal must be counterbraced or else intersecting diagonals must be placed in the panels, as shown in* Fig. 1, *near the middle of the truss.*

The value of *n* given by

$$R < n(W + w). \quad \cdots \quad (7).$$

2.)

will show the position of the head of the train when all panels between it and the centre must contain intersecting diagonals. All the other panels will need but one each, sloping upward and toward the end of the truss, as shown in Fig. 1.

Since this method is independent of the direction of approach of the train, it is only necessary to consider one-half of the truss.

It is seen in (6), that d is given in terms of d_1, hence the latter must be known in order to find d.

The centre depth is arbitrary and may be assigned at will. The depths between the centre and that point indicated by n, in inequality (7), may also be assigned at will; consequently d_1, next to the first "d" to be computed, will be known. The first "d" computed will be the "d_1" for the next "d," etc., etc., to the end of the truss.

As a margin of safety it will be well to make d a little greater than given by the second member of (6).

In long spans it would be well to make the truss depth constant for a number of panels near the centre, perhaps, even between the points given by (7). This would make a considerable number of diagonals and panels uniform in length, which would otherwise lack uniformity. Thus the construction would be simplified and cheapened.

The loads have been taken uniformly, but precisely the same methods would hold if they were not uniform.

Example.

Let the following example (the truss shown in Fig. 1) be taken :

$$\text{Span} = l = 9p = 108 \text{ feet} ; \quad \therefore \quad p = 12 \text{ feet.}$$
$$\text{Centre depth} = 16 \text{ feet.}$$

$$W = 8.00 \text{ tons.} \qquad w = 18.00 \text{ tons.}$$
$$(W + w) = 26.00 \text{ tons.}$$
$$R^1 = 4W = 32.00 \text{ "}$$

From Eq. (1):

$$R = 32 + 18\left(n - \frac{n^2 + n}{18}\right) \quad . \quad . \quad . \quad (8).$$

If $n = 3$, by (8) and (7):

$$R = 32 + 18\left(3 - \tfrac{3}{2}\right) = 74 < n(W + w) = 78.$$

Hence the diagonals De and Ed, in the panel in front of the head of the train, at d, must both be introduced.

If $n = 2$, by (8) and (7):

$$R = 62 > n(W + w) = 52.$$

Hence Cd is the only diagonal needed in the panel $CDdc$, and $d = Cc$ is to be computed from Eq. (6).

The centre depth $= Ee = Ff$ was taken at 16 feet; let Dd be taken at 15.5 feet.

Since $n = 2$; $R - n(W + w) = 10$, and $R - \dfrac{n(W + w)}{2} = 36.$

Substituting these values, and $d_1 = 15.5$, in Eq. (6):

$$d = 0.91 \; d_1 = 14.11 \text{ feet. Hence let}$$

$$d = 14.5 \text{ feet} = Cc \text{ (Fig. 1).}$$

Next, let the head of the train be at b, *i.e.*, let $n = 1$. Then by Eq. (8):

$$R = 32 + 16 = 48.$$

Also; $\quad R - n(W + w) = 22$, and $\quad R - \dfrac{n(W + w)}{2} = 35.$

For this position of load, $d_1 = 14.5$ feet. Hence, by Eq. (6):

$$d = 14.5\left(1 - \frac{22}{70}\right) = 9.94 \text{ feet. Hence let}$$

$$d = 10.00 \text{ feet} = Bb \text{ (Fig. 1).}$$

Fig. 1 represents the truss, drawn to scale, with the various depths given or computed as above, *i. e.*,

$$Ee = Ff = 16.0 \text{ feet.}$$

$$Dd = Gg = 15.5 \text{ "}$$

$$Cc = Hh = 14.5 \text{ "}$$

$$Bb = Kk = 10.0 \text{ "}$$

In the three panels adjacent to each end of the truss only two main diagonals are thus seen to be necessary, and in those no compression will ever exist. In each of the three middle panels, however, two intersecting diagonals will be necessary, since no diagonal must sustain compression.

As is evident, the expression $R - n(W + w)$ is the vertical shear at the head of the train. Hence the limiting case of the inequality (7):

$$R = n(W + w),$$

gives the two points, in the two halves of the truss, at which the vertical shear at the head of the train is zero. Between these points intersecting diagonals or counterbraces are needed, and only between them.

After the truss depths are fixed by the preceding method, the stresses in the individual members are to be found in the usual manner—as in any other bowstring truss—as shown in Chapter III.

APPENDIX IV.

REACTIONS AND MOMENTS FOR CONTINUOUS BEAMS.

Case I.—Three Spans with Two Intermediate Points of Support and Two End Supports.

The notation of Art. 35 will be used and reference will be made to Fig. 1 of that Art.

$$M_3 = \left[l_1^2 l_2 \overset{1}{\Sigma} P\left(1 - \frac{z^2}{l_1^2}\right)\frac{z}{l_1} - 2 l_3^2 (l_1 + l_2) \overset{3}{\Sigma} P\left(1 - \frac{z^2}{l_3^2}\right)\frac{z}{l_3} \right] \div$$
$$\{ 4 (l_2 + l_3)(l_1 + l_2) - l_2^2 \}. \quad \cdot \quad \cdot \quad \cdot \quad \cdot \quad \cdot \quad \cdot \quad \cdot \quad (1).$$

$$M_2 = - \left\{ M_3 l_2 + l_1^2 \overset{1}{\Sigma} P\left(1 - \frac{z^2}{l_1^2}\right)\frac{z}{l_1} \right\} \div 2 (l_1 + l_2). \quad \cdot \quad \cdot \quad (2).$$

$$R_1 = \overset{1}{\Sigma} P\left(1 - \frac{z}{l_1}\right) + \frac{M_2}{l_1} \quad \cdot \quad \cdot \quad \cdot \quad \cdot \quad \cdot \quad \cdot \quad (3).$$

$$R_2 = \overset{1}{\Sigma} P\frac{z}{l_1} - \frac{M_2}{l_1} - \frac{M_2 - M_3}{l_2}. \quad \cdot \quad \cdot \quad \cdot \quad \cdot \quad (4).$$

$$R_3 = \frac{M_2 - M_3}{l_2} + \overset{3}{\Sigma} P\frac{z}{l_3} - \frac{M_3}{l_3}. \quad \cdot \quad \cdot \quad \cdot \quad \cdot \quad (5).$$

$$R_4 = \overset{3}{\Sigma} P\left(1 - \frac{z}{l_3}\right) + \frac{M_3}{l_3} \quad \cdot \quad \cdot \quad \cdot \quad \cdot \quad \cdot \quad (6).$$

In ordinary swing-bridges where $l_1 = l_2 = l$ and a single weight P rests on l_1:

$$R_1 = P\left\{ \left(1 - \frac{z}{l}\right) - \left(1 - \frac{z^2}{l^2}\right)\frac{z}{l} \cdot \frac{2l(l + l_2)}{4(l + l_2)^2 - l_2^2} \right\} \quad \cdot \quad \cdot \quad (7).$$

$$R_4 = P\left(1 - \frac{z^2}{l^2}\right)\frac{z}{l} \cdot \frac{l l_2}{4(l + l_2)^2 - l_2^2} \quad \cdot \quad \cdot \quad \cdot \quad \cdot \quad (8).$$

These formulæ are in no wise changed for a single weight P resting on l_3, except that R_1 and R_4 are interch...:ged. Also:

$$R_3 = \frac{(R_1 - R_4)\, l - P\,(l - z)}{l_2}\; \cdots \cdots \cdots \quad (9).$$

$$\therefore R_2 = P - R_1 - R_4 + R_3 \cdot \cdots \cdots \cdots \cdots \quad (10).$$

Case II.—Two Spans with One Intermediate Support.

Reference will be made to the notation and Fig. 2 of Art. 35.

$$M_2 = - \left\{ l_1^2 \overset{1}{\Sigma} P \left(1 - \frac{z^2}{l_1^2}\right) \frac{z}{l_1} + l_2^2 \overset{2}{\Sigma} P \left(1 - \frac{z^2}{l_2^2}\right) \frac{z}{l_2} \right\} \div 2\,(l_1 + l_2)\,(11).$$

$$R_1 = \overset{1}{\Sigma} P \left(1 - \frac{z}{l_1}\right) + \frac{M_2}{l_1} \cdot \cdots \cdots \cdots \quad (12).$$

$$R_2 = \overset{1}{\Sigma} P \frac{z}{l_1} - \frac{M_2}{l_1} + \overset{2}{\Sigma} P \frac{z}{l_2} - \frac{M_2}{l_2} \cdot \cdots \cdots \quad (13).$$

$$R_3 = \overset{2}{\Sigma} P \left(1 - \frac{z}{l_2}\right) + \frac{M_2}{l_2} \cdot \cdots \cdots \cdots \quad (14).$$

If $l_2 = l_1 = l$:

$$M_2 = -\frac{l}{4} \left\{ \overset{1}{\Sigma} P \left(1 - \frac{z^2}{l^2}\right) \frac{z}{l} + \overset{2}{\Sigma} P \left(1 - \frac{z^2}{l^2}\right) \frac{z}{l} \right\} \cdot \cdots \quad (15).$$

These formulæ are based on the supposition that there may be negative reactions at A and C of Fig. 2, Art. 35. If no negative reactions are possible, and if the load is on one arm only, that arm will be a non-continuous beam for such load, and the reactions will be found by the simple principle of the lever.

Fig. 1.

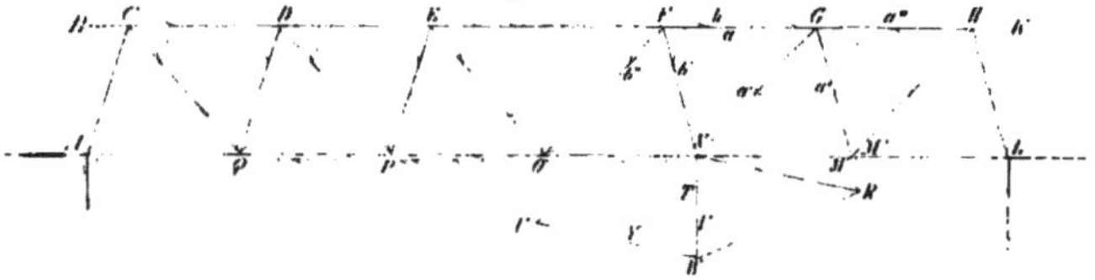

Fig. 2.

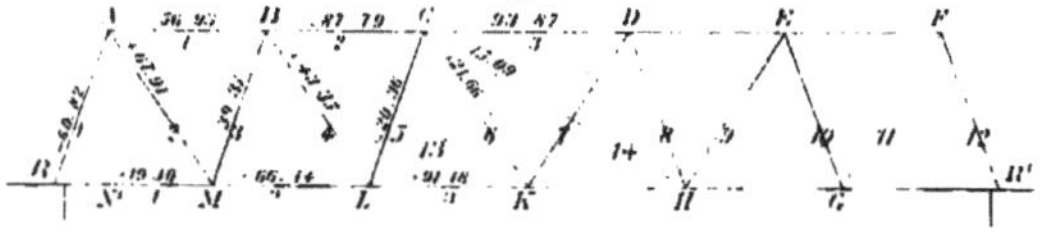

Fig. 3.

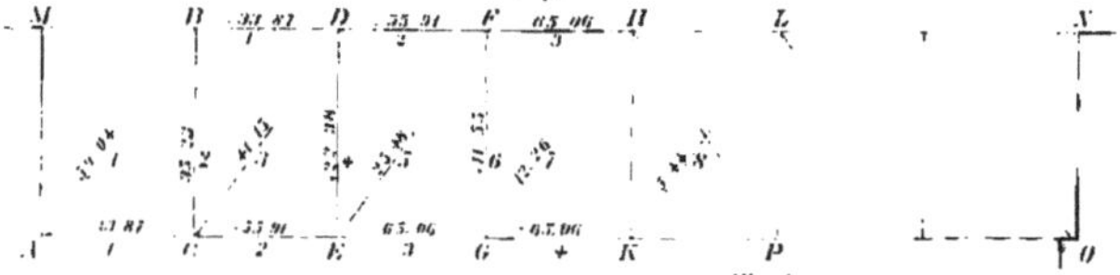

Fig. 4.

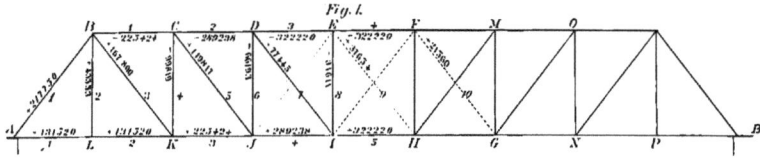

Fig. 1.

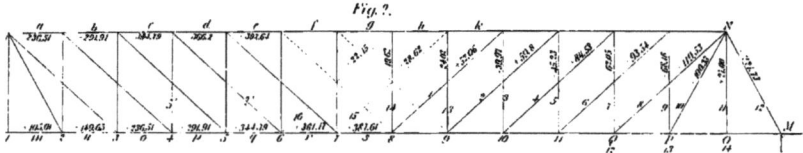

Fig. 2.

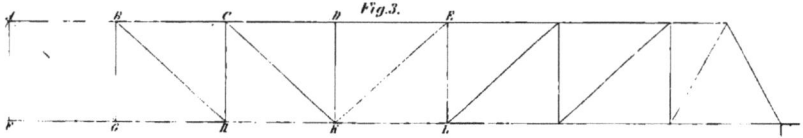

Fig. 3.

Fig. 4.

Pl. III.

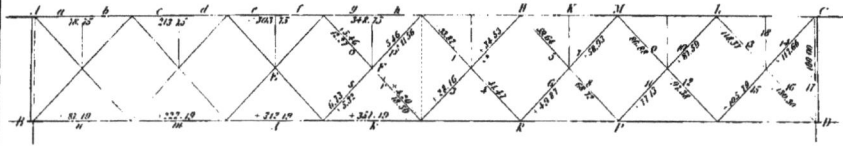

Fig. 1

Fig. 2

Fig. 3

Fig. 4

Fig. 5

Fig. 6

Pl. IV.

Scale   20 tons per inch.

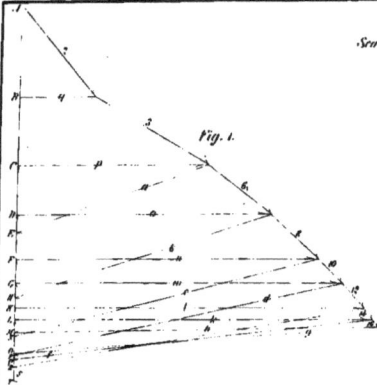

Fig. 1.

Fig. 2.

Fig. 3.

Fig. 4.

Pl. V.

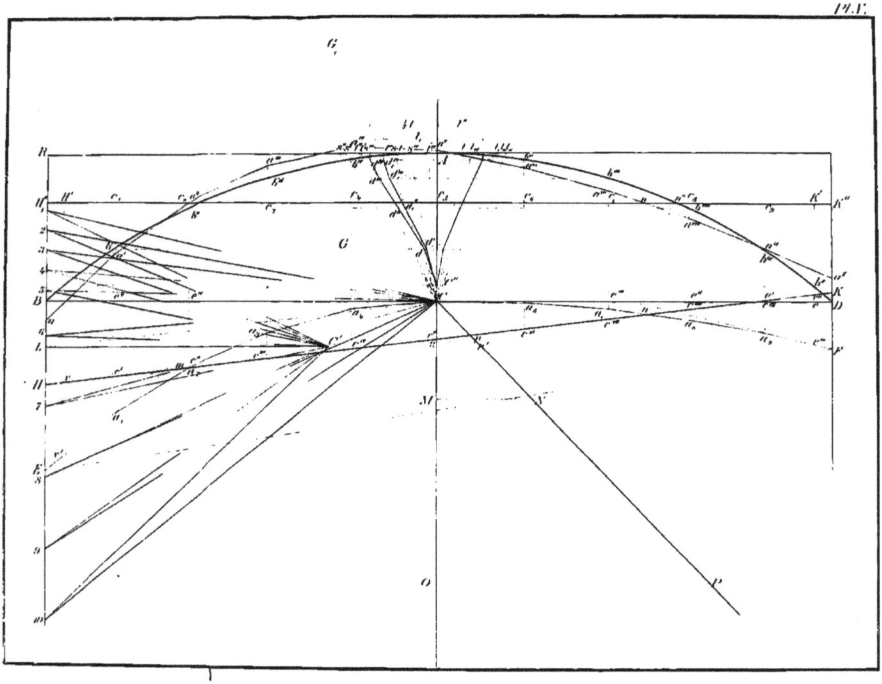

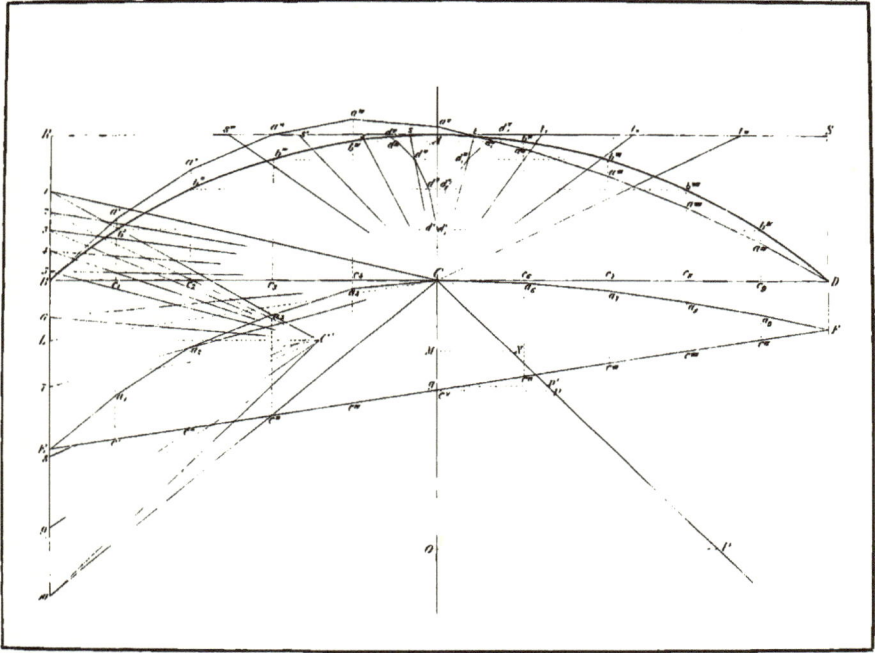

Pl. VII.

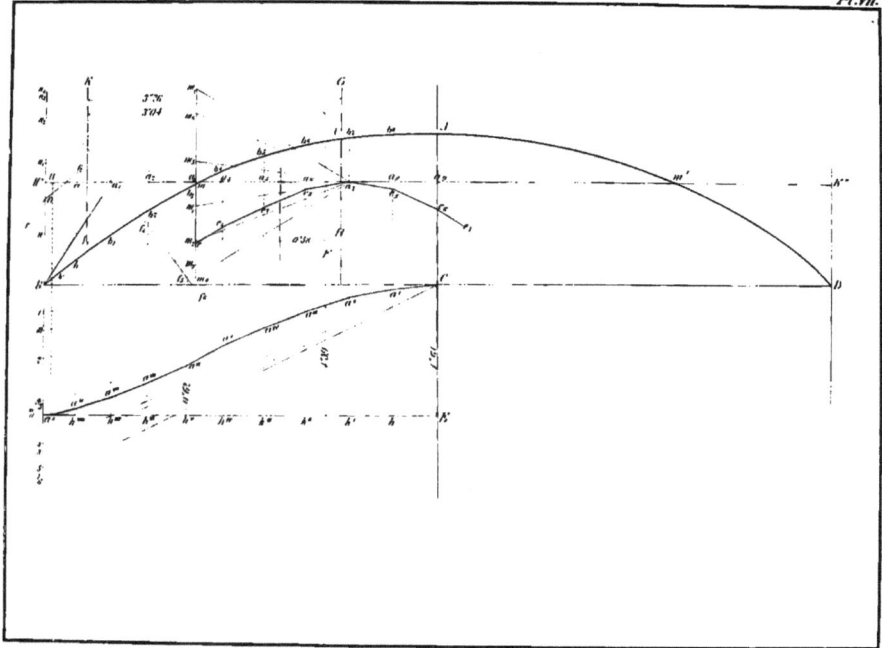

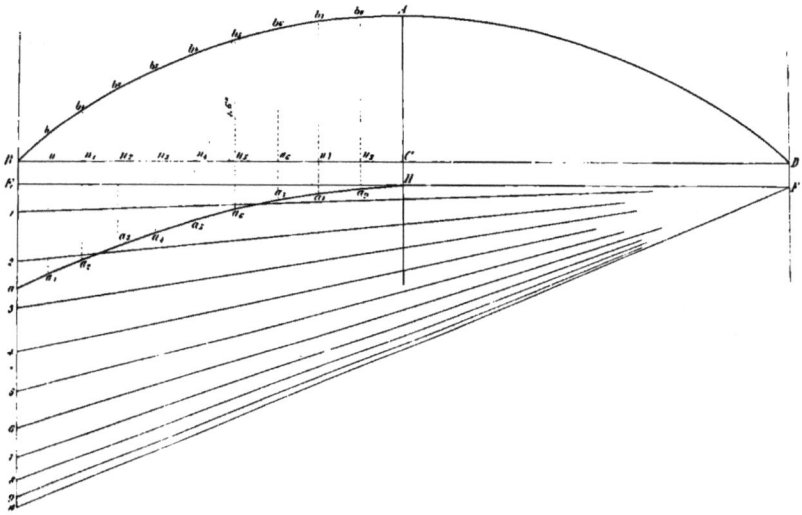

Pl. IX.

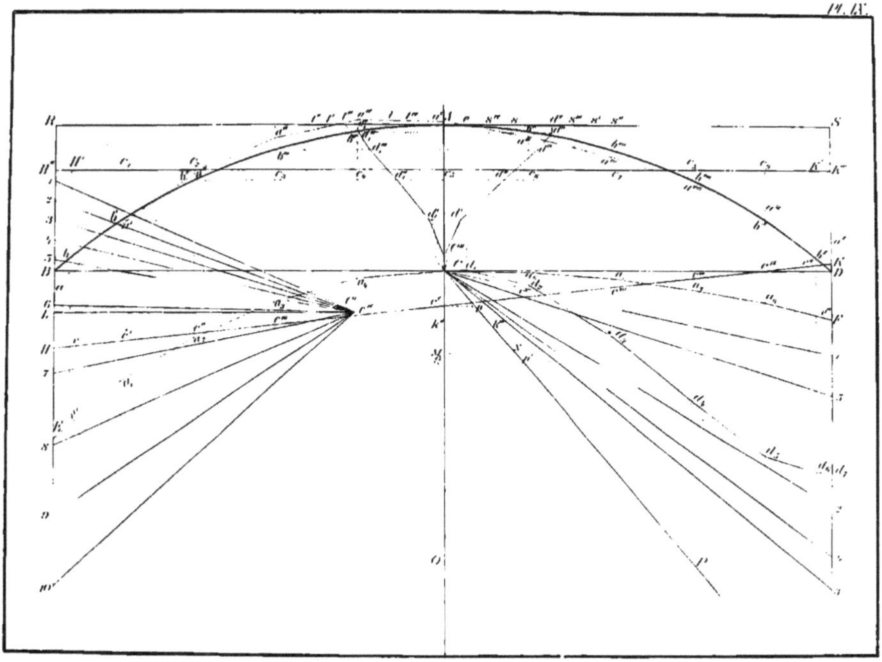

FIG. 1.

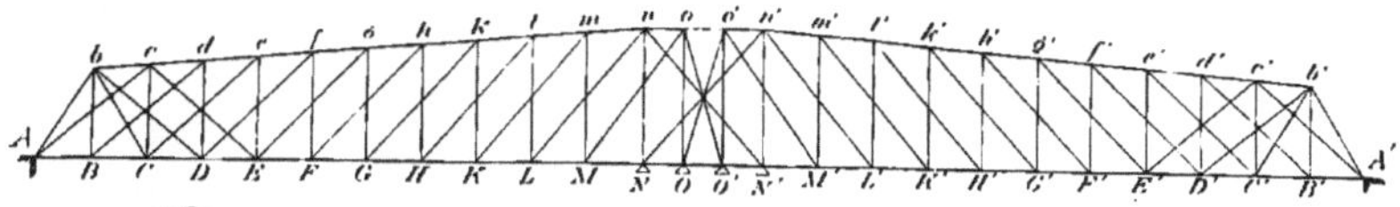

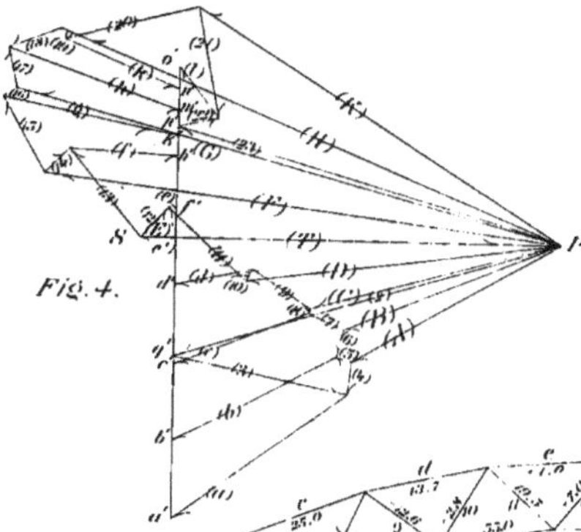

FIG. 6.

FIG. 4.

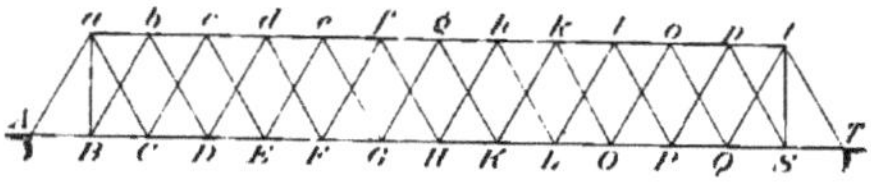

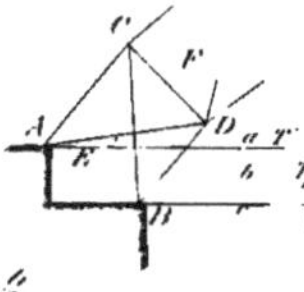

FIG. 5.

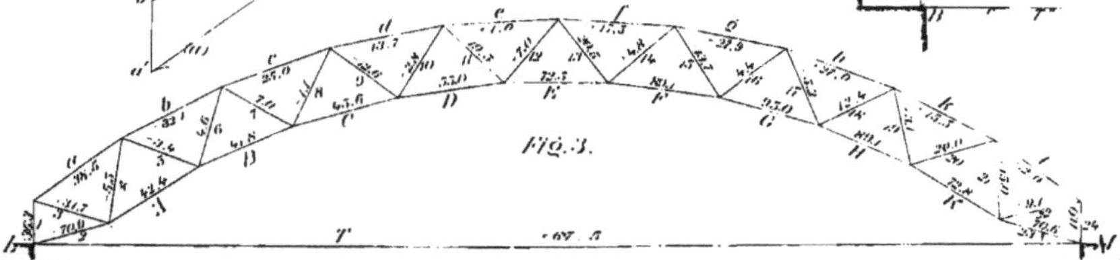

FIG. 3.

Pl. XI

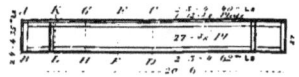

Fig. 1

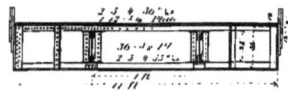

Fig. 2

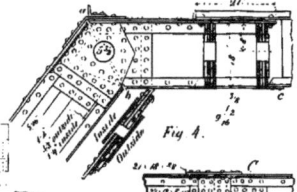

Fig. 4

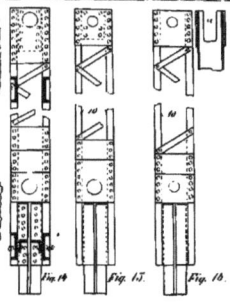

Fig. 14.   Fig. 15.   Fig. 16.

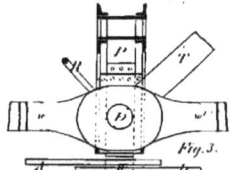

Fig. 3.

Fig. 7

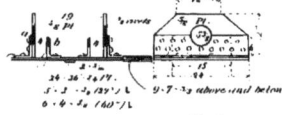

Fig. 13.

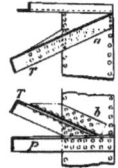

Fig. 12.

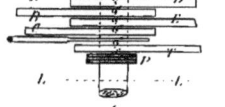

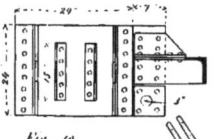

Fig. 8.   Fig. 9.

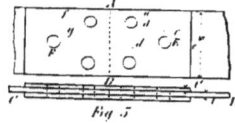

Fig. 5

Fig. 10.

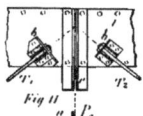

Fig. 11

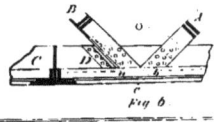

Fig. 6

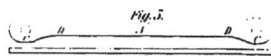

Fig. 1.

Fig. 2.

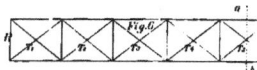

Fig. 4.

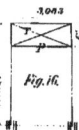

Fig. 16.

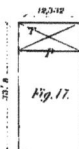

Fig. 17.

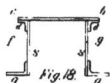

Fig. 18.

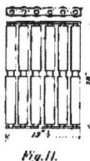

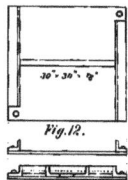

Fig. 6.

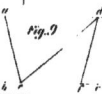

Fig. 11.

Fig. 13.

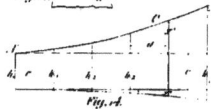

Fig. 14.

Fig. 12.

Fig. 8.

Fig. 9.

Fig. 10.

www.ingramcontent.com/pod-product-compliance
Lightning Source LLC
Chambersburg PA
CBHW020859210326
41598CB00018B/1722